# Visual C++ MFC
# Programming by Example

*John E. Swanke*

CMP Books
Lawrence, Kansas 66046

CMP Books
CMP Media LLC
1601 West 23rd Street, Suite 200
Lawrence, Kansas 66046
USA

Cover art created by John Freeman and Robert Ward.

**Distributed in the U.S. and Canada by:**
**Publishers Group West**
**1700 Fourth Street**
**Berkeley, California 94710**
**800-788-3123**

ISBN: 0-87930-544-4

VH #5  4-03

CMP**Books**

*To the one I love,*
*my wife,*
*Cathy Krinitsky*

# Acknowledgments

First, I would like to thank Paul Swanke for getting me into this topic. As my only programming brother, he has always surprised me with some new programming tidbit he's uncovered from his travels. He's also been there for me in other ways, from New Jersey to Black Rock.

I would also like to thank John "Brad" Bradberry for constantly raising the bar. Ever since our days together making beer cans for third world countries, I have fed on his acceptance, respect, and unspoken dares. It has always been a thrill and an honor to go head-to-head with the man from MIT.

I would like to thank Brian Berenbach for showing me there can be structure. What I thought was supposed to be obtuse, he showed could be organized and streamlined. Brian also stood by me once when I crashed and burned — for that I honor him as a stand up guy.

I would like to thank Berney Williams, my acquisitions editor at CMP Books, for leading me through the brutal process of writing a book. I would also like to thank Liza Niav for translating this book into English and Michelle Dowdy for making hundreds of pages of mishmash look like a book.

I would like to thank my family and friends who play an important role in my life and I apologize for all the lost time together. My friend Paul Clowmitzer once told me that writing a book was unforgivable — that you could never come up with a suitable excuse to society to explain your absence while writing it. I can only hope that my friends, family, society, and Paul can forgive me.

And finally, I would like to thank my wife, Cathy Krinitsky. Someone once said that there is a book in everyone, that all you need is the discipline to write it. I think discipline is secondary to desire. Cathy gave me that desire and then left me alone to work. Along with everything else that she means to me, Cathy helped me to live yet another one of my dreams. Thanks, Kas.

# Table of Contents

# Section II   User Interface Examples . . . . . . . **125**

# Chapter 5   Applications and Environments . . . . . **127**

# Chapter 6   Menus . . . . . . . . . . . . . . . . . . . . . . . . . **175**

# Chapter 7   Toolbars and Status Bars . . . . . . . . . **199**

# List of Figures

# Introduction

When I'm learning a new application, studying an example of what I want to do helps me be more productive in less time then if I had started from scratch reading some stale manual — or worse, a disjointedly produced CD or online documentation.

With an example, I can get a clearer sense of how the creators of a language or a class intended their software to be used. When should I use a modeless dialog box, a dialog bar, or a mini frame window? Do I go with MDI, SDI, or just a plain Dialog Box application? What do I do if I change my mind?

This book provides numerous examples of applications and features using Microsoft's Visual C++ and the Microsoft Foundation Classes (MFC). Unlike the examples you get with MFC, these are examples of what you'll find in the best-selling Windows 95/NT applications available today. For those of you who want to create the best-selling software of tomorrow, each example also includes an explanation of how it works and some tips on how to take it from there.

## What's In Store

The examples in this book are organized into the steps one might go through when writing an application, from the creation of an application's project to adding the finishing touches to its interface. Each example, however, also stands on its own so that you might use them in any order.

Similar examples are combined into chapters, which are then presented in one of the following sections.

# The Basics

Although I tried to make this book example-oriented, I found that knowing just a few of the basics up front can help make understanding an example later much easier. You can certainly skip this section if you like, but don't be surprised if I refer to it.

# User Interface Examples

The examples in this section concentrate on the user interface of an application. Because MFC is interface intensive, the vast majority of the examples in this book will be in this section. The topics have been arranged in the order in which they should be considered when creating an MFC application. They include application types, menus, toolbars and status bars, views, dialog boxes and bars, control windows, and drawing. This section is topped off with some example applications, including text editors and wizards.

# Internal Processing Examples

The examples in this section, while still applying to most MFC applications, are more representative of the unseen part of MFC than the user interface. The examples include messaging, files, lists, maps, arrays, cutting and pasting, and time.

# Packaging Examples

This last section contains examples of packaging your software in other than a standard executable, such as Dynamic Link Libraries (both MFC and nonMFC) and resource libraries.

# About the CD

Included with this book is a CD containing a working Visual C++ v5.0 project for every example in this book. If you want to find the project that corresponds to a particular book example, just locate its number among the directory names on the CD. For the most up-to-date CD-ROM data and an added v6.0 project, download the files from the publisher's ftp site at: `ftp://ftp.mfi.com/pub/rdbooks/errata/visualC.zip`.

Except where noted, most of the examples on the CD were created as an MDI application using all of the AppWizard defaults and a project name of "Wzd".

# About the SampleWizard

Also on the CD is the SampleWizard utility, which can help you add the examples in this book directly to your applications. This utility guides you through a catalog of examples which, if selected, detail the instructions and code necessary for including the example in your project. You will also be given the opportunity to substitute the example's project name ("Wzd") with your own.

The SampleWizard can be found in the `\SWD` directory on the CD. It makes use of the `\Wizard` subdirectory found in each example on the CD and contains all of the particulars for that example.

Simply execute `SW.EXE`. The rest should be intuitive.

# About the Website

Errata as well as information on my other "By Example" books may be found on my website at `http://home.earthlink.net/~jeswanke/`. Check it out.

# Section I

# The Basics

An application created with Visual C++ and MFC will automatically generate most of its own windows, handle its own messages, and do its own drawing. Microsoft has taken great pains to hide the inner workings to make creating a generic application easier. However, when something doesn't work the way you thought it should, knowing about the inner workings can help eliminate frustration programming. ("Let me try all caps this time. Well, let me try a zero this time. Why did it stop working?") More importantly, knowing how to perform tasks such as putting a window anywhere, getting a message from anywhere, and drawing anything anywhere can help distinguish your application for its user ease from an application confined by the windows, messages, and drawing provided automatically by Visual C++ and MFC.

There are four main basics to Visual C++ applications: creating a window, knowing the other MFC classes provided, sending messages to a window, and drawing inside a window. There are certainly several other basics, and we will definitely be looking at those as they come up, but these four are a pretty good start.

# Windows

In Chapter 1, we will first look at creating a window with and without MFC to get a clearer idea of just what MFC is doing. We'll find out that MFC windows are created with both a C++ class belonging to MFC and a nonC++ Windows Class that predates and exists outside of MFC. We'll take a closer look at Window Classes and review those that are supplied by the Windows operating system. And finally, we'll take a look at what goes into an MFC application.

# Classes

In Chapter 2, we'll take a broad look at the functionality provided by the Microsoft Foundation Classes. We'll examine the three MFC classes from which most of MFC is derived (`CObject`, `CWnd`, and `CCmdTarget`). We'll also take a look at the MFC classes that make up an application, the MFC classes that support the Windows interface, the classes used for drawing, accessing files, maintaining data and databases, and the MFC classes that access the Internet.

# Messaging

In Chapter 3, we'll look at how an MFC application communicates with the outside world and within itself (through messages). We'll review the four message types and follow a message through the receiving class's member functions. We'll also explore redirecting messages from this path.

# Drawing

In Chapter 4, we'll look at several aspects of drawing inside your windows, including drawing tools, drawing in colors, drawing to the screen or printer, drawing bitmaps and icons, drawing rectangles and circles, and animation.

# Windows

In this chapter, we will look at the basic element of an MFC user interface: the window. I'll compare API windows and MFC windows and describe how to create a window, destroy a window, and control communications between the Windows operating system and your window.

## Windows and the API World

A window is a rectangular area of the screen where an application displays data and looks for mouse clicks. The user interface of a Windows application can consist of dozens of windows, each with different characteristics but all interconnected (Figure 1.1).

**Figure 1.1** **Windows application user interfaces consist of windows.**

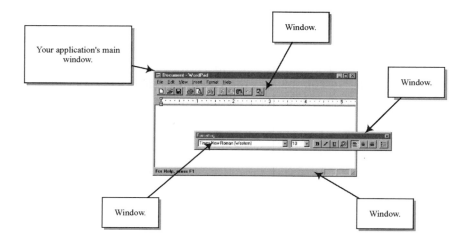

## The Three Types of Windows

Within those dozens of windows, there are only three basic types: Overlapped, Popup, and Child. Moreover, there isn't much internal difference between them, although they can be made to look different using window styles (Figure 1.2).

**Figure 1.2** **Window styles can be used to differentiate between the three window types.**

|  | Overlapped | Popup | Child |
|---|---|---|---|
| Most Basic Form | Example | Note: in its most basic form the system draws nothing. The window is all Client area. | Note: in its most basic form the system draws nothing. The window is all Client area. |
| Typical Look | Example<br>File Edit View | Hello World<br>OK | OK<br>Note: This is all Client area. A child window usually does its own drawing. |
| With Identical Settings | Example<br>File Edit View | Example<br>File Edit View | Example<br>Note: A Child window cannot have a menu. |

**Overlapped** windows are usually reserved for your application's main window and, in fact, are sometimes called *Main* windows or *Frame* windows.

**Popup** windows are usually used to communicate with the user in the form of Dialog boxes and Message boxes.

**Child** windows are usually used in views such as the text display in a text editor and in controls such as the OK button in a Dialog box. Another name for a Child window that looks like a button or any other control is a *Control* window.

The main difference between an Overlapped window and a Popup window is that a Popup window can appear without a caption (also known as a title bar). The main difference between a Child window and an Overlapped or Popup window is that a Child window can only appear inside another window with any excess removed or clipped by that other window. In addition, a Child window is the only window that can't have a menu bar.

See Figure 1.3 for a Windows application populated with Overlapped, Popup, and Child windows.

**Figure 1.3    Windows applications are made up of Overlapped, Popup, and Child windows.**

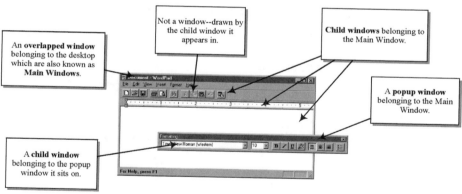

## Client and Nonclient Areas

Each window has a *nonclient area*, drawn by the system, and a *client area*, drawn by your application. The system can draw one or all of the features shown in Figure 1.4 or it can leave all the drawing to you.

**Figure 1.4   The nonclient areas of a window can optionally be drawn by the system.**

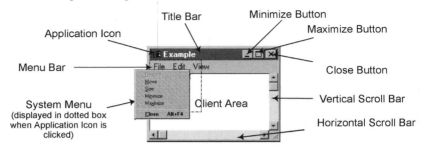

Each window is represented by a Window Object in memory that tells the Windows operating system where to draw the window and what routine to call in response to a mouse click, a key press (assuming the window has input focus), a timer expiration, etc. A Window Object is itself an instance of a Window Class. This is not a Visual C++ class, but a Microsoft Windows proprietary class that predates and exists outside of Visual C++. However, just like a C++ class, a Window Class defines several characteristics of every window created from it, such as background color and where to send messages (Figure 1.5).

**Figure 1.5   Window Objects are created from a Microsoft Windows proprietary Window Class.**

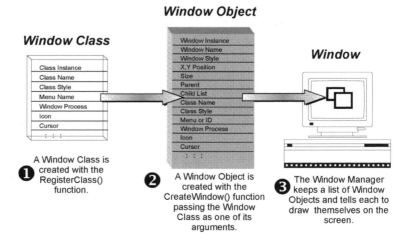

The Windows operating system provides an extensive Application Program Interface (API) to allow you to create and maintain these Window Objects. You can create a Window Object by calling `CreateWindow()`. You can change the characteristics defined by your window's class with `SetWindowLong()`. You can move your window with `MoveWindow()`. You can kill a window with `DestroyWindow()`. So what does MFC do for you?

# Windows and the MFC World

A Microsoft Foundation Classes (MFC) window is a hybrid of C++ and Windows API calls. In effect, an MFC window gives you a C++ wrapper to a lot (but not all) of the Windows API, eliminating some of the drudgery of writing a Windows application and providing some new services.

An MFC window does not have any more direct control over a Window Object then you do in the API world. If you can't do something in the API world, it can't be done in the MFC world. For example, the MFC library has the class `CWnd` for creating a window, but it's simply a wrapper around the API calls you would use in the API world.

Microsoft could have put the logic for creating and maintaining a window in the MFC library for true C++ encapsulation and control. However, this approach would have made the MFC library all but unportable, caused severe redundancies, and created more bugs.

Creating an MFC window can be tricky. First, you create an instance of `CWnd`. Then you call a member function of `CWnd`, which calls `CreateWindow()` in the API. The returned window handle (which is just an indirect pointer to the Window Object) is stored in the `CWnd` member variable `m_hWnd`.

---

**NOTE:** Because windows are created in memory that's constantly being churned, its address may constantly change. Therefore, a window handle, instead of pointing directly to the Window Object, points to another pointer that keeps track of the Window Object location.

---

Destroying a window can be equally tricky. You must be sure to destroy the Window Object and the instance of `CWnd` that wrapped it. Although the `CWnd` object knows about the Window Object, the Window Object doesn't know about the `CWnd` object (Figure 1.6).

## Figure 1.6    MFC windows are created with two objects.

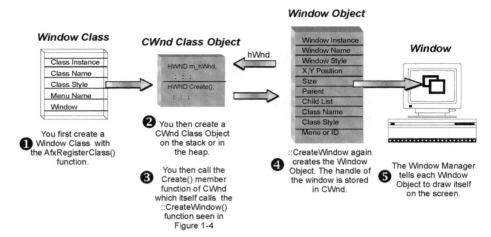

Even though a Windows application's user interface is made up of dozens and in some cases hundreds of windows, most are still created from less than ten different Window Classes. And even if there were a thousand windows in your application, each window would again still be one of only three basic types: Overlapped, Popup, or Child.

# How to Create a Window Using MFC

You can create a window using the MFC class CWnd.

```
CWnd wnd;
BOOL b=wnd.CreateEx(ExStyle, ClassName, WindowName, Style,
     x, y, Width, Height, Parent, Menu, Param);
```

The first line creates a CWnd object, and the second line creates the actual window by calling the Windows API CreateWindowEx().

```
HWND hwnd=::CreateWindowEx(ExStyle, ClassName, WindowName,
     Style, x, y, Width, Height, Parent, Menu, Instance, Param);
```

Because the CWnd class simply wraps the Windows API for creating a window (CreateWindowEx()), the arguments necessary to create a window are essentially the same between the API world and the MFC world:

- The Style and ExStyle arguments determine the "look" and type (Overlapped, Popup, or Child) of the window.

- The `ClassName` argument determines which Window Class to use when creating the window.

- The `WindowName` argument determines what, if anything, will be in the window's title.

- The `x`, `y`, `Width`, and `Height` arguments determine the location and size of the window.

- The `Parent` argument is a pointer to which window, if any, owns this window.

- The `Menu` argument points to an object in memory that this window will use for its menu — unless you are creating a Child window, in which case, this argument is an `IDnumber` that helps a Parent window identify its children.

- The `Instance` argument identifies which application this window belongs to so that messages sent to this window can be sent to the correct application's message queue. The `CWnd` class fills in `Instance` for you.

- The `Param` argument is a pointer to an optional structure of additional information used by a Window Class when creating a window.

The returned `hwnd` parameter is a pointer to the created Window Object or NULL if no window was created. This window handle is automatically saved in the `CWnd` class's `m_hWnd` member variable, as previously seen in Figure 1.6.

Now that we know the basics, let's look at the rules for filling in these arguments.

## The Rules

### Window Name

This argument is a zero-terminated text string, in which you specify what will appear in your window's title bar. If your window won't have a title bar, you can make this argument zero (0).

However, some Common Controls also use this argument. For example, a Button Control puts this argument in its button face. You can use `CWnd`'s `SetWindowText()` to change the name in a window's title bar after it's been created.

## Styles and Extended Styles

These two 32-bit arguments allow you to specify what type of window to create. You can select more than one style, as in the following example.

```
,WS_CHILD|WS_VISIBLE,
```

Window styles include:

- The styles used to create the three window types. You use WM_CHILD to create a Child window, WM_POPUP to create a Popup window

- and WM_OVERLAPPED to create an Overlapped window. If you specify none of these styles, WM_OVERLAPPED is assumed.

- The styles used to add nonclient features to your window. For example, you can use WS_VSCROLL to add a vertical scrollbar to your window. See Figure 1.7 for other nonclient area window styles.

**Figure 1.7**   **Nonclient window styles can be used to add features to your window.**

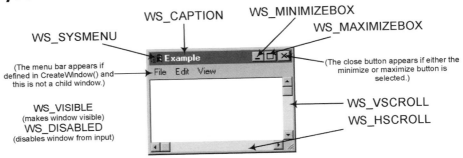

- The styles defined by each Common Control. For example, the BS_FLAT button style tells a Button Control to draw a 2D button.

- The styles used to make a window visible, enabled, and/or maximized initially.

- The styles used to mark the start of control group or to indicate which controls are eligible to receive input focus when the user is tabbing around the window.

For more examples of styles, see Appendix A.

You can use CWnd's ModifyStyle() and ModifyStyleEx() to change your window's styles after it has been created. Some styles will require you to redraw the window. You can automatically invoke CWnd's SetWindowPos() to do just that by adding a third argument to ModifyStyle().

```
CWnd wnd;
    wnd.ModifyStyle(0,WS_BORDER,SWP_NOZORDER);
```

Actually, any value you add as the third argument to ModifyStyle() or ModifyStyleEx() will cause the following SetWindowPos() options to be added: SWP_NOZORDER, SWP_NOMOVE, SWP_NOACTIVATE, and SWP_SHOW.

---

**NOTE:** Even a redrawn window might not incorporate a new style. This is especially true with Common Control windows because the individual window determines what styles it will use to redraw a window. In those cases, you will simply have to destroy and recreate the window, having first saved the style and other parameters of the existing window.

---

## X and Y Position

These two 32-bit arguments allow you to specify the location of your window in pixels. When creating Overlapped or Popup windows, X and Y are relative to the upper-left corner of the desktop window. For a Child window, X and Y are relative to the upper-left corner of the client area of the Parent window. If you set both the x and y arguments to CW_USEDEFAULT, the system will automatically pick a location for your window. The system cascades these new windows (Figure 1.8).

**Figure 1.8**  CW_USEDEFAULT **allows the system to automatically position windows.**

First Window          Next Window          Next Window

However, a Child window created with both the x and y arguments set to CW_USEDEFAULT will always be created at 0,0.

You can use CWnd's MoveWindow() to move a window after it has been created.

## Width and Height

These two 32-bit arguments allow you to specify the size of your window in pixels. If you specify CW_USEDEFAULT for both the Width and Height arguments, the system will automatically pick a size for your window based on the size of the desktop. For a Child window, however, the system will create the window with a size of 0,0. A window style of WS_MINIMIZE or WS_MAXIMIZE will override whatever values you set here.

You can use CWnd's MoveWindow() to resize a window after it has been created.

## Z-Order

Z-order determines which window appears on top of which when they all occupy the same area of the screen. The "Z" in Z-order comes from the Z in an X-Y-Z axis, where Z extends outward from the screen towards you. Your window will appear at the top of this Z-Order when initially created or when selected. However, your window will never appear above a top-most window unless it's also a top-most window. A *top-most* window is created using the WS_EX_TOPMOST window style and will appear above all other nontop-most windows, even when it isn't the currently selected window.

You can use CWnd's SetWindowPos() to change your window's Z-order after it's been created.

## Parent or Owner

This argument is a pointer to a CWnd object that identifies either a Parent window or an Owner window depending on what type of window you are creating:

- If you are creating a Child window, use this argument to identify its Parent window (the window it sits inside and is cropped by). It cannot be NULL. Child windows can only appear within their Parent window. They're destroyed when the Parent is destroyed and hidden when the Parent is hidden or minimized.

- If you are creating an Overlapped or Popup window, use this argument to identify an Owner window. If NULL, the desktop window becomes the Owner. Owned windows are always displayed above their Owners and destroyed when their Owners are destroyed. Owned windows are hidden when their Owners are minimized, but are not hidden when their Owner is hidden.

A Child window can be the Parent window of another Child window, but it can never be an Owner window. If you attempt to make a Child window an Owner, the system will simply make the top-most window of that Child window the Owner. See Figure 1.9 for an overview of this relationship.

**Figure 1.9    Owner, Parent, and Child windows have a hierarchical relationship.**

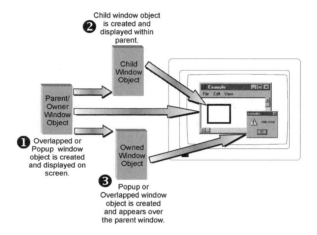

You can use CWnd's SetOwner() to change the owner of a window that already exists. You can use CWnd's SetParent() to do the same for your window's parent.

## Menu or Control ID

This argument can either identify a Menu handle or a Control ID depending on whether you're creating a Child window or an Overlapped or Popup window.

- If you are creating a Child window, use this argument to identify a Control ID. Control IDs are used to help a Parent window identify its Child windows. Because this argument is looking for an HMENU argument, you will need to use an HMENU type override to specify your ID. For example, if your Child window's Control ID is 102, you should specify it as

```
…,(HMENU) 102,…
```

- If you are creating an Overlapped or Popup window, use this argument to specify this window's menu. If NULL, the menu will default to whatever menu is specified in this window's Window Class. If the Window Class has no menu specified either, this window will be created without a menu. You can load a menu for this argument with

```
HMENU hMenu=::LoadMenu(AfxGetInstanceHandle(),
    MAKEINTRESOURCE(xx));
```

where xx is a Menu ID in your application's resources.

You can use CWnd's SetDlgCtrlID() to change the ID of a Child window that already exists. You can use CWnd's SetMenu() to change the menu of an Overlapped or Popup window.

## Instance

As mentioned previously, the CWnd class fills this argument for you. CWnd gets the Instance by calling AfxGetInstanceHandle(). An application's Instance essentially identifies which program it is in memory. AfxGetInstanceHandle() is one of several static functions provided by MFC.

## Parameter

This 32-bit argument is optional. Typically, it's a pointer to a structure or a class object you need to supply in order to create some types of windows. For example, when creating a window with the MDICLIENT Window Class, you need to supply a pointer to a CLIENTCREATESTRUCT structure here.

## Class Name

This argument is a zero-terminated string identifying which Window Class to use when creating this window. I will be discussing Window Classes and Window Processes in much more detail later in this chapter. You can't leave this argument NULL. To create a very generic MFC window, use AfxRegisterWndClass(0) to fill this argument.

# How to Destroy a Window Using MFC

As mentioned previously, deleting an MFC window can be somewhat tricky. There are two objects involved, which you must delete in the following order.

```
pWnd->DestroyWindow();    // destroys Window Object
delete pWnd;              // destroys CWnd Object
```

You could also simply delete the CWnd object, since DestroyWindow() is called from CWnd's destructor, but this isn't recommended. Destroying a window without first calling DestroyWindow() doesn't allow some destruction messages to be handled by any derived classes of CWnd. You will rarely need to destroy a window yourself. Typically, the user or the system will do this for you.

If you would like your CWnd object to be destroyed when its window is destroyed, you should add the following override to your derivation of the CWnd class.

```
CYourWnd::PostNcDestroy()
{
    delete this;
}
```

PostNcDestroy() is the last member function called before a window is destroyed. You will rarely need to override this function, however, since your CWnd and derivative classes will typically be either embedded in another class or created on the stack.

# Attaching to an Existing Window

If a window was created before or outside of your application using the Windows API and you'd like to wrap it in a CWnd class, you can use

```
CWnd wnd;
wnd.Attach(hwnd);
```

where hwnd is the handle of the existing window. Attach() simply sets CWnd's m_hWnd member variable to hwnd.

You could also use CWnd::FromHandle(hwnd), which will look to see if any CWnd objects already wrap this window handle and return a pointer to that CWnd object. If no such CWnd object exists, however, a temporary one is created. But don't hold onto this pointer — if it's temporary, the object it points to will be deleted the next time your application is idle.

# Window Classes

As mentioned previously, a Window Class is not a C++ class, but instead is a proprietary class that predates and exists outside of C++. A Window Class acts as a template from which you can create other windows that share some characteristics, including those described in the following list.

**Class Style** consists of several minor characteristics you can give your window.

**Class  Icon** defines which icon to draw in the upper-left corner of your window, if your window has an icon.

**Class Menu** defines what menu should be drawn in this window, if your window has a menu.

**Class Cursor** defines which mouse cursor should appear when the cursor moves across the client area of this window.

**Background Color** defines what color to use when erasing the background of a window. Unless you draw over it, this color will appear in the client area of your window.

**Window Process** defines what function should be called to process any messages sent to this window. Defining the Window Process is perhaps the single most important reason for the Window Class.

# Window Processes

A window interacts with its environment by sending and receiving window messages. If the system wants your window to draw itself, it sends a WM_PAINT message to it. If the system wants your window to destroy itself, it sends a WM_DESTROY message. These messages are processed by your window's Window Process, whose address is defined in the Window Class. The exact same Window Process handles the messages sent to any window created from the same Window Class. How does the same Window Process keep track of all of those windows? How does it know that it should draw window A at 10,34 and window B at 56,21? Simply by using the window's Window Object.

For example, all windows created using the BUTTON Window Class will use the same BUTTON Window Process. If a WM_PAINT message is sent to any of these windows, the BUTTON Window Process will draw the appropriate button using the size, position, and style specified in each window's Window Object (Figure 1.10).

**Figure 1.10    The BUTTON Window Process uses Window Objects to identify which window to manipulate.**

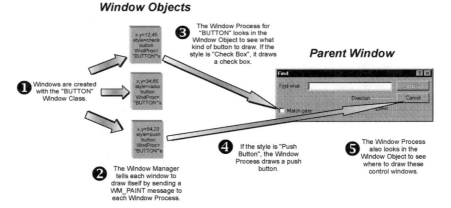

## How to Create a Window Class with MFC

When you create a Window Class, you're actually just registering a WND-CLASS structure in one of three operating system lists. The system maintains a list for each of three Window Class types:

**System Global Classes** are registered by the system on startup, cannot be unregistered, and are available to all applications.

**Application Global Classes** are registered by your application and are only available to your application and its threads.

**Application Local Classes** are registered by your application and are only available to the application or DLL that registered them.

When looking for a Window Class, the system starts with the Application Local Classes, then searches the Application Global Classes, and finally the System Global Classes.

To create your own Window Class, you can start by creating your own instance of the `WNDCLASS` structure and then registering it with MFC's `AfxRegisterClass()`. Or your can use MFC's `AfxRegisterWndClass()` to create a `WNDCLASS` object for you based on a few calling arguments.

# Using `AfxRegisterWndClass()` to Register a Window Class

The `AfxRegisterWndClass()` function is fairly automated. Several arguments you would normally need to provide are filled in for you. Even the name of the new Window Class is automatically generated for you.

```
LPCTSTR lpszClassName = AfxRegisterWndClass(
    UINT nClassStyle, HCURSOR hCursor = 0,
    HBRUSH hbrBackground = 0, HICON hIcon = 0);
```

The rules for these arguments are as follows.

## Class Name

The name generated for your new Window Class is based on the arguments you pass to this function. If you pass the same exact parameters, you will be creating the same exact Window Class. If you need to create a new Window Class, use `AfxRegisterClass()` instead.

## Style

Window Class styles can be applied as a series of `OR`'ed flags from the following list of choices.

| Class Style | Description |
| --- | --- |
| CS_OWNDC | Allocates a unique Device Context for each window created with this Window Class. See Chapter 4 for more on Device Contexts. |
| CS_PARENTDC | Retrieves a Device Context from the system cache and then sets the clipping region of that Device Context to incorporate the Parent window, such that a Child window can draw on its Parent. |

| Class Style | Description |
|---|---|
| CS_CLASSDC | Allocates one Device Context for use by all windows created from this Window Class. |
| CS_SAVEBITS | The video memory under any window created from this class will be saved, such that when the window is moved or closed, the underlying windows don't need to be redrawn — which may not be a concern with a fast machine. |
| CS_BYTEALIGNCLIENT CS_BYTEALIGNWINDOW | These styles were useful when graphic cards and CPUs were slower. With these styles the graphic cards could more easily move windows when they were on byte boundaries. |
| CS_GLOBALCLASS | If set, this is an Application Global class. Otherwise it's an Application Local class. |
| CS_VREDRAW CS_HREDRAW | If the vertical style is set and the vertical size of a window changes, the entire window is redrawn. Ditto for the horizontal. |
| CS_NOCLOSE | Disables the Close command on the System menu. |
| CS_DBLCLKS | If you don't set this and you double-click on a window created with this Window Class, the double-click won't be sent to your application — only two successive mouse clicks instead. |

## Icon

This argument is the handle to an icon that will appear in the upper-left hand corner of a window, but only if a window is also using the WS_SYSMENU window style. The icon of the Main window in your application also appears in the task bar. If you set this argument to NULL and the WS_SYSMENU style is set, a default icon will be supplied by the system.

In the MFC world, icons are handled for you in most cases. To change the icon of a window that already exists, you can use CWnd's SetIcon().

## Cursor

This argument is the handle of a mouse cursor that will appear when the mouse moves over the client area of your window. If you set this argument to NULL, the cursor will become the familiar default arrow cursor. You can load a cursor for use here with

```
HICON hIcon=AfxGetApp()->LoadCursor(xx);
```

where xx is either the name or ID of a cursor in your application's resources.

The cursor specified here is meant to be the default cursor for this window. If you would like to change your cursor shape dynamically you should process the WM_SETCURSOR messages to this window and use SetCursor() to change the cursor shape. (See Example 33 in Chapter 8.)

## Background Color

When the system creates a window, it first erases the background by drawing a rectangle over which the window will appear. This argument specifies the handle of a brush to use when filling this rectangle. (See Chapter 4 for more about brushes.) A brush object created for a Window Class will be automatically destroyed when the class is unregistered.

Rather than creating a brush object, you can instead specify one of the following system colors

```
COLOR_ACTIVEBORDER
COLOR_ACTIVECAPTION
COLOR_APPWORKSPACE
COLOR_BACKGROUND
COLOR_BTNFACE
COLOR_BTNSHADOW
COLOR_BTNTEXT
COLOR_CAPTIONTEXT
COLOR_GRAYTEXT
COLOR_HIGHLIGHT
COLOR_HIGHLIGHTTEXT
COLOR_INACTIVEBORDER
COLOR_INACTIVECAPTION
COLOR_MENU
COLOR_MENUTEXT
COLOR_SCROLLBAR
COLOR_WINDOW
COLOR_WINDOWFRAME
COLOR_WINDOWTEXT
```

However, you must typecast the color to a brush handle type and add the number one (1).

```
(HBRUSH) (COLOR_WINDOW+1)
```

Setting this argument to NULL will cause the system not to erase the screen before drawing a window. The nonclient area will be drawn as usual but the client area will retain whatever was on the screen before the window was drawn. If you set this argument to NULL, make sure your window either draws its entire client area or else processes the WM_ERASEBKGND message to erase the background then.

## Using the AfxRegisterClass() Function to Create a Window Class

If you want to take full control of creating a new Window Class, such as being able to specify your own Window Class name, you should use

```
BOOL AFXAPI AfxRegisterClass(WNDCLASS* lpWndClass);
```

where the WNDCLASS structure is as follows.

```
typedef struct _WNDCLASS {
    UINT style;                 // style of the class
    WNDPROC lpfnWndProc;        // function called by system when
                                // it has a message for a window
                                // created with this class
    int     cbClsExtra;         // extra bytes to add to the
                                // WNDCLASS structure when
                                // registering
    int     cbWndExtra;         // extra bytes to add to the Window
                                // Objects created with this class
    HANDLE  hInstance;          // instance that owns this class
    HICON   hIcon;              //icon to be used when window
                                // displays an icon
    HCURSOR hCursor;            // cursor to use when mouse is over
                                // a window created with this class
    HBRUSH  hbrBackground;      // background color to use when
                                // erasing the background area under
                                // a window created with this class
```

```
    LPCTSTR lpszMenuName;        // menu name to be used when
                                 // creating a menu for a window
                                 // created with this class
    LPCTSTR lpszClassName;       // the class name that identifies
                                 // this WNDCLASS for the system


} WNDCLASS;
```

The rules for filling in the Class `Style`, `Icon`, `Cursor`, and `Background` arguments can be found in the previous section. The rules for the remaining arguments are as follows.

## Class Name

The `ClassName` argument is a null-terminated character string that is used to identify your new Window Class. You can name your Window Class anything, but don't name it after an existing Window Class name unless you want to override that class. As mentioned previously, the system looks for a class name match in three lists starting with the Application Local Classes. If you register a class in this list with the same name as a System Global Class, your application will use your class instead.

## Menu Name

This argument points to the name of a menu in your application's resources. In the MFC world, menus are loaded for you in most cases. You can specify your own menu name here, or you can use the following if using a resource ID.

```
MAKEINTRESOURCE(IDR_MENU);
```

This will be the default menu for this window. If you specify a `NULL` here, the system will use the menu specified in `CWnd`'s `CreateEx()`. If no menu is specified there, this window will have no menu.

This argument is ignored for Child windows, since they never have a menu.

## Window Process and Instance

These two arguments are arguably the most important of a Window Class. The Window Process was discussed previously and Instance merely identifies what application contains the Window Process. In the MFC world, the default Window Process is `AfxWndProc()`, which you can retrieve for use in this argument with

```
AfxGetAfxWndProc();
```

You can use `AfxGetInstanceHandle()` to fill in the `Instance` argument.

### Class Extra and Window Extra

The `ClassExtra` and `WindowExtra` arguments provide a way of allowing your application to store proprietary data inside the Window Object or registered Window Class themselves. Class Extra specifies the number of extra bytes to allocate at the end of a class when it's registered. Window Extra specifies the number of extra bytes to add to the end of a Window Object when it's created.

In both cases, these extra bytes are used by your application to store information pertaining to a window or Window Class. However, since a `CWnd` object is also associated with a window and a `CWnd` object can save just as much information if not more, there really isn't much use for these extra byte arguments. So just set them to zero (0).

## How to Destroy an MFC Window Class

Window Classes are destroyed by unregistering them. But if you used either `AfxRegisterWnd()` or `AfxRegisterWndClass()` to register a class, these classes will be automatically unregistered when your application terminates. And even if your application didn't unregister your Window Classes, the operating system will.

## Factory Installed Window Classes

Your application will be running in an environment that will already have several registered Window Classes. Some are provided by the operating system (Windows 3.1/95/98/NT). Some are provided by MFC. The Control windows that are supplied by the system are also known as Common Controls.

## Some Important Window Classes

### Windows 3.1 and Above

| Class | Window Created |
|---|---|
| #32768 | Popup Menu Windows (Popup Menus are menus that sit in and entirely fill a Popup window.) |
| #32769 | Desktop Window |
| #32770 | Dialog Boxes |
| MDIClient | MDI Child Windows Area |

## Some Important Common Control Window Classes

### Windows 3.1 and Above

| Class | Window Created |
|---|---|
| BUTTON | Button Control Windows |
| STATIC | Static Control Windows |
| EDIT | Edit Control Windows |
| LISTBOX | List Box Control Windows |
| SCROLLBAR | Scroll Bar Control Windows |
| COMBOBOX | Combo Box Control Windows |
| ComboLBox | List Box Control Window (For the list box that appears below a Combo Box Control Window.) |

## Some Important Common Control Window Classes

### Windows 95/NT and Above

| Class | Window Created |
|---|---|
| RICHEDIT | Rich Edit Control Windows |
| SysListView32 | List View Control Windows |
| ComboBoxEx32 | Extended Combo Box Control Windows |
| SysAnimate32 | Animation Control Windows |
| msctls_trackbar32 | Slider Control Windows |

## Some Important Common Control Window Classes

### Windows 95/NT and Above (Continued)

| Class | Window Created |
|---|---|
| SysTreeView32 | Tree View Control Windows |
| msctls_updown32 | Spin Button Control Windows |
| msctls_progress32 | Progress Control Windows |
| SysHeader32 | Header Control Windows (Header Controls usually reside at the top of List View Controls.) |
| SysTabControl32 | Tab Control Windows |
| SysMonthCal32 | Month Calendar Control Windows |
| SysDateTimePick32 | Date/Time Selection Control Windows |
| msctls_hotkey32 | Hot Key Control Windows |
| tooltips_class32 | Tool tip Control Windows |
| msctls_statusbar32 | Status Bar Windows |
| ToolbarWindow32 | Tool Bar Windows |
| ReBarWindow32 | Rebar (a.k.a., CoolBar) Windows |

## Some Important MFC Window Classes

| Class | Window Created |
|---|---|
| AfxWnd | CWnd Windows |
| AfxFrameOrView | MFC Frame and View Windows |
| AfxMDIFrame | The MDI Frame Window |
| AfxControlBar | MFC Control Bar Windows |

# Other Kinds of Windows

Windows only come in three basic types — Overlapped, Popup and Child — but they're applied in several different ways in an MFC application. Except where noted, the MFC classes that wrap these windows are all derived from CWnd.

**Control Window**   A Child window that entirely draws itself to resemble a control. Examples include buttons, static text controls, and list boxes.

**Dialog Box** A Popup window with the functionality needed to fill itself with the Control windows specified in a resource file and then to process those Control windows.

**Message Box** A Popup window used by an application to prompt a user for a response.

**Toolbar** A Child window that draws its own buttons.

**Dialog Bar** A Dialog box that remains open and acts like a Toolbar.

**Status Bar** A Child window usually found at the bottom of your application's Main window and is used to display help on the command you're using.

**Frame Window** An Overlapped window that usually acts as the Parent and Owner window for all the other windows in your application. A Frame window can also direct user commands through your application's objects, which I will discuss in-depth in the next chapter.

**Document/View** This is actually made up of two MFC objects and one Child window. MFC applications are "document-centric" meaning the application lives to load, view, edit, and save documents whether those documents are text files, graphic shapes, or binary configuration files. First, an MFC Document class object is created that proceeds to load a document from disk into its member variables. Then one or more MFC View class objects are created that display those member variables. If more than one View class object is created, a document will have more than one view. Because the MFC Document class does not have an associated window, it is not derived from CWnd.

**Document Template** This actually has no window but instead is used by your application to determine what MFC document class object and what MFC view object to create when opening a document. There can theoretically be several Document Templates per application allowing an application to handle several types of documents, however a vast majority of applications simply have one template. The MFC Document Template class is not derived from CWnd. Please see Chapter 2 for more on Document Templates.

# The Desktop Window

Our last stop in this chapter is the desktop window, on which all other windows are displayed and eventually belong. The desktop window is itself a Popup window to which all other top-level windows belong. A list of top-level windows is maintained by the Windows Manager and therefore is called the Windows Manager List. The Windows Manager uses this list to populate the desktop with windows (Figure 1.11).

**Figure 1.11  The Window Manager maintains the desktop window by using the Windows Manager List.**

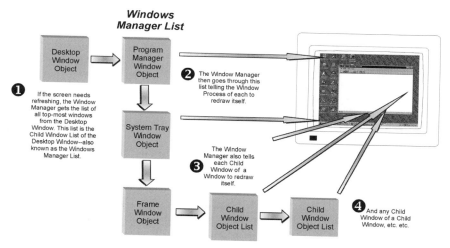

Some other interesting facts about the desktop follow.

- The desktop window is created from the #32769 Window Class (not the number "32769", but the text string #32769).

- To get a handle to the desktop window, you can use ::GetDesktopWindow().

- You can set the background image on the desktop with SystemParametersInfo() using SPI_SETDESKWALLPAPER.

- Shell_NotifyIcon() puts an icon in the Shell Tray — the collection of small icons you find in the task bar next to the time.

- You can search the windows on the desktop with ::FindWindow().

- You can find the current window at a point coordinate on the desktop with `WindowFromPoint()`.
- You can get the size of the screen with `::GetSystemMetrics (SM_CXSCREEN)` and `::GetSystemMetrics (SM_CYSCREEN)`.

# Summary

In this chapter, we have seen:

- an MFC application's method of using an MFC class and a Microsoft proprietary Window Class (not a C++ class) to create user interface windows;
- the steps to create and destroy windows and Window Classes using MFC;
- the Window Classes provided by the system;
- the three types of windows — Overlapped, Popup, and Child — of which the variety of windows you find in a Windows application interface are made;
- the desktop window and what kind of catastrophic things can be done there.

In the next chapter, we'll be looking at the classes provided by MFC.

# 2

# Classes

So far we have only looked at MFC's CWnd class. In this chapter, we will review the other important classes provided by MFC, which fall into the following categories:

- Classes that access the user interface, which includes CWnd.
- Classes that help you draw.
- Classes that provide the functionality needed to run an application.
- Classes that handle data arrays and lists.
- Classes that access databases.
- Classes that maintain files.
- Classes that allow your application to communicate over a network or the Internet.
- A few miscellaneous classes that help synchronize and debug your application.

This chapter is not intended as an MFC reference guide, but rather as an overview of what MFC has to offer. For any classes not reviewed in this chapter or for a more detailed description of a particular class, please refer to your MFC documentation.

Most OLE classes have also been omitted from this chapter because they are outside of the scope of this book.

# Base Classes

Most MFC classes are derived from three base classes: CObject, CCmdTarget, and CWnd. CCmdTarget is derived from CObject, and CWnd is derived from CCmdTarget. Classes derived from CObject have the ability at run time to get their object size and name. Classes derived from CCmdTarget can process command messages. Classes derived from CWnd can control their own window.

## CObject

The CObject class itself provides precious little functionality. Six other companion macros do the brunt of the work. Together, CObject and these macros allow classes derived from CObject to get their class name and object size at run time, create an object of their class without knowing the class name, and store and retrieve an instance of their class to an archive device without knowing its class name.

The following macros allow an instance of a class to know its class name and object size at run time.

```
DECLARE_DYNAMIC(CYourClass)     // in the .h file
IMPLEMENT_DYNAMIC(CYourClass, CYourBaseClass)
    // in the .cpp file
```

You can use CObject::GetRuntimeClass() to get the particulars of a class at run time that uses these macros.

These next macros include the functionality of the preceding macros, but also allow a class instance to be created without knowing its class name.

```
DECLARE_DYNCREATE(CYourClass)     // in the .h file
IMPLEMENT_DYNCREATE(CYourClass, CYourBaseClass)
    // in the .cpp file
```

You can use CRuntimeClass::CreateObject() to create an instance of a class that uses these macros without knowing its class name.

These next macros include all of the preceding functionality, but also allow a class instance to be saved to disk without knowing its class name.

```
DECLARE_SERIAL (CYourClass)     // in the .h file
IMPLEMENT_SERIAL (CYourClass, CYourBaseClass, schema)
    //in the .cpp file
```

See Example 70 for an example using these last two macros.

## CCmdTarget

Classes derived from the `CCmdTarget` can receive and process command messages from your application's menu or toolbar(s). The `CCmdTarget` class is discussed in much more detail in Chapter 3.

## CWnd

As discussed in Chapter 1, the member functions of `CWnd` encapsulate the Windows API responsible for creating and maintaining windows. The `CWnd` class is derived from `CCmdTarget` and, therefore, can also receive and process command messages.

---

**NOTE:** The following letters are used in this chapter to designate from which of the preceding base classes an MFC class is derived.

- O indicates the class is derived from `CObject`.
- O/C indicates a class is derived from `CObject` and `CCmdTarget`.
- O/C/W indicates a class is derived from `CObject`, `CCmdTarget`, and `CWnd`.

---

# Application, Frame, Document, and View Classes

When you create an MFC application using the Developer Studio's AppWizard, there are up to four base classes from which your application will be derived.

`CWinApp` is your application's Application Class and is responsible for initializing and running your application.

`CFrameWnd` is your application's Frame Class and is responsible for displaying and routing user commands.

`CDocument` is your application's Document Class and is responsible for loading and maintaining a document. A document can be anything from a manuscript to the settings of a network device.

`CView` is your application's View Class and is responsible for providing one or more views into the document.

> NOTE: We will be using the terms Application Class, Frame Class, etc., throughout this book to refer to your derivation of a class from these four base classes.

Which of these base classes the AppWizard includes in your application depends on what type of application you're creating.

**Dialog Applications** simply have a dialog box for their user interface and no frame, document, or view class. Dialog Applications use just a derivation of the Application Class, CWinApp. The dialog box is created using MFC's CDialog class, which is discussed in the following section.

**Single Document Interface (SDI) Applications** can load and edit one document at a time and use all four base classes mentioned previously.

**Multiple Document Interface (MDI) Applications** can load and edit several documents at once and use all four base classes plus two derivations of CFrameWnd called CMDIFrameWnd and CMDIChildWnd.

Your derivations of both the CDocument and CView classes are responsible for the Document/View described in Chapter 1. For an example using the AppWizard, see Example 2.

# CWinApp (O/C/W)

The Application Class is the first object created when your application launches and the last to execute before it terminates. At startup, the Application Class is responsible for creating the rest of the application.

- For a Dialog Application, the Application Class creates a dialog box using CDialog.
- For an SDI application, the Application Class creates one or more Document Templates (see below) and then opens an empty document using that template.
- For an MDI application, the Application Class creates one or more Document Templates and then opens an empty document using that template all within a Main Frame Class.

As we'll see in the next chapter, the Application Class also interfaces with the operating system to relay mouse messages and keyboard input to the rest of your application.

Your Application Class is derived from `CWinApp` and is named `CXxxApp` by the AppWizard, where `Xxx` is the name of your application.

## Document Templates

A Document Template defines what Frame, Document, and View class to create when your application opens a document. To create a Document Template, you create an instance of either the `CSingleDocTemplate` class for SDI applications or the `CMultiDocTemplate` class for MDI applications, and initialize it with three class pointers.

```
pDocTemplate = new CMultiDocTemplate(
    IDR_APPTYPE,
    RUNTIME_CLASS(CAppDoc),         // Your Document Class
    RUNTIME_CLASS(CChildFrame),     // Your Frame Class
    RUNTIME_CLASS(CAppView));       // Your View Class
```

The `RUNTIME_CLASS()` macro seen here returns a pointer to a class's `CRuntimeClass` structure, which is added to a class using the `DECLARE_DYNCREATE` and `IMPLEMENT_DYNCREATE` macros. A Document Template opens a document by creating an instance of all three of these classes using their `CRuntimeClass::CreateObject()` functions.

## Threads

The `CWinApp` class is itself derived from `CWinThread`. The `CWinThread` class wraps the Windows API that creates and maintains application threads in the system. You can, in fact, multitask within your own application by creating another instance of this `CWinThread` class. The `CWinApp` class represents the primary thread of execution in your application.

# CFrameWnd (O/C/W)

The Frame Class is the next object created when your application runs and is responsible for displaying and directing user commands to the rest of your application.

For an SDI application, your Frame Class is derived from `CFrameWnd`. The AppWizard automatically assigns it the name `CMainFrame`.

For an MDI application, your Frame Class is derived from `CMDIFrameWnd`. The AppWizard again assigns it the name `CMainFrame`. MDI applications also have a Child Frame Class for each document that is opened. Each Child Frame Class is derived from `CMDIChildWnd`. The AppWizard automatically assigns a Child Frame Class the name `CChildFrm`.

There is no Frame Class for a Dialog Application. As mentioned earlier, a Dialog Application is composed of an Application Class and a Dialog Class.

## CDocument (O/C)

The Document Class is usually the next object created by your application, either to open a new document or to open an existing document. The Document Class is responsible for loading a document into its member variables and allowing the View Class to edit those member variables. A document can consist of anything from a graphic file to the settings of a programmable controller.

Your Document Class is derived from `CDocument`. The AppWizard automatically assigns it the name `CXxxDoc`, where `Xxx` is your application's name.

## CView (O/C/W)

After an instance of the Document Class is created, an instance of the View Class is created. The View Class is responsible for depicting the contents of the Document Class. It might also allow your user to edit the document.

A window splitter class, `CSplitterWnd`, allows your document to have more than one view at a time. These views can be created from several instances of the same View Class or from different View Classes entirely.

The AppWizard allows you to derive your View Class from one of several base classes including `CTreeView`, `CEditView`, `CRichEditView`, `CListView`, etc. Each of these base classes imparts a different set of functionality in your application. See Example 1 for more examples of these classes. All of these classes are derived from `CView` class. The AppWizard automatically assigns the name `CXxxView`, where `Xxx` is your application's name, to your derivation of whatever base class you choose.

As mentioned previously, you can create three types of MFC applications from these four base classes: Dialog, SDI, and MDI. We will now review these application types in more detail.

# A Dialog Application

A Dialog Application is made up of an Application Class, which is derived from CWinApp, and a dialog box, which is created using a class derived from the CDialog class (Figure 2.1).

**Figure 2.1    A Dialog Application is created with an Application Class and a Dialog Class.**

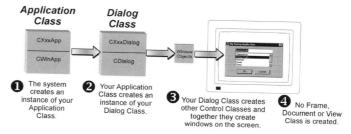

# An SDI Application

An SDI Application is made up of an Application Class derived from CWinApp, a Frame Class derived from CFrameWnd, a Document Class derived from CDocument, and one or more View Classes per document derived from one of several CView-derived View classes (Figure 2.2).

**Figure 2.2    A SDI Application is created with Application, Frame, Document, and View Classes.**

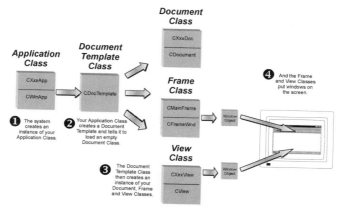

## An MDI Application

An MDI Application is made up of an Application Class derived from CWi-nApp, a Frame Class derived from CMDIFrameWnd, one or more Child Frame Classes derived from CMDIChildWnd, a Document Class per child frame derived from CDocument, and one or more View Classes per document derived from CView (Figure 2.3).

**Figure 2.3**   **An MDI Application is created with Application, Frame, Child Frame, Document, and View Classes.**

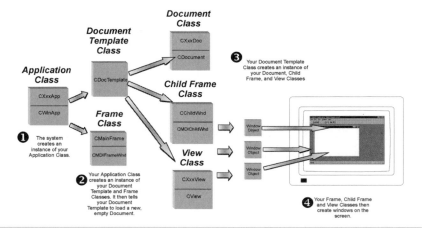

## Which Application Type Should You Choose?

Dialog Applications are generally used when there's a limited requirement for user interaction. Personally however, I would prefer to use an SDI application with a form view (with CFormView) rather than a Dialog Application. You get the simplicity of a Dialog Application but with the document features of an SDI application.

When should you go with an MDI approach over an SDI approach? If your application will only be involved with one document in its entire lifetime, then go with an SDI approach. However, if your application can generate and work on several documents — even if not all at once — then go with the MDI approach, even if you don't initially see the potential for modifying more than one document at a time. An MDI application isn't that much more complicated than an SDI application and your user gets the convenience of being able to at least view more than one document at a time.

# Other User Interface Classes

Along with Frame and View Classes, MFC provides several other classes that support your user interface.

**Control Window Classes** wrap the common controls.

**The Menu Class** does for the menu what the CWnd class does for a window.

**Dialog Classes** wrap the dialog box and the Common Dialogs.

**Control Bar Classes** wrap the control bars (toolbars, dialog bars, and status bar).

**Property Classes** wrap the Property Sheet and Page.

## Common Control Classes (O/C/W)

The Common Control Classes encapsulate the functionality of the common controls (e.g., buttons, list boxes, etc.). These classes are derived from CWnd to inherit window member functions like ShowWindow() and MoveWindow(). When these classes create a window, they use one of the common control Window Classes. For example, when you use the CButton Common Control Class to create a button, it uses the BUTTON Window Class to create the actual window.

```
Create(_T("BUTTON"), lpszCaption, dwStyle, rect,
    pParentWnd, nID);
```

The following table lists the Common Control Classes, the controls they create, and the Window Classes they use.

### Common Control MFC Classes and Their Windows Classes

| MFC Class | Common Control | Windows Class |
|---|---|---|
| CAnimateCtrl (O/C/W) | Animation Control | SysAnimate32 |
| CButton (O/C/W) | Button Control | BUTTON |
| CComboBox (O/C/W) | Combo Box Control | COMBOBOX |
| CEdit (O/C/W) | Edit Control | EDIT |
| CHeaderCtrl (O/C/W) | Header Control | SysHeader32 |

## Common Control MFC Classes and Their Windows Classes

| MFC Class | Common Control | Windows Class |
| --- | --- | --- |
| CListBox (O/C/W) | List Box Control | LISTBOX |
| CListCtrl (O/C/W) | List Control | SysListView32 |
| CProgressCtrl (O/C/W) | Progress Control | msctls_progress32 |
| CScrollBar (O/C/W) | Scroll Bar Control | SCROLLBAR |
| CSliderCtrl (O/C/W) | Slider Control | msctls_trackbar32 |
| CSpinButtonCtrl (O/C/W) | Up/Down Button Control | msctls_updown32 |
| CStatic (O/C/W) | Static Control | STATIC |
| CTreeCtrl (O/C/W) | Tree Control | SysTreeView32 |
| CTabCtrl (O/C/W) | Tab Control | SysTabControl32 |
| CDateTimeCtrl (O/C/W) | Date/Time Picker Control | SysDateTimePick32 |
| CMonthCalCtrl (O/C/W) | Month Calendar Control | SysMonthCal32 |
| CHotKeyCtrl (O/C/W) | Hot Key Control | msctls_hotkey32 |
| CToolTipCtrl (O/C/W) | Tool Tip Control | tooltips_class32 |

Not all MFC Common Control Classes simply wrap a common control Window Class. Three MFC classes actually provide functionality not found in the common controls. The following table shows these classes, the MFC class from which they're derived, and the additional support they provide.

## Derived MFC Classes and Their MFC Base Class

| MFC Class | MFC Class Derived From | Functionality Added |
| --- | --- | --- |
| CBitmapButton | CButton | Better support for bitmaps on buttons. |
| CCheckListBox | CListBox | Check boxes in a list box. |
| CDragListBox | CListBox | User-draggable items in a list box. |

## The Menu Class (0)

The CMenu class wraps the Windows API that creates and maintains menus. CMenu also has two member functions, Attach() and Detach(), that allow you to wrap an existing menu object the same way a CWnd object can wrap an existing window.

## Dialog Classes

The CDialog class wraps the Windows API that creates dialog boxes. Dialog boxes are Popup windows that, when created, can populate themselves with the Control windows defined in a dialog template.

### Common Dialog MFC Classes

The MFC library also has six classes that wrap the Windows APIs that create the Common Dialogs. Common Dialogs are dialog boxes that come pre-populated with controls to prompt your user for some commonly requested information, such as filenames for loading and saving, colors, fonts, and printing parameters. Common Dialogs save you the job of writing these dialogs yourself while also presenting your user with a familiar dialog box.

The following table shows the purpose of the Common Dialogs, the Windows APIs that provides them, and the MFC Common Dialog Classes that wrap the Windows APIs.

#### Common Dialog MFC Classes and Their Use

| Common Dialog Box | Window API Call | MFC Class |
| --- | --- | --- |
| Select Color | ::ChooseColor() | CColorDialog |
| Open/Save File | ::GetOpenFileName()<br>::GetSaveFileName() | CFileDialog |
| Find or Replace Text | ::FindText()<br>::ReplaceText() | CFindReplaceDialog |
| Select a Type Font | ::ChooseFont() | CFontDialog |
| Setup Print Page | ::PageSetupDlg() | CPageSetupDialog |
| Print | ::PrintDlg() | CPrintDialog |

# Control Bar Classes (O/C/W)

The Control Bar Classes wrap the Windows API that provides your application with toolbars, status bars, dialog bars, and rebars.

- The CToolBar (O/C/W) and CToolBarCtrl (O/C/W) classes help you to create and maintain toolbars. The CToolBarCtrl class encapsulates the functionality of the system-provided Window Class ToolbarWindow32 that does the actual creation and maintenance of a toolbar. The CToolBar class, on the other hand, is mostly MFC. CToolBar provides more functionality than CToolBarCtrl, but at the price of more memory required to support it.

- The CStatusBar (O/C/W) and CStatusBarCtrl (O/C/W) classes create and maintain status bars. The CStatusBarCtrl class encapsulates the system-provided Window Class msctls_statusbar32 that does the actual creation and maintenance of a status bar. The CStatusBar class, on the other hand, is mostly MFC. CStatusBar provides more functionality than CStatusBarCtrl, but at the price of more memory required to support it.

- The CDialogBar (O/C/W) class creates and maintains dialog bars. A *dialog bar* is a hybrid of a dialog box and a toolbar. It displays a dialog box template, but it floats and docks to your Main Frame window like a toolbar.

- The CRebar (O/C/W) and CRebarCtrl (O/C/W) classes create and maintain Rebars. *Rebars* are window controls that can contain a toolbar, a dialog bar, or any other child window. They provide a *grabber bar* (a double line at the left side) that allows them to be more easily moved. They also more easily allow a bitmap to be drawn in their backgrounds. The CReBarCtrl class encapsulates the functionality of the system-provided Window Class ReBarWindow32 that does the actual creation and maintenance of a Rebar. The CReBar class, on the other hand, is mostly MFC. CReBar provides more functionality then CReBarCtrl, but at the price of more memory required to support it.

# Property Classes

The Property Classes wrap the Windows API that can provide your application with Property Pages and Property Sheets. One or more Property Pages appear in a Property Sheet to create the tabbed view familiar to Windows users. This view is typically used to select program options.

- The CPropertySheet (O/C/W) class creates a Property Sheet. Although CPropertySheet is not derived from CDialog, it's quite similar.
- The CPropertyPage (O/C/W/CDialog) class creates a Property Page. CPropertyPage is derived from CDialog.

For an example of a Property Sheets and Pages, see Example 8.

# Drawing Classes

## Device Contexts Classes

The CDC (O) class wraps the Windows API that allows you to draw. This class also maintains a *device context*, which is an object in memory that contains all of the characteristics of a device on which you intend to draw. Those characteristics include the maximum dimensions of the device and which way is up. Devices can be screens or printers. For a much more detailed discussion of device contexts and the member functions of the CDC class, please see Chapter 4.

There are four MFC classes derived from the CDC class that provide additional functionality.

CClientDC is typically used to conveniently create and destroy a device context for you. CClientDC is usually created on your stack. Its constructor creates a device context for the client area of your window by calling CDC::GetDC(). Then when your routine returns, CClientDC's destructor destroys that context by calling CDC::ReleaseDC(). No muss, no fuss, no forgotten device contexts to release and cause resource leaks.

CWindowDC does for the nonclient area of your window what the CClientDC class does for your client area.

CPaintDC calls CWnd::BeginPaint() when it's constructed to get the device context. The device context in this case only allows you to draw to the area of your window's client area that has been invalidated — as opposed to drawing to the entire client area. The CPaintDC class also calls CWnd::EndPaint() when it's deconstructed.

CMetaFileDC is used to create a Microsoft metafile. A metafile is a disk file that contains all of the drawing actions and modes necessary to draw a figure. You can create a metafile by opening a metafile device context and then using your drawing tools to draw to it as if it were a screen or printer device. The generated file can then be reread to create the figure at a future point to one of the other devices. Please see Chapter 4 for more on metafiles.

## Graphic Object Classes

A device context isn't sufficient to contain all of the drawing characteristics a drawing function needs. In addition to the device context, Windows has several other graphic objects it uses to store drawing characteristics. Those additional characteristics include everything from the width and color of drawn lines to what font to use when drawing text. The Graphic Object Classes wrap all six of these graphic objects.

The following table lists the MFC class, the graphic object it wraps, and the drawing characteristics stored in that object.

## Graphic Object Classes and the Handles They Encapsulate

| MFC Class | Graphic Object Handle | Graphic Object Purpose |
| --- | --- | --- |
| CBitmap | HBITMAP | Bitmap in memory. |
| CBrush | HBRUSH | Brush characteristics — when filling a shape, the color and pattern to be used. |
| CFont | HFONT | Font characteristics — when writing text, the font to be used. |
| CPalette | HPALETTE | Palette colors. |
| CPen | HPEN | Pen characteristics — when drawing an outline, the line thickness to be used. |
| CRgn | HRGN | A region's characteristics — including the points that define it. |

# File Classes

The CFile() class wraps the Windows API for creating and maintaining a flat file. Three MFC classes have been derived from CFile to provide additional functionality:

CMemFile allows you to create a file in memory instead of on disk. When you construct a CMemClass object, the file is immediately opened and you can use its member functions to read and write to it as if it were a disk file. For an example of a memory file, see Example 65.

CSharedFile is similar to the CMemFile class except that it's allocated on the global heap, which makes it available for you to share it using the clipboard and DDE. For an example of a shared memory file, see Example 76.

CStdioFile allows you to read and write text strings terminated with carriage control and line feed characters. For an example of using CStdioFile, see Example 64.

Another file class of interest is CFileFind (O) which locates local files on disk. The CFileFind class is the base class for two other MFC classes, CFtpFileFind and CGopherFileFind, which can help locate files on the Internet. These classes will be discussed later in this chapter.

## CArchive **and Serialization**

The CArchive class uses the CFile class to save the class objects of your document to disk in a process called serialization. With *serialization*, the member variables in your classes and whole class objects can be stored to an archive device in sequence so that they can be restored in the exact same sequence later. See Chapter 13 for several examples of serialization.

# Database Classes

The MFC library has classes to supports two types of database.

**Open Database Connectivity** (ODBC) classes wrap the ODBC API that most database vendors support. If your application uses MFC's ODBC classes, it can support any Database Management System (DBMS) that supports the ODBC standard.

**Data Access Objects** (DAO) classes support a newer database API that has been optimized for use with the Microsoft Jet database engine. You can still access ODBC compliant database systems and other data sources through the Jet engine.

## ODBC Classes

There are three main ODBC classes.

CDatabase (O) opens a DBMS database using the ODBC API. After you've constructed a CDatabase object you can use its OpenEx() member function to establish a connection to a database. Calling the Close() member function of CDatabase closes the connection.

CRecordset is used to store and retrieve records through a database connection.

CDBVariant represents a column in a record set without concern for the data type.

See Example 72 for more on accessing an ODBC database.

## DAO Classes

The DAO classes include three classes similar to the ODBC classes.

CDaoDatabase (0) opens a DAO database.

CDaoRecordSet (0) holds records.

COleVariant represents a record column.

The DAO Classes also include three more classes.

CDaoWorkSpace (0) manages a database session allowing transactions to be Committed (stored to the database) or Rollbacked (undone).

CDaoQueryDef (0) represents a query definition.

CDaoTableDef (0) represents a table definition including the field and index structure.

See Example 73 for more on accessing a DAO database.

# Data Collection Classes

The Data Collection Classes maintain and support arrays, lists, and maps of data objects.

CArray and its derivatives support an array of data objects. An array is made up of one or more identical data objects (e.g., integers, classes, etc.) that are contiguous in memory and can therefore be accessed with a simple index. The CArray class can grow or shrink its size dynamically. There are several derivations of CArray (e.g., CByteArray, CDWordArray, etc.) that allow you to create a type-safe array. However, there's also a CArray<type,arg_type> template class that allows you to make any array type-safe.

CList and its derivations support a linked-list of data objects. A linked-list is made up of one or more identical data objects (e.g., integers, classes, etc.) that aren't continuous in memory and are doubly-linked so that you can go forwards and backwards through the list. There are several derivations of CList (e.g., CPtrList, CObList, etc.) that allow you to create a type-safe list. However, there's also a CList<type,arg_type> template class that allows you to make CList type-safe for any type.

CMap and its derivations support a dictionary of data objects. A data dictionary stores one or more identical data objects (e.g., integers, classes, etc.) under a binary or text key. You can use this key to retrieve a data item. For example, since Windows doesn't keep track of which MFC CWnd objects belong to which window, your application uses a CMap object to associate window handles with their companion CWnd objects. There are several derivations of CMap (e.g., CMapWordToPtr, CObToString, etc.) that allow you to create a type-safe dictionary. However, there's also a CMap<class KEY,class ARG_KEY,class VALUE,class ARG_VALUE> template class that allows you to make any map type-safe.

For an example of each of these classes, see Examples 78, 79, and 80.

# Other Data Classes

The MFC library also includes several miscellaneous data classes.

- The CTime and CTimeSpan classes have member functions that both read in and format time strings for the time_t data type (Example 81).

- The `COleDateTime` and `COleDateTimeSpan` classes have member functions that read in and format time and date strings for the OLE `DATE` data type (Example 81).

> ## CTime **or** COleDateTime
>
> Which you should use, `CTime` or `COleDateTime`, depends on how fancy you want to get. The `time_t` data type that `CTime` supports is easy to store and pass around because it's just a 32-bit integer. However, `CTime` doesn't have some of the nice member functions of `COleDateTime`, like getting the day of a year. And of course, the `time_t` data type will have its own Y2K meltdown in the next century because it represents the number of seconds since 1/1/70.

- The `COleCurrency` class has member functions that read in and format currency numbers for the OLE `CURRENCY` data type.
- The `CString` class has member functions that manipulate text strings. The data type that `CString` manipulates is `TCHAR`.
- The `CPoint` class provides member functions to manipulate a `POINT` structure, which defines the X and Y coordinates of a point.
- The `CSize` class provides member functions that manipulate a `SIZE` structure, which defines the width and height of a size.
- The `CRect` class provides member functions that manipulate a `RECT` structure, which defines the four points of a rectangle.

# Communication Classes

The MFC library contains classes that allow your application to communicate over a network or over the Internet.

## Network Classes

`CASyncSocket` (0) allows your application to communicate with other applications over a network. `CASyncSocket` wraps the Window Socket API which is based on the UNIX® sockets implementation in the Berkeley Software Distribution (BSD, release 4.3) from the University of California at Berkeley. The `CASyncSocket` class gives you access to the bit and byte minutia of network communication.

CSocket is derived from the CAsyncSocket class and allows you to send and receive files over a network without worrying about the minutia. You can create a CSocketFile object and associate it with a CSocket, then create a CArchive object and associate it with the CSocketFile object. Sending a file is now just a matter of serializing it out over the network.

# Internet Classes

MFC has two sets of Internet classes: one that allows your application to become a client (e.g., a web browser) and another to add functionality to your Internet HTTP server software (e.g., a web site).

## Client Classes

MFC's Internet client classes encapsulates the Win32 Internet Extensions (WinInet) allowing you C++ access to the Internet using one of four protocols: File Transfer Protocol (FTP), Hypertext Transfer Protocol (HTTP), gopher, and file.

The CInternetSession (0) class has two ways to interact with files on a web site. For simple reading or downloading of files, you can use the OpenURL() member function passing it the Universal Resource Locator (URL) of the web site file. OpenURL() returns a pointer to another MFC Internet class depending on the URL you used.

| URL Prefix | Internet Class Pointer Returned |
|---|---|
| file:// | CStdioFile* |
| http:// | CHttpFile* |
| gopher:// | CGopherFile* |
| ftp:// | CInternetFile* |

Each of these Internet classes is derived from CStdioFile. You can then read from, but not write to, this file object.

To write to a file or perform some other protocol specific operation, you must use one of the other member functions of CInternetSession to open a connection first. The connection is returned within yet another MFC Internet class.

| Protocol | CInternetSession Member Function Called | Internet Class Pointer Returned |
|---|---|---|
| FTP | GetFtpConnection() | CFtpConnection* |
| HTTP | GetHttpConnection() | CHttpConnection* |
| Gopher | GetGopherConnection() | CGopherConnection* |

You can then interact with a web site using the member functions of CFtpConnection, CHttpConnection, or CGopherConnection.

Three other MFC Internet client classes of interest are CFtpFileFind, CGopherFileFind, and CGopherLocator. CFtpFileFind and CGopherFileFind are derived from the CFindFile class mentioned earlier and can locate files over the Internet. The CGopherLocator class gets a gopher "locator" from a gopher server and makes the locator available to CGopherFileFind.

## Internet Server Extension Classes

MFC has several classes that allow you to add functionality to your Internet server. For these classes to work, however, you must have an ISAPI-compliant HTTP server, which Microsoft does not supply with MFC. One such compliant server is Microsoft's Internet Information Server.

- The CHttpServer class can be created by your server to handle a specific request from a client browser.

- The CHttpFilter class can be used to look for and act on certain communications that you can specify through your server.

# Other Classes

## Debug Classes

The CMemoryState class can help you pinpoint a memory leak. You create a CMemoryState object and call its Checkpoint() member function to take a snapshot of your memory. At a later point, you take another snapshot using another CMemoryState object and compare your results. The DEBUG_NEW macro makes use of this class to automatically locate memory leaks in the debug version of your application.

The CException (0) classes are MFC's implementation of C++ exception handling. There are eleven classes derived from CException handling the exceptions listed in the following table.

| MFC Class | Exception |
|---|---|
| CArchiveException | Errors using the CArchive class. |
| CDaoException | Errors using the CDaoDatabase classes. |
| CDBException | Errors using the CDatabase classes. |

| MFC Class | Exception |
|---|---|
| CFileException | Errors using the CFile class. |
| CInternetException | Any Internet class errors. |
| CMemoryException | Any memory errors. |
| CNotSupportedException | Attempting to use an unsupported function. |
| CResourceException | Attempting to locate or allocate a resource. |
| CUserException | Usually used to indicate that the user did something exceptional. |

To handle exceptions, you would use the following template.

```
TRY
{
    // attempt something risky
}
CATCH(CUserException, e)
{
    //handle exception
}
AND_CATCH(CMemoryException, e)
{
    //handle another exception
}
CATCH_ALL(e)
{
    //handle all exceptions
}
END_CATCH
```

# Synchronization Classes

The MFC Synchronization Classes are used to prevent data objects from being corrupted in a multithreaded application. As mentioned previously, an MFC application can have several threads running at once. If more than one of these threads is modifying the same data object at the same time, they might corrupt that data by storing to the same memory location at the same time.

Four MFC classes and two MFC helper classes help to synchronize data access in a multithreaded application:

CMutex is used to prevent more than one thread from accessing a data object at the same time. To use CMutex, you add it to the member variables of your data class. Then you construct another MFC class, CSingleLock, referencing the CMutex class to any member functions that access these member variables. You then call the Lock(int timeout) member function of CSingleLock. If another thread is already accessing this data, the Lock() function will not return until the other thread calls Unlock() or you time out. The CMultiLock class allows you to specify several CMutex objects so that you can wait on several accesses at once (Figure 2.4).

**Figure 2.4    You can use CMutex and CSingleLock to synchronize data access.**

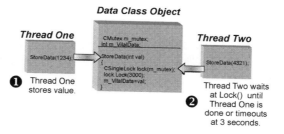

CSemaphore is identical to the CMutex class except that it can allow one or more simultaneous accesses at one time. CSingleLock and CMultiLock are used as before.

CCriticalSection is identical to the CMutex class except it's constructed right on the stack and is faster than CMutex.

CEvent allows you to synchronize your data access based on any event — not just access by another thread. Threads will wait until you allow them to continue by calling the SetEvent() and ResetEvent() member functions of CEvent.

## Wait Cursor Class

The CWaitCursor class can be used to conveniently change the mouse cursor to an hourglass to indicate the system is busy. CWaitCursor is usually constructed on the stack just before a time-consuming task during which input is prohibited. Then when the function returns, CWaitCursor automatically deconstructs and the wait cursor goes away (Example 34).

# Summary

In this chapter, you have seen some of what the MFC class libraries have to offer your application, including:

- Classes from which your application is built.
- Classes that create your user interface.
- Classes that help your application to draw.
- Classes that handle data collections, databases, and flat files.
- Classes that allow communication over networks and the Internet.

We also touched briefly on how classes communicated with each other using messages and the CCmdTarget class. In the next chapter, we will explore this subject in much more depth.

# Messaging

In Chapter 1, we looked at the basic elements of an MFC user interface: windows, Window Classes, and CWnd. In Chapter 2, we looked at the other classes that make up the MFC library and specifically what classes form the core of an MFC application. In this chapter, we'll discover how MFC classes and their windows communicate with one another. We will find out that there are three types of messages (Windows, Command, and Control Notification) and that messages are either sent or posted. We'll follow a message as it's being processed by the MFC Window Process. And finally, we'll look at ways to redirect messages for some powerful results.

## Sending or Posting a Message

As mentioned in Chapter 1, every window has a Window Process that handles messages delivered to it. Messages can come from the system, your own application, or another application. Messages can tell your Window Process to perform a task (e.g., to initialize itself, draw or destroy a window, etc.) or notify it of an event (e.g., that a mouse just clicked on a window).

There are two ways for the system or your application to deliver a message: by sending it or by posting it.

## Sending a Message

When you send a message, you are calling the Window Process of a window directly. Communication is immediate. Your application doesn't continue until after the Window Process returns a result to your calling function.

## Posting a Message

When you post a message, you are putting your message in the message queue of the application that owns that window. The application then looks in its message queue whenever it's idle and processes the messages there by removing them and sending them on to their intended window. Communication is delayed until the target application gets the time to process your message. Your calling function returns immediately after it posts the message, but the result simply indicates whether or not the message was posted successfully — not the result of the called Window Process (Figure 3.1).

**Figure 3.1**   **Communication is immediate when sending a message, but may be delayed when posting a message.**

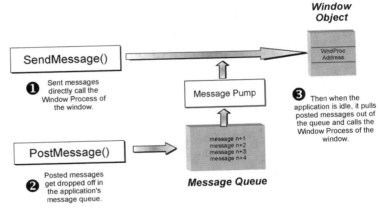

## Sending Versus Posting a Message

Mouse and keyboard messages are usually posted. All other messages are usually sent. Posted messages receive special mouse and keyboard processing in the message queue. In general, you should always send a message unless you want your action delayed until all mouse and keyboard messages have been processed.

I

3

# How to Send a Message Using MFC

To send a message with MFC, you should first get a pointer to the CWnd class object to which you want to send the message. Then you should call the SendMessage() member function of CWnd.

```
LRESULT Res= pWnd->SendMessage(UINT Msg, WPARAM wParam,
    LPARAM lParam);
```

The pWnd pointer points to the targeted CWnd class object. The Msg argument is the message and the wParam and lParam arguments contain the message's parameters, such as where a mouse clicked or what menu item was selected. The target window returns the result of your message in the Res argument.

To send a message to a window that doesn't have a CWnd class object, you can call the Windows API directly using a handle to the target window

```
LRESULT Res =::SendMessage(HWND hWnd, UINT Msg, WPARAM wParam,
    LPARAM lParam);
```

where hWnd is the handle of the target window.

# How to Post a Message Using MFC

Posting a message with MFC is almost identical to sending a message except you use PostMessage() instead of SendMessage(). The value returned by Res is also different. Rather than containing a value returned by the target window, Res is simply a Boolean value that indicates whether or not the message was successfully placed in the message queue.

## Retrieving a Posted Message

Normally once a message has been posted, you let your application send it on in the background. However, in special circumstances, such as when you want to stop your application until it receives a certain message, you may want to remove a message yourself. There are two ways to remove a message from your application's message queue, but neither involves MFC.

- The first method allows you to peek in the queue to see if a message is there without disturbing anything.

```
BOOL res=::PeekMessage(LPMSG lpMsg, HWND hWnd, UINT wMs
    FilterMin, UINT wMsgFilterMax, UINT wRemoveMsg);
```

- The second method actually waits until a new message arrives in the queue and then removes the message and returns with it.

```
BOOL res=::GetMessage(LPMSG lpMsg, HWND hWnd,
    UINT wMsgFilterMin, UINT wMsgFilterMax);
```

In both cases, the hWnd argument specifies the window for which you want to intercept messages, although if you use NULL, messages for all windows will be intercepted. The wMsgFilterMin and wMsgFilterMax arguments correspond to the Msg argument in SendMessage() and indicate the range of messages for which you're looking. If you use 0,0, all messages will be intercepted. If you use WM_KEYFIRST, WM_KEYLAST or WM_MOUSEFIRST, WM_MOUSELAST, all keyboard or mouse messages will be intercepted.

The wRemoveMsg argument specifies whether PeekMessage() should actually remove the message from the queue. (GetMessage() always removes the message.) This argument can be either:

- PM_REMOVE, which causes PeekMessage() to remove the message, or
- PM_NOREMOVE, which causes PeekMessage() to leave the message in the queue and return with a copy of it. Of course, if you leave a message in the message queue and then call PeekMessage() again looking for the same type of message, you'll return with the exact same message.

The lpMsg argument is a pointer to a MSG structure that contains the retrieved message.

```
typedef struct tagMSG {
    HWND    hwnd;      // window handle message is intended for
    UINT    message;
    WPARAM  wParam;
    LPARAM  lParam;
    DWORD   time;      // the time the message was put in the queue
    POINT   pt;        // the location of the mouse cursor when the
                       // message was put in the queue
} MSG;
```

# The Three Types of Messages

There are three types of messages floating around an MFC application: Window Messages, Command Messages, and Control Notifications.

# Window Messages

Window Messages are generally concerned with the inner workings of a window, such as creating it, drawing it, and destroying it. They're usually sent from the system to a window or from window to window.

When sending a Window Message using `SendMessage()` or `PostMessage()`, the `Message`, `wParam`, and `lParam` arguments are formatted as follows.

| Message | wParam | lParam |
|---------|--------|--------|
| WM_XXX | command defined | command defined |

`WM_XXX` can be one of dozens of Window Messages, such as:

- `WM_CREATE`, which tells a window to initialize itself,
- `WM_PAINT`, which tells a window to draw itself, or
- `WM_MOUSEMOVE`, which tells a window that the mouse just moved across it.

For some common Window Messages, see Appendix B. For a complete list of Window Messages, refer to your MFC documentation.

# Command Messages

Command Messages are generally concerned with processing a user request. They're generated when the user clicks on a menu item or a toolbar and are sent to the class object that can handle the request (e.g., load files, edit text, save options, etc.).

When sending a Window Message using `SendMessage()` or `PostMessage()`, the `Message`, `wParam`, and `lParam` arguments are formatted as follows.

| Message | wParam | | lParam |
|---------|--------|------------|--------|
| WM_COMMAND | 0 | Command ID | 0 |

The Command ID is either the ID of the menu item selected or the toolbar button that was clicked. Notice that the Command ID can be no larger than one word. Any attempt to make it larger is ignored — the system simply stuffs a zero (0) in the high order word. Also note that `lParam` must be zero (0). Some Control Notifications also use the `WM_COMMAND` message and the only way to differentiate Command Messages from Control Notifications is if `lParam` is NULL.

## Control Notifications

Control Notifications are generally sent by Control windows to their parents when something significant happens, such as when a combo box is about to be opened. Control Notifications give Parent windows the opportunity to further control their windows when necessary. For example, when a combo box opens, the Parent can stuff it with information that wasn't available when the combo box was first created.

Control Notifications have gone through an evolution. Consequently, the Message, wParam, and lParam arguments used with SendMessage() have three formats.

### First Control Notification Format

The first Control Notification format is just a subset of Windows Messages.

| Message | wParam | lParam |
|---------|--------|--------|
| WM_XXX | command defined | command defined |

WM_XXX can be any of the following messages.

- WM_PARENTNOTIFY signifies that a control window has either been created or destroyed or that the mouse has clicked on the window.
- WM_CTLCOLOR, WM_DRAWITEM, WM_MEASUREITEM, WM_DELETEITEM, WM_CHARTOITEM, WM_VKEYTOITEM, or WM_COMPAREITEM are all messages sent to a Parent that draws its own Control window.
- WM_HSCROLL or WM_VSCROLL are sent by a scroll bar control to tell its Parent to scroll the window.

### Second Control Notification Format

The second Control Notification format uses the WM_COMMAND message, which it shares with Command Messages.

| Message | wParam | | lParam |
|---------|--------|---|--------|
| WM_COMMAND | XN_XXX | Control ID | Window Handle |

The lParam argument is used to differentiate between Command Messages and Control Notifications. A Control Notification has a valid window handle in lParam, which identifies the control making the notification. A Command Message has a NULL value in lParam.

The XN_XXX value differs depending on what control is making the notification. For example, an XN_XXX value of EN_CHANGE tells a Parent window that the displayed text in an Edit Box Control has just changed. Several other examples are listed in Appendix B.

## Third Control Notification Format

The third and most flexible Control Notification format uses the WM_NOTIFY message.

| Message | wParam | lParam |
|---------|--------|--------|
| WM_NOTIFY | Control ID | Pointer to NMHDR |

The lParam value points to a structure that can contain everything you'd ever want to know about the control making the notification without being confined by space or type. This structure is called NMHDR.

```
typedef struct tagNMHDR {
    HWND hwndFrom;       // Window handle of Control Window
                         // making the notification.
    UINT idFrom;         // Control ID of Control Window
                         // making the notification.
    UINT code;           // notification code ex: the user
                         // has clicked the Control Window
} NMHDR;
```

NMHDR stands for Notification Message Header. Why Header? Because some controls send an even larger structure with NMHDR as the header. Even functions that don't know about the larger structure are still be able to process the notification header.

# How MFC Receives a Posted Message

The only significant difference between MFC processing of a posted message and a sent message is that a posted message spends some time in the message queue of your application. There it sits until the message pump pulls it back out.

## The Message Pump

The message pump in an MFC application is in the Run() member function of CWinApp. Run() is called when your application first starts. Run() then divides its time between performing background processing, such as getting rid of temporary CWnd objects, and checking the message queue. When a new message comes in, Run() pumps it — that is, uses GetMessage() to grab the message out of the queue, runs two translation functions on the message, and then calls the Window Process for which the message was intended by using the DispatchMessage() function (Figure 3.2).

**Figure 3.2     The message pump performs background processing and checks the message queue.**

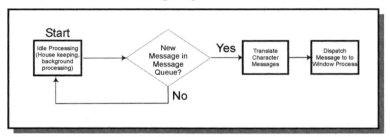

The two translation functions the message pump calls are PreTranslate-Message() and ::TranslateMessage.

PreTranslateMessage() gives the MFC class of the target window an opportunity to translate the message before the message is sent to it. For example, CFrameWnd uses PreTranslateMessage() to convert accelerator keys (e.g., Ctrl+s to save your file) into Command Messages. Messages that cause a translation are usually thrown away and their translation, if any, is reposted to the queue.

::TranslateMessage is a Windows function that translates raw key codes into key characters.

Once a message is dispatched with DispatchMessage(), MFC processes it as if it were a message sent with SendMessage().

# How MFC Processes a Received Message

The goal of processing a received message is simple: direct the message to a function that can handle it using the message identifier in the message.

NonMFC windows accomplish this goal with a simple case statement where each case performs some function or calls some other function.

```
MainWndProc(HWND hWnd, UINT message, W PARAM wParam,
            LPARAM lParam)
{
    switch(message)
    {
    case WM_CREATE:
        :   :   :
            break;
    case WM_PAINT:
        :   :   :
            break;
    default:
        return(DefWindowProc(hWnd, message, wParam,
               lParam));
    }
    return(NULL);
}
```

Any left over messages are passed on to a default message handler. Unfortunately, a case statement doesn't lend itself very well to C++ and encapsulation. In a C++ world, you want messages to be handled by the member functions of a class that specialize in that type of message.

So MFC replaces the case statement with something much more complicated and convoluted, but which allows you to handle messages in the privacy of your own class as long as you do three things:

1. derive your class from the CWnd class object that will receive the message (or CCmdTarget for Command Messages);
2. write a member function in that class that handles the message; and
3. define a lookup table (called a Message Map) in your class with an entry for your member function along with the identifier of the message it handles.

MFC then directs an incoming message to your handler by calling the following functions in the order shown.

1. `AfxWndProc()` receives the message, finds the `CWnd` object to which the message belongs, and then calls `AfxCallWndProc()`.

2. `AfxCallWndProc()` saves the message (message identifier and parameters) for future reference and then calls `WindowProc()`.

3. `WindowProc()` sends messages on to `OnWndMsg()` and then, if unprocessed, `DefWindowProc()`.

4. `OnWndMsg()` calls either `OnCommand()` for `WM_COMMAND` messages or `OnNotify()` for `WM_NOTIFY` messages. Everything left over is a Windows message. `OnWndMsg()` searches your class's Message Map for a handler that will process any Windows message. If `OnWndMsg()` can't find a handler, it returns the message to `WindowProc()`, which sends the message on to `DefWindowProc()`.

5. `OnCommand()` checks to see if this is a Control Notification (`lParam` is not NULL); if it is, `OnCommand()` tries to reflect the message to the control making the notification. If it isn't a Control Notification — or if the control rejected the reflected message — `OnCommand()` calls `OnCmdMsg()`.

6. `OnNotify()` also tries to reflect the message to the control making the notification; if unsuccessful, `OnNotify()` calls the same `OnCmdMsg()` function.

7. `OnCmdMsg()`, depending on what class received the message, will potentially route Command Messages and Control Notifications in a process known as Command Routing. For example, if the class that owns the window is a Frame Class, Command and Notification Messages are also routed to the View and Document Classes looking for a Message Handler for the class.

## Why Message Maps?

Why Message Maps? This is, after all, C++. Why doesn't `OnWndMsg()` simply call a predefined virtual function for each Windows Message? Because it would be too CPU-intensive. As it is, `OnWndMsg()` does the unthinkable and drops into assembler when scanning a Message Map to speed up the process.

---

**NOTE:** You can override `WindowProc()`, `OnWndMsg()`, `OnCommand()`, `OnNotify()`, or `OnCmdMsg()` to modify this process. Overriding `OnWndMsg()` allows you to cut into the process before the Window Messages have been sorted. Overriding `OnCommand()` or `OnNotify()` allows you to cut into the process before a message can be reflected.

---

Message Reflection and Command Routing are discussed in much more detail in the following section.

I

3

# The Functions that Process a Message

We will now follow step by step as a Window Process receives and deciphers a message.

### AfxWndProc()

The Window Process of all MFC windows is

```
LRESULT AfxWndProc(HWND hWnd, UINT nMsg, WPARAM wParam,
   LPARAM lParam)
```

If a message is sent, `SendMessage()` essentially calls `AfxWndProc()` directly. If a message is posted, the message pump calls `AfxWndProc()` through `DispatchMessage()`.

The first thing `AfxWndProc()` does is find the `CWnd` object of the target window. Although a window doesn't know anything about its `CWnd` object, your application keeps track of the pairings in a map.

Once the `CWnd` object is found, `AfxCallWndProc()` is called.

### AfxCallWndProc()

The prototype for `AfxCallWndProc()` is

```
LRESULT AfxCallWndProc(CWnd *pWnd, HWND hWnd, UINT nMsg,
   WPARAM wParam, LPARAM lParam)
```

`AfxCallWndProc()` saves the message for future reference. In fact, even if you change the parameters in a message later on, this Window Process will refer to the parameters it saves here when it does default processing with the `Default()` member function. Once saved, `WindowProc()` is called.

### WindowProc()

The prototype for `WindowProc()` is

```
LRESULT WindowProc(UINT nMsg, WPARAM wParam, LPARAM lParam);
```

`WindowProc()` next calls `OnWndMsg()`, which attempts to find a handler for this message in your class. Any unhandled messages returned to `Window-Proc()` are passed along to `DefWindowProc()`. `DefWindowProc()` is the repository of all unwanted messages. If you didn't have a handler for a message, that doesn't mean it isn't an important message. For example, you will rarely want to process the WM_NCPAINT message, but it is important because DefWindowProc() uses this message to draw the nonclient area of your window.

---

**NOTE:** The MFC CControlBar class, which is the base class for toolbars, dialog bars, and status bars, overrides WindowProc(). If any of the following messages go unprocessed in a Control Bar, they are sent back to its Parent (which is usually the Main Frame Class).

| | | | |
|---|---|---|---|
| WM_NOTIFY | WM_COMMAND | WM_DRAWITEM | WM_MEASUREITEM |
| WM_DELETEITEM | WM_COMPAREITEM | WM_VKEYTOITEM | WM_CHARTOITEM |

You can automatically process the buttons in a toolbar, status bar, or dialog bar in its Parent window, too. However, you should try to encapsulate as much of this functionality in a toolbar, status bar, or dialog bar class as you can. The CMainFrame class can get pretty massive as it is.

---

### OnWndMsg()

The prototype for OnWndMsg() is

```
BOOL OnWndMsg(UINT message, WPARAM wParam, LPARAM lParam,
    LRESULT* pResult);
```

OnWndMsg() calls either OnCommand() if the message is WM_COMMAND or OnNotify() if the message is WM_NOTIFY. All other messages are assumed to be Window Messages and OnWndMsg() proceeds to search your class's Message Map for a handler.

## Message Maps

OnWndMsg() locates a message handler for a Window Message by searching the Message Maps in its class and derivations. Message Maps are static tables of data that contain an entry for every message handler in a class. Every derivation of CWnd can have a Message Map (Figure 3.3).

**Figure 3.3** `OnWndMsg()` **searches Message Maps to locate message handlers for Window Messages.**

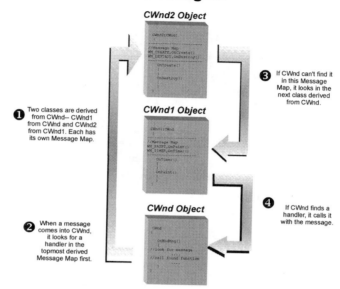

Each Message Map is bracketed by two macros.

```
BEGIN_MESSAGE_MAP(CXxx,CYyy)
    :    :    :
    Message Map entries
    :    :    :
END_MESSAGE_MAP
```

CXxx is this class derivation's name and CYyy is the name of next class derivation you want OnWndMsg() to search, usually the class from which CXxx is derived. The BEGIN_MESSAGE_MAP macro in the next class identifies the class from which it was derived, and so on, until the Message Maps in all class derivations have been searched. Other than marking the location of the Message Maps for OnWndMsg(), you can ignore how these macros work.

Each Message Map entry uses the following structure.

```
struct AFX_MSGMAP_ENTRY
{
    UINT nMessage;
    UINT nCode;
    UINT nID;
    UINT nLastID;
    UINT nSig;
    AFX_PMSG pfn;
};
```

- `UINT nMessage` identifies the particular `WM_xxx` message.
- `UINT nCode` is the Control Notification code, which is zero (0) for Window Messages. As we'll see in the following text, the functions that handle Command Messages and Control Notifications use this same Message Map.
- `UINT nID` is the Command ID, which is zero (0) for Windows Messages.
- `UINT nLastID` is the last Command ID, which is zero (0) for Windows Messages, in a range of Command IDs that start with `nID`. Functions that handle Command Messages can process a range of IDs.
- `UINT nSig` defines what the message handler's calling arguments should be. There are over sixty predefined values for `nSig` in `AFXMSG_.H` For example, a `nSig` value of `iww` will cause `OnWndMsg()` to format `wParam` and `lParam` as two `UINT` arguments before calling the message handler. The returned value is an integer.
- `AFX_PMSG pfn` is the address of message handler.

So when `OnWndMsg()` gets a Windows Message, it looks in its class derivations for the top Message Map. `OnWndMsg()` then scans the entries there for a `nMessage` match; if found, it formats `wParam` and `lParam` into the calling arguments specified by `nSig` and calls the function specified by `pfn`.

If `OnWndMsg()` doesn't find an entry in the top Message Map, it checks the `BEGIN_MESSAGE_MAP` macro for the next Message Map to scan, and so on until all the Message Maps in an object are checked. If `OnWndMsg()` scans all the Message Maps in an object and finds no message handler, the Window Message is returned to `WindowProc()`, which then delivers it to `DefWindow-Proc()`.

## Message Map Macros

Not only does AFXMSG_.H contain the predefined values for nSig, it also contains several macros that help simplify creating a Message Map entry. The macros come in two formats:

- ON_MESSAGE(), which handles any WM_XXX message, and
- a macro for every Windows Message in the form ON_WM_XXX, where WM_XXX is any currently defined Windows Message.

As a rule, you should use the ClassWizard to add these macros automatically to your class's Message Map (Example 59). However, not all of these macros can be added automatically. You can add macros manually by referring to the complete list of macros in Appendix B.

### OnCommand()

We now continue our story with WM_COMMAND messages, which when we last saw them were being sent to OnCommand() for processing.

```
LRESULT OnCommand(WPARAM wParam, LPARAM lParam);
```

Even at this point, a WM_COMMAND message can be either a Command Message or a Control Notification. OnCommand() checks to see if lParam is a valid window handle; if it is, OnCommand() treats the message as a Control Notification, where lParam is the control's handle, and reflects the message to the window that sent it.

## Message Reflection

As mentioned before, Control Notifications are sent to a parent window when the state of a control changes. Control Notifications therefore give a parent window the limited ability to customize or supplement the way it controls work. As an example, when a combo box drops down, a parent window can react by modifying the contents of the drop down list.

Putting this functionality in the parent window however counters the goal of object-oriented programming, where every object should contain all of its own functionality. In the example above, every time that combo box is moved to a new parent window, the drop down logic must be moved to the new parent window as well. It is therefore preferable to keep Control Notifications processing with the control. Since each control already has an MFC class that you can override to add other functionality, it would be really nice if you could process Control Notifications to that list. Then every time you moved the control to a new parent, you wouldn't have to worry about what code you have to copy into the new parent. MFC supports this functionality in a process called *Message Reflection*.

Whenever an MFC window realizes it has just received a Control Notification, it knows that it has a control window somewhere and potentially an MFC-derived class controlling it. It therefore tries to "reflect" that notification back to the MFC-derived class to give it the opportunity to process it (Figure 3.4). In this way, you can put all of the functionality for any of your controls in one neat class package.

## Figure 3.4   Message Reflection offers a Control Notification back to its Control Window.

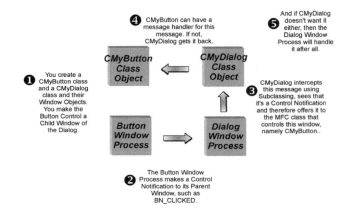

If the Control doesn't want the message back, OnCommand() treats the message just like a Command Message and calls OnCmdMsg().

### OnNotify()

WM_NOTIFY messages, you'll recall, are sent to the OnNotify() function.

```
BOOL OnNotify(WPARAM wParam, LPARAM lParam, LRESULT &lResult);
```

Any message sent here is automatically assumed to be a Control Notification and the message is reflected. If the control class doesn't want the message back, OnNotify() calls OnCmdMsg().

OnCommand() and OnNotify() both call ReflectLastMsg() to offer a Control Notification back to its Control Window. ReflectLastMsg() gets the window handle of the Control Window and looks in its Message Map for WM_COMMAND+WM_REFLECT_BASE or WM_NOTIFY+WM_REFLECT_BASE and the notification code.

## Message Map Macros

There are three sets of predefined macros — one for each Control Notification format — that you can use in a Control Window's Message Map to process its reflected messages:

- `ON_WM_XXX_REFLECT()` for the first Control Notification format,
- `ON_CONTROL_REFLECT()` for the second Control Notification format, and
- `ON_NOTIFY_REFLECT()` for the third Control Notification format.

Please refer to Appendix B for a complete list.

### OnCmdMsg()

If a Control Notification isn't wanted by its Control Window, both `OnCommand()` and `OnNotify()` handle it just like a Command Message by calling

```
BOOL OnCmdMsg(UINT nID, int nCode, void* pExtra,
    AFX_CMDHANDLERINFO* pHandlerInfo);
```

where

- `UINT nID` is the Command ID.
- `Int nCode` is the Notification Code. For a Command Message, a value of `CN_COMMAND` is stuck here (which equates to zero (0)).
- `Void *pExtra` depends on what type of message is being processed.

| When processing: | pExtra contains: |
|---|---|
| `WM_NOTIFY` Control Notifications | A pointer to the `AFX_NOTIFY` structure which points to `NMHDR` |
| Menu item and toolbar updates (enabling, disabling, etc.) | A pointer to a derivation of `CCmdUI` (discussed below) |
| Everything else | NULL |

- `AFX_CMDHANDLERINFO* pHandlerInfo` is always NULL when `OnCommand()` or `OnNotify()` calls `OnCmdMsg()`. When this argument does contain a pointer to an `AFX_CMDHANDLERINFO` structure, `OnCmdMsg()` doesn't execute any Command Message handler it finds. Instead, `OnCmdMsg()` just returns the address of the handler it would have executed.

```
struct AFX_CMDHANDLERINFO
{
    CCmdTarget* pTarget;
        // the object the function resides in
    void (AFX_MSG_CALL CCmdTarget::*pmf)(void);
        // the function address
};
```

This additional functionality is used when enabling and disabling menu items and toolbar buttons, as discussed in the following text.

Command Messages and Control Notifications are then offered to several classes in a process known as Command Routing.

## Command Routing

`OnCmdMsg()` is actually a member function of `CCmdTarget`, not `CWnd`. This is important because it allows any class derived from `CCmdTarget` to receive a Command Message — even those without a window. For example, the Document Class has no associated window — it depends on the View Class to display its document. But a Document Class is the best place to handle a load or save command. So while a Document Class doesn't handle Windows Messages like `WM_CREATE` or `WM_DESTROY`, `OnCmdMsg()` allows it to handle Command Messages like `WM_COMMAND` and `WM_NOTIFY`. `CWnd` is itself derived from `CCmdTarget`, so it too can support Command Messages (Figure 3.5).

**Figure 3.5**   `CCmdTarget` **supports Message Maps in MFC Objects without a window.**

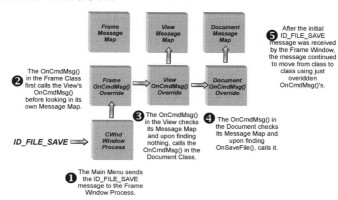

MFC implements Command Routing by overriding `OnCmdMsg()` in all of the main application classes (Application, Frame, View, and Document). For example, the View Class overrides `OnCmdMsg()` so that it can offer Command and Control Notification messages to the Document Class. And the Document Class overrides `OnCmdMsg()` so that it can offer Command messages to the Document Template Class. The following list shows how Command and Control Notifications are routed to the main classes in your application.

### `OnCmdMsg()` **Command Routing**

| Command Messages Sent to the MFC Class | Are also Routed to the Following Classes in the Following Order |
|---|---|
| `CFrameWnd` | All derivations of `CView` with the input focus<br>Its own Message Maps<br>All derivations of `CWinApp` |
| `CMDIFrameWnd` | All derivations of `CMDIChildWnd` with the input focus<br>All derivations of `CView` with the input focus<br>Its own Message Maps<br>All derivations of `CWinApp` |
| `CMDIChildWnd` | Its own Message Maps |
| `CView` | Its own Message Maps<br>All derivations of `CDocument` |

### `OnCmdMsg()` **Command Routing**

| Command Messages Sent to the MFC Class | Are also Routed to the Following Classes in the Following Order |
| --- | --- |
| CDocument | Its own Message Maps<br>All derivations of CDocumentTemplate |
| CDialog | Its own Message Maps<br>That of its parent<br>That of its thread |
| CPropertySheet | Its own Message Maps<br>That of its parent<br>That of its thread |

The effect of this routing is cumulative. A Command Message sent to CMDIFrameWnd will be offered first to the Message Maps in the active CMDIChildWnd, then the active CView, then the CDocument of the active CView, then the CDocTemplate of that document, then the CMDIFrameWnd's own Message Map, and finally CWinApp.

Command Routing stops as soon as OnCmdMsg() finds a message handler. In the previous example, if OnCmdMsg() finds a handler in CDocument, it calls that handler with the Command Message and returns with the result without going on to CDocTemplate.

## Message Map Macros

Although Command Messages and Control Notifications are routed identically, their predefined Message Map macros are different. Command Messages take the form of

```
ON_COMMAND()
```

while Control Notifications take the general form of

```
ON_CONTROL() for WM_COMMAND notifications and
ON_NOTIFY() for WM_NOTIFY notifications.
```

See Appendix B for a complete list.
See Figure 3.6 for the big picture.

**Figure 3.6**  **This illustration presents the big picture of how MFC processes a received message.**

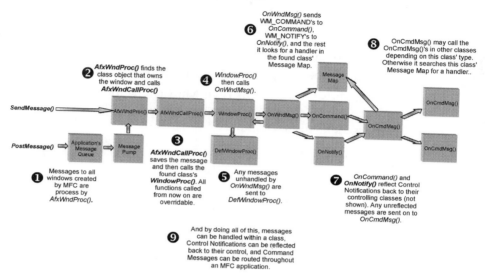

# Processing User Interface Objects

Some of the member functions we have just reviewed are also used by MFC to update the status of its user interface. Specifically, MFC uses OnCmdMsg() to enable, disable, and check all the items in a menu and all the controls in a control bar (toolbar, dialog bar, and status bar). MFC updates the status of menu items whenever one opens. Every time your application is idle, it updates the status of the controls in its control bars.

In either event, MFC enters a loop in which it calls OnCmdMsg() up to two times for every menu item or control bar control. The first time OnCmdMsg() is called, your application is given a chance to enable or disable the item or control itself. If you haven't provided a special user interface (UI) handler for the action requested, OnCmdMsg() is called a second time to check whether the item or control has a message handler at all; if not, OnCmdMsg() automatically disables the item or control itself.

---

NOTE: If you would like to turn off this automatic disabling feature, set m_bAutoMenuEnable to FALSE in your CMainFrame class. Unfortunately, this doesn't do a darn thing for the toolbar.

---

When `OnCmdMsg()` is called to allow your application to update the interface, it's given the following arguments.

```
OnCmdMsg(nID, CN_UPDATE_COMMAND_UI,CCmdUI *pCmdUI, NULL);
```

- `nID` is the Control ID of the menu item or control bar control to update.
- `*pCmdUI` points to a class that contains member functions that allow your application to update the item or control.

`OnCmdMsg()` then looks through the same Message Maps in the same order as for Command Messages until it finds a special UI message handler. The predefined macro for handling `CN_UPDATE_COMMAND_UI` is

```
ON_UPDATE_COMMAND_UI(id, Handler)
```

For a range of Control IDs, the predefined macro is

```
ON_UPDATE_COMMAND_UI_RANGE(id, idLast, Handler)
```

With either macro, your handler will have the same format.

```
void Handler(CCmdUI *pCmdUI)
{
    pCmdUI->Enable(pFlag);
}
```

The `CCmdUI` object you are passed contains the following member variables and functions of interest.

- `m_nID` is the Control ID, which is useful for the second macro format when a range of Control IDs can be updated with the same handler.
- `Enable(BOOL bOn)` causes the item or control to be disabled if set to `FALSE`.
- `SetCheck(int nCheck)` will check the item if `nCheck` is one (1).
- `SetRadio(BOOL bOn)` updates radio button groups.
- `SetText(LPCTSTR string)` sets the text in a control or menu item.

Each of these member functions is overridden, such that the code necessary to disable a menu item is called when a menu is being updated and the code for a control is called when a control bar is being updated.

When `OnCmdMsg()` is called the second time to determine whether your application even has a handler for a menu item or control, it's given the following arguments.

```
BOOL bHandler =OnCmdMsg(nID, CN_COMMAND, CCmdUI *pCmdUI,
    AFX_CMDHANDLERINFO *info);
```

As mentioned the first time we looked at `OnCmdMsg()`'s arguments, the presence of a valid pointer to a `AFX_CMDHANDLERINFO` structure causes `OnCmdMsg()` to simply look for a message handler, rather than execute one. If the returned value is `FALSE`, there is no handler and the item or control is disabled.

For examples of updating the user interface using `CCmdUI` class objects, see Examples 16, 17, and 18.

# Creating Your Own Windows Messages

There are over 70 system-defined Windows Messages from `WM_NULL` to `WM_USER`. You can also create two types of custom Window Messages for your own purposes: those that are defined statically above `WM_USER` and those that are defined dynamically by the system when given a character string identifier.

---

**NOTE:** Remember that Windows Messages are intended more for the internal workings of a window rather than processing user commands. If you simply want to tell another part of your application or another application entirely to process a user command, then use the already defined `WM_COMMAND` Windows Message and a new Command ID.

---

## Statically Assigned Windows Messages

Windows has reserved a range of integers between zero (0) and `WM_USER-1` for its own system-defined Windows Messages. A range starting with `WM_USER` and extending to `0x7fff` has been set aside for your own custom messages. You can define your own messages with a simple #define statement.

```
#define WM_MYMESSAGE1 WM_USER
#define WM_MYMESSAGE2 WM_USER+1
    :    :    :
```

You would then send or post this message just like any Windows Message with

```
SendMessage(WM_MYMESSAGE1,wParam,lParam)
```

or

```
PostMessage(WM_MYMESSAGE1,wParam,lParam).
```

You can capture this new message in your Message Map using the macro

```
ON_MESSAGE(WM_MYMESSAGE1,Handler)
```

where your handler would have the format

```
LRESULT Handler(WPARAM,LPARAM)
{
}
```

You can even get fancy and define your very own Message Map macro by modifying this macro.

```
#define ON_MY_MESSAGE1() \
    {WM_MYMESSAGE1, 0, 0, 0, AfxSig_lwl, (AFX_PMSG)(AFX_PMSGW)\
    (LRESULT (AFX_MSG_CALL CWnd::*)(WPARAM, PARAM))&OnMyMessage1},
```

where your handler now automatically has the name OnMyMessage1, but still has the same calling arguments. To change the calling arguments, look in the AFXMSG_.H file for a more appropriate nSig value to replace AfxSig_lwl. Then modify the last line in the macro to reflect the new arguments.

For an example, see Example 62.

## Dynamically Assigned Windows Messages

For sending messages between applications, you should instead create a new Windows Message based on a descriptive character string with

```
UINT wm_MyMessage1=::RegisterWindowMessage(LPCSTR Identifier);
```

- Identifier is the descriptive character string.
- wm_MyMessage1 is a dynamically allocated Windows Message between 0xc000 and 0xffff.

To handle registered messages in your Message Map you would use

```
ON_REGISTERED_MESSAGE(wm_MyMessage1, Handler)
```

passing it the value assigned by `RegisterWindowMessage()`.
The Handler to this message looks like

```
LRESULT Handler(WPARAM,LPARAM)
{
}
```

A dynamically assigned Windows Message makes maintaining Windows Messages between applications easier because you don't have to be concerned that every application knows that `WM_MYMESSAGE1` is a particular `WM_USER+ n` value. Instead, all applications can use a string like "My Message 1" to identify the message and `RegisterWindowMessage()` figures out what integer value to use.

For example, if an application registers a message with the identifier "Program Button" for the first time, `RegisterWindowMessage()` might assign a value of `0xc012`. Another application registering the same identifier will instead be returned the preassigned value of `0xc012`. Both applications can now communicate Windows Messages to each other, even if they were written at different times and with different sets of `WM_USER` definitions.

# Redirecting Messages

Now that we know how messages get from point A to point B, let's look at the advantages of intercepting and redirecting these messages. We have already reviewed one method of redirecting messages.

- Message Reflection was discussed previously and offers a way to process Control Notifications in the control's own class object. Message Reflection allows any functionality you add to a control's class to be easily ported between applications.

There are six other powerful techniques for redirecting messages.

- Subclassing and Superclassing offer a way of adding functionality to a Window Class by intercepting messages to its Window Process and processing some of those messages in your very own Window Process.
- Overriding `OnCmdMsg()`.
- Using `SetWindowsHookEx()`.
- Using `SetCapture()`.
- Creating your own localized message pump function.

# Subclassing and Superclassing

As mentioned earlier, every window has a Window Process that's defined in its Window Class. This Window Process handles every message sent to that window. A WM_PAINT message tells this process to draw the window. A WM_DESTROY message tells the Window Process to destroy the window. If you are writing the Window Process yourself, modifying its behavior is no problem. To draw the window differently, simply change your WM_PAINT handler. However, if this Window Process belongs to someone else, there's no source code to change. For example, every button in your application is created with a system-provided BUTTON Window Class that has its very own Window Process. If you wanted to change the look of this window, you can't change its WM_PAINT handler because it's unavailable.

So how do you change the look of a button, short of rewriting the control from scratch? By simply replacing the address of the original Window Process in your window object with the address of your own Window Process. You aren't forced to write the Window Process from scratch by doing this because you only need to handle the messages you want to handle in your Window Process. Anything you don't want to handle can be sent on to the original Window Process. For example, you can intercept the WM_PAINT message and pass WM_DESTROY directly to the original Window Process. You could even let WM_PAINT through first and when the original Window Process returns, you can add any additional drawing to the original Window Process drawing.

Changing the address of a Window Process in a window object that's already been created is called *subclassing*.

Changing the address of a Window Process in a Window Class before any window has been created from it is called *superclassing*.

The advantage superclassing has over subclassing is in allowing you to process WM_NCCREATE messages. These messages are unavailable to you with subclassing, since subclassing involves modifying a window that's already processed the WM_NCCREATE message (has already been created). In practice, however, superclassing is rarely used. To use superclassing, you would need to create a new Window Class for every type of window you wanted to create and you wouldn't be able to change the characteristics of an existing window. If you need to use the WM_NCCREATE message to change the way a window is created, create the window as hidden and change it after it's been created.

See Figure 3.7 for an example of subclassing and Figure 3.8 for an example of superclassing.

**Figure 3.7**   **Subclassing allows you to intercept specific messages and forward the rest by changing the address of a Window Process.**

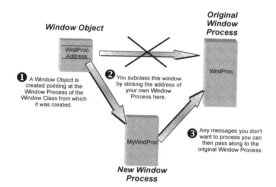

**Figure 3.8**   **Superclassing is similar to subclassing, but also allows you to change the address of a Window Process before any windows have been created.**

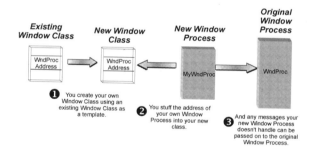

## Subclassing Windows using MFC

You can subclass a window using CWnd's SubclassWindow(). You pass the handle of the window you want to subclass to SubclassWindow(). SubclassWindow() then sticks the address of your class's Window Process into this window. You can then intercept and handle any message to this window from the safety of your own class member functions. Any leftover messages are sent on to the original Window Process. See Figure 3.9 for an example of subclassing a button control using CButton class. Working examples may be found in Examples 47 and 48.

**Figure 3.9** CWnd's SubclassWindow() **allows you to subclass using MFC.**

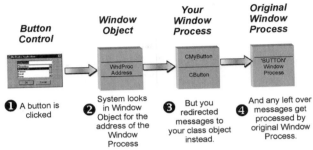

**Overriding** OnCmdMsg()

As mentioned earlier, Command Messages are automatically offered to your Frame, Document, Application, and View Classes. You can also offer Command Messages to other classes by overriding OnCmdMsg(). Start by making sure the new class is derived from CCmdTarget. Then from a class that currently processes Command Messages, such as the CMainFrame Class, call the OnCmdMsg() member function of the new class.

For example, a modeless dialog box does not automatically receive any Command Messages from the main menu or toolbars. If you would like to process a main menu command in a modeless dialog box, you need to override your CMainFrame's OnCmdMsg() and there call OnCmdMsg() of the modeless dialog box. Modeless dialog boxes, as well as all classes that control a window, are already derived from CCmdTarget.

For an example of overriding OnCmdMsg(), see Example 61.

---

NOTE: You can also override the OnWndMsg() function of any class derived from CWnd to redirect Window Messages. However, Window Messages are usually irrelevant outside their own window, making this a seldom-used technique.

---

**Using** SetWindowsHookEx()

The ::SetWindowsHookEx() Windows API function allows you to intercept messages before they make it to their intended Window Process. You can even intercept a message sent with SendMessage() before it calls its target Window Process. The syntax for SetWindowsHookEx() is as follows.

```
HHOOK SetWindowsHookEx(
    int msgId,
    HOOKPROC hookProc
    HINSTANCE procInstance,
    DWORD threadId)
```

- int msgId is the type of message you're looking for, such as:
  - WH_CALLWNDPROC, which looks for any message about to be processed by a Window Process;
  - WH_JOURNALRECORD, which looks for any keyboard or mouse message posted to the system message queue; and
  - WH_CBT, which looks for messages that indicate if any window is about to be created, destroyed, moved, etc.
- HOOKPROC hookProc is a function in your application that's called when one of these messages is found.
- HINSTANCE procInstance is your application's instance handle. It should be zero if you are looking for messages from your own application.
- DWORD threadId is the ID of the thread to look for messages. If zero, every process in the system is examined.

SetWindowsHookEx() puts a hook procedure that looks for your message into a chain of procedures. Your hook procedure can optionally call the next procedure in the chain with CallNextHookEx() or simply discard it. You must also unhook your procedure with UnhookWindowsHookEx() before you terminate.

SetWindowsHookEx() has been used in functions that learn keystrokes for a keyboard macro and in applications that teach someone how to use an application.

## Using SetCapture()

Normally when you use the mouse to click on a window, a window message is sent to that window. In some applications, however, especially those that require you to examine other windows, you want to be able to intercept mouse messages no matter where the user clicked. To do this, you can use CWnd's SetCapture(). Upon calling SetCapture(), all mouse messages will be posted to your window until you call ReleaseCapture(). The capture is also canceled automatically for you after your user clicks outside of your window once. Therefore, you need to continually call SetCapture() when you receive a mouse click message.

### Localized Message Pumps

To suspend the execution of your application until it receives a particular posted message, you can write your very own message pump. Message boxes and dialog boxes both have their own message pumps that suspend program execution until the user is finished with them. You might use a message pump to help port an application that previously operated in an environment that always required the program to suspend itself for user input.

Writing your own message pump isn't as hard as it sounds. Your application currently uses one inside the Run() member function of CWndApp, for which Microsoft has provided the source.

## Summary

In this chapter, we discovered that:

- MFC classes and their windows communicate with each other by sending or posting messages.

- Posted messages travel through a message queue before being delivered to a Window Process, while sent messages essentially call the Window Process directly.

- There are three types of messages: Windows Messages, Command Messages, or Control Notifications.

- Messages are processed through Message Maps, instead of case statements, with five overridable functions: WindowProc(), DefWindowProc(), OnWndMsg(), OnCommand(), OnNotify(), and OnCmdMsg().

- MFC uses the message processing routines to also allow your application to enable and disable menu items and toolbar buttons.

- And finally, there are seven ways to redirect messages with powerful results: subclassing, superclassing, Message Reflection, overriding OnCmdMsg(), using SetCapture() or SetWindowsHookEx(), and writing your own message pump.

In the next chapter, we're going to look at the MFC classes for drawing.

# 4

# Drawing

In the last few chapters, Windows did all the drawing for us — from painting the nonclient area of a window to drawing the Common Controls. In this chapter, we will find out how to draw our own controls and views using the MFC class CDC. We will discover:

- the types of drawing tools Windows and MFC support;
- the drawing modes and characteristics Windows and MFC support;
- two ways, metafiles and paths, to store drawing commands for future display;
- how to draw in dithered and nondithered colors; and
- when exactly Windows draws its windows.

## Device Contexts

To write your own routine to draw a line on the screen, you would need the following types of calling arguments:

- the line coordinates (the start $x,y$ to the end $x,y$),
- the line color, and
- the line thickness.

To get really sophisticated, you would also need:

- the ability to specify the measurement units in which to draw in (pixels, inches, or centimeters) and
- the ability to draw your line on the printer or to a graphic file.

For portability, you'd probably also want to be able to draw to any display or any printer without worrying about the video card or printer type.

If you take all of these attributes into consideration, your routine could look something like this

```
DrawLine(x1,y1,x2,y2,color,width,device,limits,units)
```

Although certainly straightforward, this function has a hidden cost — every time you draw a line, all of these arguments have to be pushed onto the stack. This function doesn't allow room for enhancements, either. To add arguments, you would have to edit any reference to the old function.

So how would you lower the number of calling arguments needed and allow room for expansion? One method would be to write a C++ class containing most of the arguments your routine needed, which you could then pass as one argument. You could even have this class construct itself with some valid default values so you could occasionally use it "as-is". You could also keep the class around for as long as you're drawing to save time reconstructing it.

Windows uses this exact method with an object called a Device Context — except that this object is created without a C++ class, since Windows predates and exists outside of C++. Device Context objects are created with some default values so that you can use them "as-is". To draw a line using Windows, all you need is a pointer to a Device Context object, called a Device Context Handle (HDC), and the line coordinates.

```
::MoveTo(HDC hdc,int x,int y);
::LineTo(HDC hdc,int x,int y);
```

To draw an entire shape, the ::MoveTo() function is called only once. Thereafter, the ::LineTo() function is called. Not one extra argument is pushed onto the stack (Figure 4.1).

**Figure 4.1** **Device Contexts allow you to lower the number of calling arguments needed and allow room for expansion.**

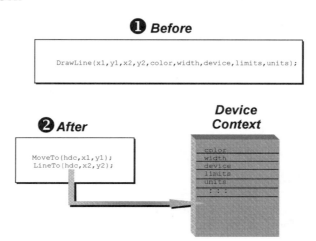

**❶ Before**

```
DrawLine(x1,y1,x2,y2,color,width,device,limits,units);
```

**❷ After**

```
MoveTo(hdc,x1,y1);
LineTo(hdc,x2,y2);
```

**Device Context**

```
color
width
device
limits
units
: : :
```

---

**NOTE:** Because device contexts are created in memory that's constantly being churned, its address may be constantly changing. Therefore, a device context handle, instead of pointing directly to the Device Context object, points to another pointer that keeps track of the Device Context location.

---

## Device Independence

A Device Context also enables *Device Independence*, which allows you to write software without specifying a device. All your software has to know is that it's writing to a Device Context, whether it be a screen or an Epson printer. The Device Context can later be assigned to a printer, the screen, a bitmap, or a file.

# Creating a Device Context in the MFC World

MFC wraps a Device Context object with the CDC class. The m_hdc member variable of CDC points to the Device Context it owns. There are four types of Device Contexts you can create:

- for the screen,
- for a printer,
- for just the informational part of a printer, and
- for a bitmap.

## Screen

To create a Device Context for the screen, you would use

```
CDC *pDC=GetDC();
```

where GetDC() is a member function of the CWnd class and returns a Device Context that allows you to draw to the client area of its window. If you would also like to draw in the nonclient area, use

```
CDC *pDC=GetWindowDC();
```

If you would like to draw to the entire screen, like a screen saver or game might do, use

```
CDC *pDC=CDC::FromHandle(::GetDC(NULL));
```

Device Contexts for the screen come from one of three sources: Private Device Contexts, Class Device Contexts, and Common Device Contexts.

### Private Device Contexts

Private Device Contexts are created when your window is created and exist for the life of your window. To make your window create a Private Device Context, you must set the CS_OWNDC flag in the Window Class of the window in which you're drawing. GetDC() will then return a pointer to the Private Device Context, rather than creating a Device Context. Any changes you make to the Private Device Context will be maintained for the life of that window. Private Device Contexts are used in graphic-intense applications where constantly allocating a Device Context can become CPU-intensive.

## Class Device Contexts

Class Device Contexts are created when you register a Windows Class and exist for the life of that Window Class. To make a Windows Class create its own Class Device Context, you must set the CS_CLASSDC flag. GetDC() simply returns a pointer to this Class Device Context. Again, the Device Context retains its values for the life of the Window Class. As noted in Chapter 1, however, care must be taken not to allow two windows created from the same Window Class to draw at the same time. Also, note that Microsoft considers a Class Device Context to be obsolete.

## Common Device Contexts

A Common Device Context is created when you use GetDC() in a common resource heap. To get a Common Device Context, don't set the CS_OWNDC or CS_CLASSDC flags in your Windows Class. With earlier versions of Windows, there were just five preallocated Common Device Contexts within the operating system itself — thus, the name GetDC() as opposed to CreateDC(). You had to make sure you released a Device Context after using it or your system would grind to a halt. Although Device Contexts are now created, resource memory is still limited. After using a Common Device Context, you must still release it with

```
ReleaseDC(pDC);
```

---

**NOTE:** Private and Class Device Contexts don't need to be released.

---

You will most likely be using Common Device Contexts to do your drawing. Any settings you make to a Common Device Context are lost each time you release them. However, you can save and restore a Device Context's settings with these two CDC member functions.

- SaveDC() saves the current state of the device context.
- RestoreDC() restores the device context to a previous state saved with SaveDC().

## Printer

To create a Device Context for the printer, you must first create a CDC class object and then use its CreateDC() member function.

```
CDC dc;
dc.CreateDC(LPCTSTR lpszDriverName,
    LPCTSTR lpszDeviceName,
    LPCTSTR lpszOutput,
    const void* lpInitData)
```

- lpszDriverName is the name of the device driver that came with your printer (e.g., "HPDRV").
- lpszDeviceName is the name of the device to which you are printing (e.g., "HP Laserjet III"). Device drivers can support more than one type of device.
- lpszOutput specifies the device's port name (e.g., "LPT1:").
- lpInitData is any device-specific initialization data.

After using a Printer Device Context, you must destroy it with the DeleteDC() member function of the CDC class. However, if you created the CDC class object on the stack, the Device Context will be automatically deleted when your routine returns.

```
CMyClass::Drawing(...)
{
    CDC dc;
    dc.CreateDC(...);
}        // CDC object and Device Context destroyed
```

## Memory

To create a Device Context that will allow you to draw directly into a bitmap, you must first create a CDC class object with its CreateCompatibleDC() member function.

```
CDC dc;
dc.CreateCompatibleDC(HDC hdc);
```

CreateCompatibleDC() initializes the Device Context object that it creates using the settings in the Device Context that you pass it. Thus the name, Compatible DC. You should pass a Device Context from the device on which your created bitmap will appear (e.g., screen or printer).

To actually draw to a bitmap, you must also create an empty bitmap object and associate this Device Context with that bitmap.

```
CDC dc;                              // create CDC object
CBitmap bitmap;                      // create CBitmap object
dc.CreateCompatibleDC(pDC);          // create Device Context object
bitmap.CreateCompatibleBitmap(pDC, ICON_WIDTH, ICON_HEIGHT);
                                     // create Bitmap object
dc.SelectObject(&bitmap);            // point memory Device Context
                                     // to this bitmap
```

After drawing to the bitmap, you must delete this Device Context using

```
dc.DeleteDC();
```

For a complete example, see Example 58.

## Informational

The CDC class provides one last type of Device Context for just the informational part of a printer.

```
CDC dc;
dc.CreateIC(…);
```

where the calling arguments are the same as for CreateDC().

An Information Device Context contains only the device characteristics of a printer or other nonscreen device, such as color support. An Information Device Context is faster to create than a regular Printer Device Context, since it contains only the characteristics of a device.

You must delete this context after using it, as described in the previous section.

---

**NOTE:** Just as with the CWnd class and the window it controls, the CDC class can attach itself to an existing Device Context with Attach(), and detach itself with Detach(). To create a temporary CDC object that wraps an existing Device Context, you can use FromHandle(). However, this temporary class object is deleted the next time the application goes into the idle loop.

---

# Drawing Routines

The CDC class also wraps the entire Windows graphics API into its member functions. These member functions include:

- functions to draw a point,
- functions to draw lines,
- functions to draw shapes,
- functions to fill and invert shapes,
- a function to scroll the screen,
- functions to draw text, and
- functions to draw bitmaps and icons.

## Point Drawing

Pixel drawing is nothing more than changing the color of individual pixels.

| SetPixel() | Try to draw a pixel in the specified color. Returns the actual color drawn. |
|---|---|
| SetPixelV() | Same as above, except faster because it does not return the actual color drawn. |

## Line Drawing

Line drawing is nothing more than changing the color of a series of pixels on the screen.

| MoveTo() | Starts line, arc, and poly drawing by moving to an initial position. |
|---|---|
| LineTo() | Draws a line from an initial position to another point. |
| Arc() | Draws an arc. |
| ArcTo() | Draws an arc, then updates initial position. |
| AngleArc() | Draws a line, then an arc, and updates initial position. |
| PolyDraw() | Draws a series of lines and Bézier splines. |
| Polyline() | Draws a series of lines. |
| PolyPolyline() | Draws multiple series of line. |

## Shape Drawing

A shape here is considered to be a series of closed lines.

| | |
|---|---|
| Rectangle() | Draws a rectangle. |
| RoundRect() | Draws a rectangle with rounded corners. |
| Polygon() | Draws a polygon. |
| PolyPolygon() | Creates two or more polygons. |
| Ellipse() | Draws an ellipse. |
| Pie() | Draws a pie-shaped wedge. |
| Draw3dRect() | Draws a three-dimensional rectangle. |
| DrawEdge() | Draws the edges of a rectangle. |
| DrawFrameControl() | Draw a frame control. |

## Shape Filling and Inverting

Filling and inverting changes all the pixel colors inside a shape.

| | |
|---|---|
| FillRect() | Fills a rectangle. |
| InvertRect() | Inverts the colors of a rectangle. |
| FrameRect() | Draws a border around a rectangle. |
| FillSolidRect() | Fills a rectangle with a solid color. |
| ExtFloodFill() | Fills an area with the current brush. Provides more flexibility than the FloodFill() member function. |

The next three functions involve a Region object, which is essentially a list of lines forming a shape that is maintained by the Device Context. For a full discussion on Regions, see "Clipping Attributes" on page 106.

| | |
|---|---|
| FillRgn() | Fills a region. |
| FrameRgn() | Draws a border around a specific region. |
| InvertRgn() | Inverts the colors in a region. |

## Scrolling

Scrolling moves pixel colors around the screen.

| | |
|---|---|
| ScrollDC() | Scrolls a rectangle of bits horizontally, vertically, or both. In other words, it moves the screen image left, right, up, or down. |

## Text Writing

You might think that text is printed, not drawn. But in a graphical user interface, even a text character is just a picture made out of pixels.

| | |
|---|---|
| TextOut() | Writes a character string at a specified location. Nothing fancy. |
| ExtTextOut() | Writes a character string within a rectangular region. |
| TabbedTextOut() | Writes a character string at a specified location and converts any tabs in the string to spaces, based on a table passed with this function. |
| DrawText() | Draws text in the specified rectangle with a lot more options than TextOut(), including centering text and showing text on multiple lines. |

## Bitmap and Icon Drawing

A bitmap or icon is just a massive array of pixel colors. A header usually indicates how many pixels are in a row so that a drawing routine knows when to start the next row. A bitmap drawing routine usually just copies this array to video memory. An icon has the additional ability to have a transparent color. In other words, when an icon is drawn, one of its colors is substituted with whatever was originally on the screen.

| | |
|---|---|
| DrawIcon() | Draws an icon at the specified location. |
| BitBlt() | Copies a bitmap from a specified Device Context, usually loaded from disk or created in memory, as outlined previously. |
| StretchBlt() | Same as BitBlt(), except it attempts to stretch or compress the bitmap to fit the destination. |
| PatBlt() | Creates a bit pattern. |

For examples of drawing, see Chapter 11.

# Drawing Attributes

All of the functions reviewed in the previous section use the Device Context object to simplify their calling arguments. Because there is just one Device Context for all drawing functions, each function may use only use twenty percent of the attributes stored in a Device Context.

A Device Context either contains a particular drawing attribute itself or it points to yet another object that does. For example, the attributes for drawing a line (e.g., width or color) are stored in a separate Pen object, to which the Device Context points. Please see Figure 4.2 for the other objects to which the Device Context points and the MFC classes that wrap them.

**Figure 4.2** **A Device Context can point to Helper Device Context Objects that contain additional drawing attributes.**

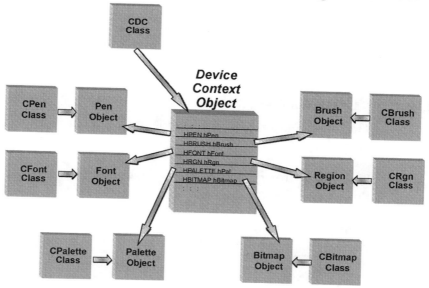

You create these additional graphic objects the same way you create a window object: first create the MFC class object, then call the Create() member function of that class. Once created, you need to tell your Device

Context to use this new object using `SelectObject()`, which simply points the Device Context to your new graphic object:

```
CPen pen;                         // create MFC object
pen.CreatePen(...);               // create Pen object
CPen *pOldPen=dc.SelectObject(&pen);
    // points the DC to your new object
    // returns a pointer to old pen object
```

Although the Device Context points to five different graphic objects, you only use one `SelectObject()`. The Device Context determines which pointer to replace based on a signature word in the object itself.

After you're through drawing with your new object, you must delete the object from memory. Since the Device Context is still pointing to your object, you must first point it to another object — usually the previous object.

```
dc.SelectObject(pOldPen);    // select another object
pen.DeleteObject();          // deletes Pen object (returning
                             // destroys the CPen object)
```

## Device Context Attributes

The following drawing attributes are found in a Device Context and its helper objects.

**Line attributes** control how line functions draw their lines (e.g., color and thickness). Most of these attributes are contained in the Pen object.

**Filling attributes** control how functions that draw shapes fill them in. Most of these attributes are stored in the Brush object.

**Text attributes** control how text is drawn. The font attributes are stored in the Font object. Color and alignment (e.g., left or centered) are stored in the Device Context itself.

**Mapping attributes** control what the $x,y$ coordinates in a drawing operation mean. In other words, are the coordinates in pixels or inches and which way is up? There is no additional object for mapping — all information is stored in the Device Context, itself.

**Palette attributes** can be used to control line and shape colors. Graphic applications in particular use the palette to control color, especially on systems with limited video memory. All palette attributes are stored in the Palette object.

**Blending attributes** control how the lines interact with the background on which they're being drawing. For example, do you simply want to blow away the background, or do you want to XOR with the background color so that you can erase the line later by simply redrawing it?

**Clipping attributes** create a region outside of which you can't draw. These attributes are stored in the Region object.

**Bitmap attributes** contain the bitmap used in bitmap drawing functions.

# Line Drawing Attributes

Line attributes control how lines are drawn. The Device Context points to a Pen object using an HPEN handle, which contains most line drawing attributes. You create a Pen object in MFC by first creating a CPen class object, then calling one of two member functions of CPen.

| CreatePen() | Creates a Pen object with the specified width, color, and pattern. The pattern can be anything from a solid line to a dashed line, for which you specify the length of the dashes and spaces. |
|---|---|
| CreatePenIndirect() | Creates a Pen object using the LPLOGPEN structure. |

Rather than create your own Pen object, there are several predefined Pen objects available, which you can select directly into your Device Context using SelectStockObject(). You identify which stock object to select with one of these flags.

| BLACK_PEN | Contains the attributes to draw a black pen. |
|---|---|
| WHITE_PEN | Contains the attributes to draw a white pen. |
| NULL_PEN | Causes drawing functions not to draw a line. For example, a rectangle draws a line and then fills the rectangle. Selecting this Pen object would cause no line to be drawn. |

## Shape Filling Attributes

The Brush object controls how shapes are filled. Your Device Context points to a Brush object using an HBRUSH handle. You create a Brush object in MFC by first creating a CBrush class object, then calling one of six member functions of CBrush.

| | |
|---|---|
| CreateSolidBrush() | Creates a Brush object with a solid color. |
| CreateHatchBrush() | Creates a Brush object with the specified hatched pattern and color. |
| CreateBrushIndirect() | Creates a Brush object using the settings in a LOGBRUSH structure. |
| CreatePatternBrush() | Creates a Brush object with a pattern specified by a bitmap. |
| CreateDIBPatternBrush() | Creates a Brush object with a pattern specified by a device independent bitmap (DIB). |
| CreateSysColorBrush() | Creates a Brush object with the default system color. |

Rather than create your own Brush object, there are several predefined objects available to you from the system. You can select one of these directly into your Device Context using SelectStockObject() and one of these flags.

| | |
|---|---|
| BLACK_BRUSH | Identifies a Brush object that contains the attributes needed to fill a shape in black. |
| WHITE_BRUSH | Contains the attributes to fill a shape in white. |
| DKGRAY_BRUSH | Contains the attributes to fill a shape in dark gray. |
| GRAY_BRUSH | Contains the attributes to fill a shape in gray. |
| LTGRAY_BRUSH | Contains the attributes to fill a shape in light gray. |
| NULL_BRUSH or HOLLOW_BRUSH | Causes drawing functions not to fill a shape. For example, a rectangle draws a line and then fills the rectangle. Selecting this Brush object would cause the rectangle not to be filled. |

# Text Drawing Attributes

The text attributes control how text is drawn. There are three types of text attributes: color, alignment, and font.

## Text Color

Two member functions allow you to set the foreground and background colors for drawing text.

| | |
|---|---|
| SetTextColor() | Sets the foreground color (i.e., text color). |
| SetBkColor() | Sets the background color on which the text is written. |

## Text Alignment

The text alignment attribute determines how your text is aligned (ex: left, right, centered). As such, the x and y arguments passed in the TextOut() function can either indicxate the left, right, bottom, etc. of the text you're drawing. With the default alignment, x and y represent the upper-left corner of the text, but there are several other variations. To change the text alignment attribute of your device context, you would use the SetTextAlign() function and one of the following flags:

| | |
|---|---|
| TA_LEFT | The x argument used in TextOut() to indicate the left side of the text drawn. TA_LEFT is the default setting. |
| TA_TOP | The y argument used in TextOut() to indicate the top of the text drawn. TA_TOP is the default setting. |
| TA_RIGHT | The y argument used in TextOut() to indicate the right side of the text drawn. A bounding rectangle is computed around the text to determine where to start the text. |
| TA_CENTER | The y argument used in TextOut() to indicate the middle of the text drawn. |
| TA_BASELINE | The y argument used in TextOut() to indicate the baseline of the text drawn. |
| TA_BOTTOM | The y argument used in TextOut() to indicate the bottom of the text drawn. |

Note that these attributes can be combined. For example, ORing TA_CENTER and TA_BOTTOM would draw a text string that is centered and on top of a point indicated by x and y.

## Text Font

Font attributes include the typeface, size, and other aspects of your text's appearance. These attributes do not define a font you can use, but rather tell the system what kind of font you would like to use. The system then looks in its list of available fonts for the closest match and uses that font.

Font attributes are contained in a Font object, which the Device Contact points to using the HFONT handle. You create a Font object in MFC by first creating a CFont class object, then calling one of four member functions of CFont.

| | |
|---|---|
| CreateFont() | Create a font object and define several characteristics of the font you would like, including point size, weight (normal or bold), and italic or not. |
| CreateFontIndirect() | Same as above, except uses a LOGFONT structure. |
| CreatePointFont() | Simplified version of CreateFont() — all you have to do is specify point size and typeface. |
| CreatePointFontIndirect() | Simplified version of CreateFontIndirect() — all you have to do is specify point size and typeface. |

Rather than create your own Font object, there are several predefined objects available which you can select directly into your Device Context using SelectStockObject(). You identify which stock object to select with one of these flags.

| | |
|---|---|
| SYSTEM_FONT | This is the default font that Windows uses to draw menus, title bars in windows, etc. |
| SYSTEM_FIXED_FONT | This is the default font that Windows used prior to v3.0. |
| ANSI_FIXED_FONT | The typeface is Courier with no proportional spacing (every character takes up the same space on a line). The special character set (characters outside of A-Z) is the ANSI standard. |

| ANSI_VAR_FONT | The typeface is MS Sans Serif with proportional spacing (the letter I takes up less space in a line then the letter w). The special character set is the ANSI standard. |
|---|---|
| DEVICE_DEFAULT_FONT | The preferred font for the device. |
| OEM_FIXED_FONT | Same as ANSI_FIXED_FONT except the special character set is OEM-specific. |

## Mapping Modes

Windows has several Mapping Modes that allow you to draw in pixels, inches, or millimeters. For most applications, however, you'll never have to worry about mapping modes. The default mode is pixel, such that x and y equate to pixel values. However, if you do want x and y to be in inches, you will need to be aware of the difference between Logical Units and Device Units.

## Logical vs. Device Units

Logical Units are the x and y values you pass to a drawing function. They can indicate inches or millimeters. Device Units, on the other hand, are the number of pixels your x and y values generated on the screen or dots on a printer. Mouse clicks are reported back to you in Device Units. If you want to know where on your image the user just clicked, you will need to convert them back to Logical Units. When working in Logical Units, you are working in the Logical Coordinate System. Device Units are in the Device Coordinate System (also called the Physical or Client Coordinate Systems).

The mapping mode is stored in the Device Context using SetMapMode() with one of the following flags.

| MM_TEXT | This is the default mapping mode. The value in x and y is exactly equivalent to one (1) screen pixel or printer dot. A positive y goes down the screen or printed page. |
|---|---|
| MM_HIENGLISH | The value in x and y is equivalent to 1/1000 of an inch on the screen or printed page. Windows determines how many pixels are required for the current screen device to equal 1/1000 of an inch. A positive y goes up the screen or printed page. |
| MM_LOENGLISH | The value in x and y is equivalent to 1/100 of an inch on the device and y goes up. |

| MM_HIMETRIC | The value in x and y is equivalent to 1/100 of a millimeter on the device and y goes up. |
| MM_LOMETRIC | The value in x and y is equivalent to 1/10 of a millimeter on the device and y goes up. |
| MM_TWIPS | The value in x and y is equivalent to 1/1440 of an inch on the device and y goes up. This is usually used for text drawing — one twip is equivalent to 1/20th of a font point. |

Two additional mapping modes allow you to set how many Logical Units equate to Device Units: MM_ANISOTROPIC and MM_ISOTROPIC. Isotropic mode makes sure that one unit in the x direction is equivalent to one unit in the y direction. In the Anisotropic mode, anything goes.

Defining two rectangles sets your preferences for Windows' conversion from the logical world to the device world. The first rectangle is in Logical Units and represents the extent of the area in which you want to draw. The second rectangle is in Device Units (i.e., pixels) and represents the area of the device you want the first rectangle to represent. In Window parlance, the first rectangle is called a Window and the second rectangle is called a Viewport (Figure 4.3).

**Figure 4.3**  **Setting the Window and Viewport rectangles defines how Windows converts from the logical to the device world.**

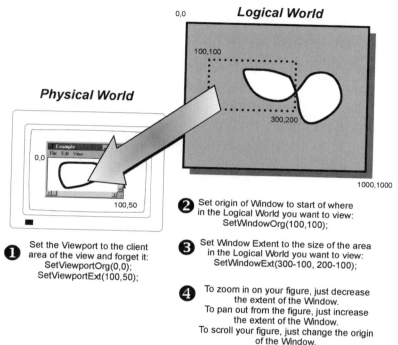

To set the Window and Viewport rectangles, you use these member functions of CDC.

| SetWindowOrg() | Sets the upper-left value of the rectangle that represents the logical world. |
|---|---|
| SetWindowExt() | Sets the size of the rectangle that represents the logical world. |
| SetViewportOrg() | Sets the upper-left corner of the rectangle that represents the device world. |
| SetViewportExt() | Sets the size of the rectangle that represents the device world. |

---

NOTE: If you use the Isotropic mapping mode, you must set the Window values before you set the Viewport values.

---

## Inverting, Zooming, and Panning

Once you're in either the Anisotropic or Isotropic mapping mode, you can get some powerful graphic effects for free.

**Inverting**   You can invert the image you draw on the screen just by inverting the Viewport values (Figure 4.4).

```
pDC->SetWindowOrg(0,0);           // 0,0 in logical world equals
pDC->SetViewportOrg(0,480);       // 0,480 in device world
pDC->SetViewportExt(640,480);     // the maximum in the
                                  // device world
pDC->SetWindowExt(640,-480);      // equals this maximum
                                  // in the logical
```

**Figure 4.4**   **Inverting the Viewport values allows you to invert the image on the screen.**

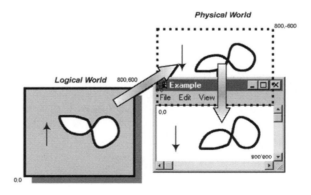

Physical World

Logical World   800,600

To make positive Y go up, first set the
Y extent of the Window to negative:
SetWindowExt(800,-600);

❷ Then, to move the figure back into
the view, set the Window origin to
the Y extent as a positive value:
SetWindowOrg(0,600);

❸ The formula to convert a logical unit
into a device unit is:

*DevUnit=((LogUnit-WindowOrg)\*DevExtent/LogExtent)+DevOrg*

**Zooming**   You can zoom or shrink an image just by making the Viewport values larger or smaller. Use `CDC::ScaleViewportExt()` to zoom by percentages.

**Panning**   All of your image may not fit on the screen at once. You can create it entirely in the logical world but only portions may be visible in the device world. You can move the viewable part of your image on the screen just by moving the Viewport rectangle. Use `CDC::OffsetViewportOrg()` to help.

> **NOTE:** You can use `SetViewportOrg()` and `SetWindowOrg()` to pan in any drawing mode. However, `SetViewportExt()` and `SetWindowExt()` are reserved for the Anisotropic and Isotropic mapping mode.

## Converting

The `CDC` class also provides two functions to convert from one world to the other.

| | |
|---|---|
| `DPtoLP()` | Converts Device Units into Logical Units. |
| `LPtoDP()` | Converts Logical Units into Device Units. |

## Dialog Units to Pixels

Dialog Units is a subject on the fringes of mapping modes. Dialog templates created with the Dialog Editor are filled with sizes and coordinates that are in Dialog Units, as opposed to pixels. Dialog Units allow a dialog template to proportionally adjust to different typefaces and font sizes. To convert from dialog units to pixels, you can use the following member function of `CDialog`:

```
MapDialogRect(LPRECT &rect);
```

where `rect` contains a set of Dialog Units and the function converts them to pixels for this instance of the dialog.

If you can't create an instance of the dialog first, you can use a static function of the `CDialogTemplate` class.

```
static void CDialogTemplate::ConvertDialogUnitsToPixels(LPCSTR typeface,
    int point, int x, int y, SIZE *pSize)
```

`typeface` and `point` are the current default typeface and font size of your dialog box. `x` and `y` are in dialog units. This function returns pixels in `pSize`.

## Palette Attributes

The palette attributes maintain the colors your application requires, so that those colors can be loaded into the system palette when your application becomes active. For most applications, you'll use your application's palette because you'll only be using the few colors already provided for you by the system. However, in some graphic-intense applications that require lots of different and unique colors, you need to start worrying about video cards that allow a maximum of only 256 unique colors.

Palette attributes are contained in the Palette object, which the Device Context points to using an `HPALLETE` handle. You create a Palette object in MFC by first creating a `CPalette` class object, then calling the `CreatePalette()` member function.

Rather than create your own Palette object, there is one predefined Palette object that you can select directly into your Device Context using

```
SelectStockObject(DEFAULT_PALETTE).
```

This is the default palette with which a new Device Context is initialized. This palette is provided more as a way to reset your color palette to its original settings than to provide an exciting set of new colors.

For much more on palettes, see "Colors and the Palettes" on page 111.

## Blending Attributes

The blending attributes control how the lines or text you draw interact with the background on which they're drawn. For most applications, you never have to worry about blending with the background. The default blending mode simply blows away what was in the background. However, for applications that draw image on top of image, you need to worry about what effect each additional image has on what went before.

The Device Context contains all the attributes needed for blending — no additional objects are required. You can change two blending aspects:

- how lines blend with the background colors on which their drawn, and

- whether to use the text background color when drawing text.

---

**NOTE:** You can also change how a bitmap blends with its background. However, this attribute is contained in the drawing function, itself, and not the Device Context.

---

## Line Blending

To set how a line blends with its background, use the `SetROP2()` member function of the CDC class. The `ROP` stands for Raster Operation, which means this function only works on Raster devices, such as your screen. Line blending doesn't work on some printers or any plotters that are vector devices.

There are several line blending modes that `SetROP2()` can set. The following are used most often.

| | |
|---|---|
| R2_COPYPEN | Default — line simply blows away color below. |
| R2_NOT | The final color of the line is the inverse of the current screen color. The Pen object color is ignored. |
| R2_XORPEN | The final line color is the XOR of the Pen object color and the screen color. |
| R2_MERGEPEN | The final line color is the OR of the Pen object color and the screen color. |
| R2_BLACK | The line is always black. The Pen object color is ignored. |
| R2_WHITE | The line is always white. The Pen object color is ignored. |

The `R2_XORPEN` mode is frequently used to allow lines to seemingly move over objects without disturbing them. An example of this is lasso or marquee selecting, in which you drag a rectangle around several objects to select them. To draw a lasso, you need to draw a rectangle becoming progressively larger without disturbing the underlying image. You can do this by drawing the first rectangle in `R2_XORPEN` mode. You can then erase this rectangle without destroying the underlying image by simply redrawing it, because the XOR of a XOR color is the original pixel color.

## Text Blending

Text blending is used to determine whether to draw text with a background or to simply draw the text on top of whatever is already there. To set the background mode, use the `SetBkMode()` member function of CDC using one of two flags.

| | |
|---|---|
| OPAQUE | This is the default. The area in which text is drawn is blown away with the color specified in the `SetBkColor()` function. |
| TRANSPARENT | This mode allows you to write text on top of an image without disturbing it. |

## Clipping Attributes

The clipping attributes define a region in your drawing area outside of which any attempt to draw is ignored. The Device Context points to a Region object using an HRGN handle, which contains a list of lines that outlines a clipping region. You create a Region object in MFC by first creating a CRgn class object, then calling one of several member functions of CRgn. These are the most common.

| CreateRectRgn() | Initializes a CRgn object with a rectangular region. |
|---|---|
| CreateRectRgnIndirect() | Initializes a CRgn object with a rectangular region defined by a RECT structure. |

Clipping regions can be used for some interesting graphic effects. For example, you can create a clipping region that forms the letters in the word "STOP", then draw a bitmap using this device context. The result is the word "STOP" filled in with portions of the bitmap you drew.

As with Palettes and Blending, most applications will never care about clipping regions.

## Bitmap Drawing Attributes

The bitmap attribute in a Device Context is simply a pointer to a bitmap object using an HBITMAP handle. You create a Bitmap object in MFC by first creating a CBitmap class object, then calling one of several member functions of CBitmap, of which these are the most common.

| LoadBitmap() | Creates a bitmap object from your application's resources, which you possibly created with the bitmap editor. |
|---|---|
| LoadOEMBitmap() | Uses a bitmap object that comes predefined from Windows. See the MFC User Documentation for a list. |
| CreateBitmapIndirect() | Creates a bitmap object from information supplied in the BITMAP structure. See the BITMAP structure in the following section. |

| `CreateBitmap()` | Same as `CreateBitmapIndirect()`, except that the information found in the `BITMAP` structure of `CreateBitmapIndirect()` is instead in the calling arguments to this function. |
|---|---|
| `CreateDiscardableBitmap()` | Same as `CreateBitmap()`, except that the created bitmap object will be automatically discarded for you if it isn't pointed to by a Device Context the next time your application is idle. |
| `CreateCompatibleBitmap()` | Creates a bitmap object from information supplied in the bitmap object pointed to by an existing Device Context. All you have to supply is the width and height. |

## Device Dependent Bitmaps Versus Device Independent Bitmaps

A bitmap is nothing more than an array of pixel colors with a header that indicates how many pixel colors make up a row before the next row is started. Each color is represented with a 32-bit RGB value. (For more information on RGB values, see "Colors and the Palettes" on page 111.)

However, rather than having an array of RGB values, most bitmaps save space by having an array of byte or word indices into another array of color values or a color table (Figure 4.5).

**Figure 4.5**  **Most bitmaps use an array of color indices to save space.**

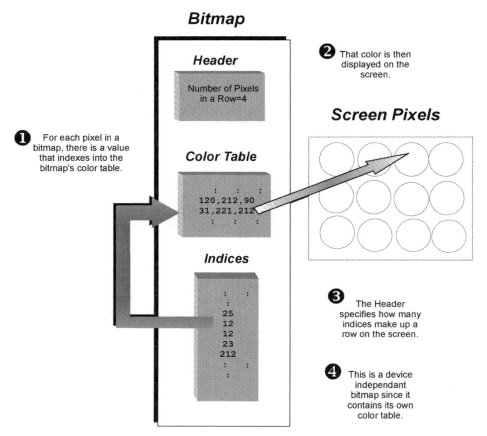

Both Device Dependent Bitmaps (DDB) and Device Independent Bitmaps (DIB) are made up of an array of indices. However, a Device Dependent Bitmap does not have its own color table. Instead, the indices in a DDB point to your application's color palette. A Device Independent Bitmap has it's own color table.

The Bitmap objects you create or load are all Device Dependent Bitmaps. They contain no color table of their own and depend on your application's palette for their color values. Bitmaps that reside on the disk are Device Independent Bitmaps. A .bmp file contains a file header, a bitmap header, a color table, and the array of pixel colors indices. When you use CBitmap's Load-Bitmap(), this DIB bitmap file is converted to a DDB. The color table in the

DIB is converted to colors in your application's palette and the pixel color array is adjusted accordingly.

## Drawing Bitmaps

To draw a bitmap on the screen or printer, you need two Device Contexts: one that contains the bitmap attribute and another to which you copy the bitmap. Why can't the Device Context that points to the Bitmap object simply draw it to the screen or printer itself? Who knows? Certainly the drawing functions provided are more versatile than one that simply copies its own bitmap to the screen.

The calling arguments for the most popular bitmap drawing function is

```
pDC->BitBlt(int x, int y, int nWidth, int nHeight, CDC* pSrcDC,
    int xSrc, int ySrc, DWORD dwRop);
```

- pDC is the Device Context to which the bitmap will be copied.
- int x and int y are the upper-left corner of the location in which to draw the bitmap.
- int nWidth and int nHeight are the size of the bitmap.
- pSrcDC is the Device Context that points to the bitmap to draw.
- int xSrc and int ySrc are the upper-left corner of the bitmap to copy.
- DWORD dwRop defines how to blend the bitmap with its background on the destination device.

This function not only copies bitmaps, but with some dwRop values, a source bitmap object isn't even used.

For examples of drawing bitmaps, see Examples 56 and 57.

# Metafiles and Paths

Windows and MFC provide two ways to store a figure to draw later: Metafiles and Paths.

**Metafiles** essentially capture any drawing to a Device Context, so that it can be replayed later or even stored to disk. As mentioned previously, bitmaps are always device dependent once loaded into memory. You can, therefore, think of Metafiles as bitmaps that are device independent in memory — however, they are a lot slower to draw.

**Paths** capture drawing functions just like Metafiles, except that Paths can draw modified versions of their shapes. You can also fill a Path figure or turn it into a clipping region.

## Metafiles

To create a Metafile using MFC, you first create a special Device Context that is a derivative of the `CDC` class called `CMetaFileDC`.

```
CMetaFileDC dcMeta;                    // creates the class object
dcMeta. CreateEnhanced(pDCRef, "myMetaFile.EMF", lpDimension,
    "description");
                                       // creates the Device Context
```

---

**NOTE:** Actually, we're creating an Enhanced Metafile here — the older version was not device independent and should be considered obsolete. However, for the purposes of simplifying this discussion, I will refer to an Enhanced Metafile as simply a Metafile.

---

After creating the Metafile Device Context, you then draw to it as you would any Device Context. When you're finished, you save your Metafile to disk by closing it.

```
HENHMETAFILE hMetaFile=dsMeta.CloseEnhanced( );
dcMeta.DeleteDC();               // delete metafile device context
::DeleteEhMetaFile(hMetaFile);   // delete metafile from memory
```

To load and play an existing Metafile, use

```
HENHMETAFILE hMetaFile=::GetEnhMetaFile("myMetaFile.EMF");
pDC->PlayEhMetaFile(hMetaFile);
```

## Paths

A Path is created just like a Metafile except that you start by drawing to a normal Device Context. To start a Path you call

```
CDC *pDC=GetDC();
pDC->BeginPath();
```

All subsequent line drawing functions will then be stored to this Path. All drawing functions that don't draw lines will be ignored. To mark the end of a Path call

```
pDC->EndPath();
```

This stops the drawing and selects the Path to the Device Context. Yes, the Device Context has yet another object similar to pens and brushes to which it points for a Path. However, the Path object is not initialized to any value when the Device Context is created.

---

**NOTE:** You cannot append to a Path. Calling `BeginPath()` again will simply destroy the path currently defined in the Device Context.

---

There are several functions you can use to draw a created and selected Path, including the following

| | |
|---|---|
| `StrokePath()` | Draws the path using the current Pen object. |
| `FillPath()` | Fills a path using the current Brush object. |
| `StrokeAndFillPath()` | Both of the above at once. |
| `SelectClipPath()` | Adds Path to the current clipping region. |

# Colors and the Palettes

For most applications, you will use the RGB macro to define a color for drawing. The syntax for the RGB macro is

```
RGB(BYTE nRed, BYTE nGreen, BYTE nBlue)
```

where nRed, nGreen, and nBlue represent color intensities. Their ranges are from 0 to 255. For example, to specify the color gray, you would use RGB(127,127,127). White would be RGB(255,255,255). Black would be RGB(0,0,0). To draw a solid blue line from 10,10 to 20,30, you would use

```
CDC *pDC=GetDC();              // create a Device Context object
CPen pen;                      // create a CPen class object
pen.CreatePen(PS_SOLID,1,RGB(0,0,255));
                               // create a Pen object with blue color
pDC->SelectObject(&pen);       // point Device Context to Pen object
pDC->MoveTo(10,10);            // draw line
pDC->LineTo(20,30);
```

## Dithered Colors

Colors created using the RGB macro are actually dithered colors. A dithered color is produced by a pattern of differently colored dots that simulate the desired color. If you look closely at a pink color drawn using RGB(255,127,127), you'll notice that it is actually made up of interspersed red and white pixels that look pink from a distance. Some standard colors will be solid when using the RGB() macro, such as red and white, but any nonstandard color you draw will be dithered. Dithering enables you to draw any color imaginable without having to worry about system palettes and video memory. However, any nonstandard color will appear a little fuzzy.

## Nondithered Colors

Unfortunately for most CAD and graphic applications, the fuzziness from dithering is unacceptable. Lines seem to bleed into the shapes they surround unless they are a standard color.

So why dither? After all, each pixel in a screen can actually display the entire spectrum of colors from black to white. Which means you can display any color imaginable — without dithering — just by setting a pixel color directly. All a pixel needs is a 32-bit RGB value to tell it what color to be.

However, since each pixel would require a 32-bit integer to fully define a color, an 800 × 600 display would need almost 2Mb of video memory to fill the screen, which is prohibitive for most systems.

You could force everyone who uses your application to buy lots of video memory, but most application developers work with the system palette, instead.

## The System Palette

The system palette allows you to have solid colors of any shade without lots of video memory. It was created based on the idea that even though there may be hundreds of thousands of pixels in your display, they may contain only a couple of hundred unique colors. Rather than store the same color value over and over again, these unique colors are instead stored in the system palette and each pixel then simply points to that palette. Consequently, each pixel may only require four or eight bits to point to a color in this palette and video memory requirements plummet (Figure 4.6).

**Figure 4.6    The System Palette enables you to lower video memory requirements.**

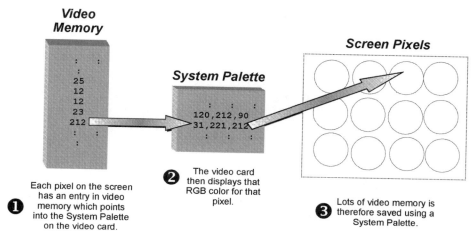

The number of bits required per pixel depends on the current screen settings for the System Palette, which are set through the Control Panel of the Windows operating system.

| System Palette Setting | Bits Required per Pixel in Video Memory | Video Memory Required for an 800 × 600 Screen |
|---|---|---|
| 16 Color | 4 bits | 234Kb |
| 256 Color | 8 bits | 468Kb |
| High Color | 16 bits | 937Kb |
| True Color | 32 bits | 1.875Mb |

Notice that setting the System Palette to "True Color" brings us right back where we started — with no need for a System Palette, but a big requirement for video memory. With True Color, every pixel gets its own unique solid color. Buying enough video memory means never having to worry about the System Palette or dithering.

## Using the System Palette

The following steps enable you to use the system palette.

### 1. Create Your Own Application Palette

You create an application palette by filling a LOGPALETTE structure with the colors you need and then creating a Palette object.

```
LOGPALETTE *lp = (LOGPALETTE *)calloc(1, sizeof(LOGPALETTE) +
        (NUM_MAP_COLORS * sizeof(PALETTEENTRY)));
lp->palVersion = 0x300;
lp->palNumEntries = NUM_COLORS;
for (color = 0; color < NUM_COLORS; color++)
{
    lp->palPalEntry[color].peRed=
        GetRValue(application_colors[color]);
    lp->palPalEntry[color].peGreen=
        GetGValue(application_colors[color]);
    lp->palPalEntry[color].peBlue=
        GetBValue(application_colors[color]);
    lp->palPalEntry[color].peFlags =   NULL*;
}
pPal = new CPalette;
pPal->CreatePalette(lp);
free(lp);
```

### 2. Copy Your Application Palette into the System Palette

Your application's palette is copied into the system palette when your application becomes active and receives the WM_QUERYNEWPALETTE message. Most applications process this message in the CMainFrame class.

```
CMainFrame::OnQueryNewPalette()
{
    CDC *pDC=GetDC();
    pDC->SelectPalette(pPal,FALSE);
//pPal is your CPalette class object
```

```
if (pDC->RealizePalette())
        pDC->UpdateColors();
    Release(pDC);
}
```

The RealizePalette() member function of CDC attempts to store the colors in your application's palette into the system palette. First, RealizePalette() tries to stuff all of your application's colors into spots that are currently unused by any other application running. If there are no currently unused spots, RealizePalette() starts blowing away colors that are used by other applications. You can see the effect this has whenever you switch from one application to another on the desktop. The colors in the application you just left sometimes suddenly go crazy. If RealizePalette() still doesn't have enough room in the system palette, your application's colors are matched up with similar colors already in the palette — sometimes they are good matches, sometimes they aren't (Figure 4.7).

The UpdateColors() member function of CDC attempts to conform the colors in your drawing area to the new system palette. Before RealizePalette() modifies the system palette, it first saves a copy of the old palette. UpdateColors() examines the color of each pixel that used to point to the old system palette, tries to find the same or similar color in the new palette, then modifies the pixel color to point to the new system palette color. UpdateColors() is a way for your application to adjust to a new system palette without having to redraw everything. After a while, though, your colors will start to drift (Figure 4.8).

**Figure 4.7**   **The** CDC **member function** RealizePalette() **allows your application's palette to work with the System Palette.**

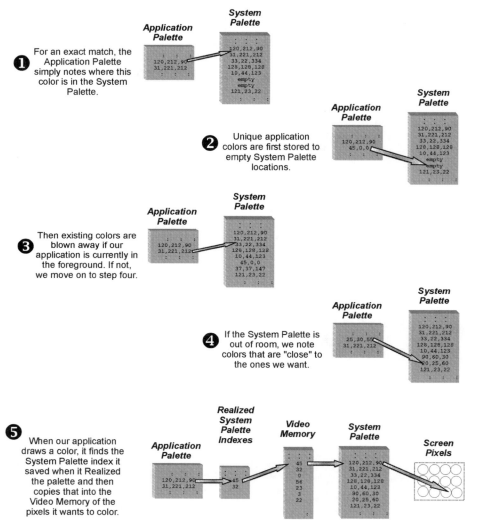

**Figure 4.8** The `CDC` **member function** `UpdateColors()` **allows your application to adjust to a new System Palette without having to redraw everything.**

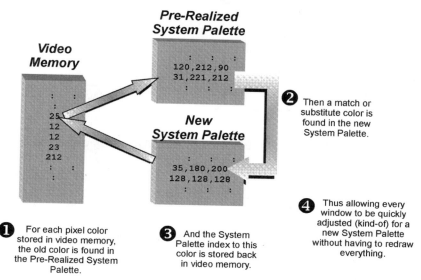

### 3. Define Colors Using `PALETTERGB()` and `PALETTEINDEX()`

Rather than use the `RGB()` macro to define colors, you must now use two other macros:

```
PALETTERGB(BYTE nRed, BYTE nGreen, BYTE nBlue)
```

or

```
PALETTEINDEX(nIndex)
```

where `nIndex` is an index to your application's palette colors.

You would use the following to create a blue Pen object.

```
pen.CreatePen(PS_SOLID,1,PALETTERGB(0,0,255));
    //create a Pen object with blue color
```

Internally, the difference between these three macros is the fourth byte of the color word created, as seen here.

| RGB() | | | |
|---|---|---|---|
| 0 | Blue | Green | Red |
| PALETTEINDEX() | | | |
| 1 | Index | | |
| PALETTERGB() | | | |
| 2 | Blue | Green | Red |

---

**NOTE:** The RGB() macro will use color dithering even if you take the time and effort to put the exact color you want in the system palette! If you go to all the bother of using the system palette, make sure you access it by using PALETTERGB() or PALETTEINDEX().

---

### 4. Fix up Colors When Your Application is not Active

WM_PALETTECHANGED informs you that another application has just changed the system palette, usually because the application just became active. We are, therefore, given an opportunity to make our colors work with this new palette, but with the understanding that our requirements are secondary to this newly active application. If your application has lots of colors, you might not want to bother processing this message and live with the fact that your application will appear psychedelic until it again has control of the system palette. However, if you want to try to update your application palette, use

```
CMainFrame::OnPaletteChanged()
{
//if we changed the palette, ignore this message
if (pFocusWnd == this)
        return;
    pDC=GetDC();
    pDC->SelectPalette(pPal, FALSE);
    if (pDC->RealizePalette())
        pDC->UpdateColors();
    Release(pDC);
}
```

Actually, the last part of this routine is identical to `OnQueryNewPal-ette()`. However, because we are now a background application, `Realize-Palette()` will not overwrite any used spots in the system palette. `UpdateColors()` will now be faced with an even tougher job of coming up with matches that are close to the original.

## Animating Colors

Every time you use `RealizePalette()`, a `WM_PALETTECHANGED` message is sent to every application in the system if a color changes. If you change your colors often, this can become rather CPU-intensive. For this reason, you should use `AnimatePalette()` instead of `RealizePalette()`.

1. Use `PC_RESERVE` when creating the application palette entry you want to change.

2. Use `AnimatePalette()` to change the color.

```
PALETTEENTRY  p_entry;
p_entry.peRed =     GetRValue(color);
p_entry.peGreen =   GetGValue(color);
p_entry.peBlue =    GetBValue(color);
p_entry.peFlags =   PC_RESERVED;
pPal->AnimatePalette(index,1,&p_entry);
```

# Controlling When and Where to Draw

Normally, you should only draw to a window when processing a `WM_PAINT` or `WM_DRAWITEM` window message. The `WM_PAINT` window message is sent by the system when it's time to draw the client area of a window. The `WM_DRAWITEM` message is sent to the owner of a user-drawn control when it's time to draw some part of that control.

You can draw at other times, but you must be aware that the next time a `WM_PAINT` message is processed, whatever you drew will be overwritten by your paint routine. The only other time you might want to draw to a window would be to move objects or lines as a result of mouse commands. An example of this is when the mouse is used to draw a lasso box around a group of items in the view.

> **NOTE:** You could use `CWnd::LockWindowUpdate()` to stop your paint routine from drawing to the window. However, the window then can't be moved.

## Processing `WM_PAINT`

The standard way to handle `WM_PAINT` messages is

```
void CMyWnd::OnPaint()
{
    CPaintDC dc(this); // creates a Device Context

    // draw to dc
    :    :    :
}
```

The `CPaintDC` class shown here is a derivative of the `CDC` class. Not only does it create a Device Context for you, but it also releases the Device Context when `OnPaint()` returns and `CPaintDC` deconstructs itself.

## Drawing Only Invalidated Areas

To prevent excessive screen flicker and speed up refresh time, the Device Context you get from `CPaintDC` does not always allow you to draw to the entire client area of a window, but instead allows you to redraw only to the parts of the window that have changed (e.g., the last line in a text document). `CPaintDC` tells the Device Context to clip any drawing outside of an invalidated area, simply by adding the invalidated area to the Device Context's Region object. If you recall, the Region object defines a clipping region outside of which you can't draw.

Normally, a region becomes invalidated when a window is new or another window that had drawn over it is now closed. When your application enters its idle loop, it sends a `WM_PAINT` message to any window with an invalidated region. You can speed this process up by invalidating an area yourself and forcing the WM_PAINT message to be sent to a window immediately:

```
InvalidateRect(rect);   // invalidates rectangular area of
                        // client window -or--
Invalidate();           // invalidates entire client area
UpdateWindow();         // sends WM_PAINT message immediately
```

---

**NOTE:** One of the more frustrating problems you might encounter when writing a drawing routine is to have it seemingly draw nothing. Everything seems as it should, but nothing appears on the screen. This clipping action is usually the problem and can be bypassed for testing by simply ignoring the Device Context you get from CPaintDC and creating your own using CClientDC.

---

## OnDraw()

When drawing to a view (a window created by the CView class or one of its derivatives), you should process WM_PAINT messages in CView's OnDraw() rather than OnPaint(). CView can print your view to the printer by simply calling OnDraw() with a printer Device Context, which it does when the user uses the standard Print command in the File menu.

Unfortunately, a lot of applications don't implement OnDraw() because they don't use CView directly. Some use the CFormView class for their view, which is made up of Common Controls and Common Controls don't use OnDraw() to draw themselves. You can still print this type of view, but more as a screen capture than a drawing.

## Processing WM_DRAWITEM

Some Common Control windows allow you to draw the control yourself. You might want to keep the functionality of a control, like a button, but with a different look instead, such as the tab on a folder. To draw your own Common Control, you would create that control using a style that makes it Owner drawn. For a Button control, that's the BS_OWNERDRAW style. The Parent of such a control would then receive a WM_DRAWITEM window message whenever it was time to draw some part of the control. Included with the WM_DRAAWITEM window message is the following structure.

```
typedef struct tagDRAWITEMSTRUCT {
    UINT    CtlType;
    UINT    CtlID;
    UINT    itemID;
    UINT    itemAction;
    UINT    itemState;
    HWND    hwndItem;
    HDC     hDC;
    RECT    rcItem;
    DWORD   itemData;
} DRAWITEMSTRUCT;
```

The hDC value in this structure is your Device Context for drawing the control. To wrap a CDC class around it, you could use

```
CDC *pDC=FromHandle(hDC);
```

---

**NOTE:** For true encapsulation, you should process the WM_DRAWITEM message in a class derived from the Common Control. For example, you would create a class called CMyButton derived from CButton that processes WM_DRAWITEM using message reflection, discussed in Chapter 3.

---

## Drawing at Other Times

As mentioned previously, you might also find yourself drawing in tandem with mouse commands, such as when using the mouse to draw a line. For these and other times you draw without WM_PAINT or WM_DRAWITEM, there are two other derivatives of the CDC class that can make your life easier.

- The CClientDC() class creates a Device Context just like CPaintDC(), except that it doesn't try to clip anything. The CClientDC class creates your Device Context when your routine starts and releases it when your routine returns.

```
CMyClass::Foo()
{
    CClientDC dc(this);


       :      :      :
}
```

---

**NOTE:** One of the hazards of using a convenience like a Device Context is that you always have to remember to release it when you're done, or else suffer the pain of having your application slowly consume all memory. Rather than use CDC *pDC=GetDC(); to create a Device Context, you might be better served to make a habit of using CClientDC, since you never have to worry about releasing your Device Context.

---

• The CWindowDC() class is identical in purpose to the CClientDC(), except that it creates a Device Context that allows you to draw to the entire window, including the nonclient area.

## Summary

In this chapter, we looked at the Device Context object and how it simplifies the calling arguments of drawing functions. We reviewed the various drawing tools and modes available to us under MFC. We discovered how convoluted drawing nondiffused colors can be when there isn't enough video memory. We also discovered that you can wait for the system to send you a message to draw or you can force a window to draw immediately.

This concludes the pure text portion of this book. In the following sections, we will look at MFC and Visual C++ from the other angle. Actual features you might want to add to your own application will be presented, along with a step-by-step guide to implementing them. And yes, there will be more notes explaining how these features are implemented, so that they might be broadened and enhanced.

# Section II

# User Interface Examples

The examples in this section concentrate on the user interface aspect of the applications you can create with the help of the Developer Studio, the Microsoft Foundation Classes (MFC), and Visual C++. As you might expect from a user interface development tool, the vast majority of the examples in this book will be in this section. The topics have been arranged in the order in which they should be considered when creating an MFC application. Example 1 outlines a battle plan for creating any MFC application with references to the examples that follow. The topics covered in this section by chapter include the following.

## Applications and Environments

Examples in this chapter include planning the implementation of applications using MFC, et al., using both the Wizards and brute force. Most of the common aspects of your application interactions with its environment are included, such as initializing its screen, displaying its icons, processing command line options, and saving its preferences.

## Menus

The next area of concern is application menus — namely, adding commands, updating with the state of your application, and experimenting with appearance. Also included is how to direct menu commands to your application's classes using the Class Wizard.

## Toolbars and Status Bars

The examples in this chapter look at creating toolbars and status bars with the Developer Studio's editors. Examples are included of updating controls for both types of bars to reflect the state of your application, as well as adding nonstandard controls to either one.

## Views

If you choose to create either an SDI or MDI application, your application's view will be your user's primary mode of interacting with your application. The type of application you're creating will determine the type of view you will want to build into it. Other aspects of the view include splitting it up, and conditionally changing the look of the mouse cursor.

## Dialog Boxes and Bars

The second mode of interaction with your application will be through dialog boxes and bars. They can be modal or modeless. They can be entirely created by your or you can customize a system-supplied dialog box.

## Control Windows

Dialog Boxes are populated with buttons and edit boxes that are collectively known as Control Windows (Child Windows supplied by the operating system). Not only can you fill your dialog boxes with them, but you can put them in views, bars, or anywhere there's a window.

## Drawing

The examples in this chapter cover a range of items, from drawing figures and text to manipulating bitmaps.

# 5

# Applications and Environments

Welcome to Applications and Environments. Here we will plan out an MFC application, determine whether it will have a Dialog, Single Document (SDI), or Multiple Document (MDI) interface, and help distinguish it from all the other MFC applications generated by the Developer Studio by giving it some pizzazz.

**Example 1    Planning Your MFC Application**   We will devise a strategy for using the Developer Studio, et al., to turn your application idea into an actual application.

**Example 2    Creating an MFC Application Using the AppWizard**
We will use the Application Wizard to generate a set of classes and resources that will become the foundation of your MFC application.

**Example 3    Creating a Class Using the ClassWizard**   We will use the Class Wizard to add classes to your application.

**Example 4     Initializing the Application Screen**   We will take control of the initial size and placement of your application's window.

**Example 5     Saving the Application Screen**   We will save the size and position of your application's window for its next execution.

**Example 6     Processing Command Line Options**   We will convert command line flags into Boolean variables we can use in our application.

**Example 7     Dynamically Changing Application Icons**   We will change your application's icon, which not only appears in the upper left corner of the application, but also appears in your system's task bar.

**Example 8     Prompting the User for Preferences**   We will prompt our user for your application's options.

**Example 9     Saving and Restoring User Preferences**   We will save your application's options in the system registry.

**Example 10     Terminating Your Application**   We will look at a way to control how your application exits.

**Example 11     Creating a Splash Screen**   We will create an initial screen for your application that presents its name and affiliation.

# Example 1   Planning Your MFC Application

## Objective

You would like to create an application using Visual C++, the MFC libraries, and the Developer Studio's wizards and editors.

## Strategy

We will start by deciding what type of MFC application would best suit your needs: Dialog, SDI, MDI, or none of the above. We will then pick the best view and document for that application. Next, we'll review other ways your user can interact with your MFC application and what Developer Studio editors you can use to add those interfaces. If the functionality in your application can be shared with other applications, we'll look at your library

Example 1   Planning Your MFC Application   **129**

choices. And finally, since every application that the Developer Studio creates looks almost alike, we'll explore several features you can add to your application to make it stand out.

# Steps

## Pick an Application Type

1. There are three types of MFC applications you can create automatically with the Developer Studio's AppWizard utility (as seen in Example 2): Dialog, SDI, or MDI. Manually, you can create any type of hybrid application. For more details on deciding just what type of application to choose, see Chapter 2. To make a quick decision now, try following these guidelines.

- If you are creating an application with a limited need for a user interface, or if you want your interface to be totally unique, then create a Dialog Application. Typical Dialog Applications include applications that configure hardware devices, screen savers, and game programs.

- If your application will be editing a document, you should pick one of the other two application types. In this context, "editing a document" is meant in the broadest possible sense. The document we refer to here can be a text file, a spread sheet file, one or more tables in a third-party database, or your own proprietary binary file. It can even be the stored settings of a vast array of hardware devices. Editing simply means adding, deleting, or modifying the data in any one of these types of documents.

  - A Single Document Interface SDI) Application allows you to edit only one document at a time. If your application can only physically edit one document at a time, as is the case of an application that monitors an array of hardware devices, then you should pick an SDI interface. Otherwise, you should create an MDI application — even if there doesn't initially appear to be any benefit to editing more than one file at a time.

  - A Multiple Document Interface (MDI) Application allows you to edit several documents at once. An MDI application isn't that much more complicated than an SDI application, and your user gets the added convenience of being able to at least view more than one document at a time.

2. If you have decided to create a Dialog application, you're done with this example. A Dialog application has no view or document, and there's really not much more to the interface. However, you might look at the rest of this example to see some of the ways to add pizzazz to your application.

## Pick a Type of View

1. If you picked an SDI or MDI application, you must choose the type of view you want before you exit the AppWizard. In the last step of the AppWizard, you can choose one of the following view classes.

- For a simple text editor application, pick CEditView.

- For an application that can edit rich text format (RTF) files, pick CRichEditView. (This selection will also cause your application to pick CRichEditDoc for your document class.)

- For a graphic application, pick CScrollView.

- For a simple monitoring or accounting application, pick CListView (Example 36).

- To start the creation of an Explorer-type interface, pick CTreeView. (You can manually add a CListView in a later step.)

- To create a view out of a dialog box template, pick CFormView. (A dialog box is a window inhabited by several other Control windows, such as buttons and edit boxes. See Chapter 1 for more on this topic.)

2. You can also indirectly pick either the CRecordView or CDaoRecordView class for your view in an earlier step of the AppWizard, in which you decide what database support to add to your application. If you pick either of the "Database View" options in the second step, this view is added to allow you to easily access the records in an ODBC or DAO database.

## Pick a Document Type

1. There are three basic types of documents your application can interact with: flat files, serialized files, or databases. The choice is usually made for you by the type of application you intend to write. Let's review the choices.

Example 1   Planning Your MFC Application   **131**

- A flat binary or text file is the simplest of documents your application can support and can usually follow any storage method you can dream up. To see what MFC classes support flat files, see Example 63.

- Serialized files represent MFC's organized way of storing binary files so that they can be easily retrieved, even when they were made by a previous version of your application. Any type of document, whether it's a text file, a spreadsheet file, or database data, can be stored this way (Examples 66 to 70).

- Your application can also work with the Microsoft Jet Engine Database Management System (DBMS) and any other third-party ODBC-compliant DBMS (Examples 72 and 73).

## Other Considerations

1. The view represents the primary method your user has to interact with their document. However, there are several other features you can add to your application to allow them to interact.

- You can add commands to the main menu using the Menu Editor. (See Examples 12 and 13 for more on this topic. For popup menus, see Example 21.)

- You can add toolbars and buttons using the Toolbar Editor (Example 22).

- You can add modal dialog boxes, which allow your user to enter detailed information while suspending your application (Example 40).

- You can add modeless dialog boxes, which are dialog boxes that don't suspend your application (Example 41).

- You can add dialog bars, which are a hybrid of toolbar and modeless dialog box, allowing your user to dock this type of dialog box (Example 45).

- You can add Property Sheets, which are Window's conventional way to allow a user to enter and save their preferences (Example 8).

2. As you start to add layer upon layer of functionality to your MFC application, you may start to wonder which class should contain what functionality. For example, which class should process what command message and what window message? You can easily access the data and functionality of one class from another, as seen in Appendix D, but where should that functionality initially reside? Here's a rough guide to help you decide.

- The Application Class, derived from `CWinApp` and controlling no window, should have precious little additional functionality itself, other than to control the creation and destruction of your application. This can include processing command line flags and providing a customized method of opening documents. The Application Class also serves some application-wide services such as background processing and superclassing.

- The Mainframe Class, derived from `CFrameWnd` and controlling your application's Main window, should be in charge of all application-wide interfaces, including the toolbars, status bar, menu, and dialog bars. However, if any of these bars has additional functionality, it should be encapsulated in its own class. Support for user preferences is also found typically in the Mainframe Class.

- The Document Class, derived from `CDocument`, should contain any data pertaining to your application's document. For true C++ encapsulation, the Document Class shouldn't allow direct access to its data — not even from the View Class — but instead should contain wrapper functions to access its data. The Document Class should also contain all of the functionality necessary to load and save a document, from simple binary files to ODBC databases. If your application does nothing but access an ODBC database, your Document Class may contain nothing more than the logic necessary to open and close that database, since the database is the main repository of your data. The Document Class should be an island unto itself, getting its information from a storage device and handing it out to the view, but rarely storing information in another class.

- The View Class, derived from `CView`, should contain all of the logic necessary to view and edit the data in the Document Class. Any menu or toolbar commands that operate specifically on the document, such as cut or paste, should be processed here. All mouse messages affecting the view should be processed here. All drawing, reporting, editing, selecting, and printing should be done here. All dialog boxes and popup menus should be spawned here. If this or any class starts to become massive, you should factor any common functionality into a new base class. Create a new `CMyBaseView` class and stuff some basic functionality there. Or you can encapsulate some functionality into its own class. The functions that select, cut, and paste in the view make a good candidate for its own class.

- Other Classes should encapsulate as much of their own functionality as possible. Dialog classes should contain everything they need to prompt

Example 1    Planning Your MFC Application    **133**

the user, as should dialog bars, toolbars, and status bar classes. A self-drawn control should be drawn from within its own class.

- Any time you factor out common functionality from your application into a base class, you can put the new base class into an MFC Extension Class. You can then move the new MFC Extension Class into a Dynamically Linked Library, from which applications can share this functionality.

---

**NOTE:** As another rule of thumb, if you find that one class is constantly accessing the functions and data of another class, it would probably behoove you to move that functionality to the other class.

---

3. And finally, there are several ways to snazz up your application to help distinguish it from the 3,700 other applications created every day using the Developer Studio.

- Add a splash screen that appears when your application first starts up (Example 11).

- Show an animation of progress rather than a "Please Wait" message (Example 43).

- Make your toolbar buttons larger and stick words in them (Examples 24 and 25).

- Put other types of status messages in the status bar (Example 31).

- Add controls other than buttons to your toolbar or use a dialog bar, instead (Examples 28 and 45).

## Notes

- Writing an application using MFC is a constant battle to decide whether to use an existing MFC feature or to write one yourself. Knowing the basics from Section I, you could, for example, easily write your own button control window. You would then have total control over the final product — there's nothing undocumented. If it doesn't perform as expected, you can follow the debugger right down into it to find out why and change it, if desired. However, the work of creating a button control has already been done and tested forever. It's also more universally found and understood and allows someone else to more easily pick up your code where you left off. I highly recommend that if the MFC method of providing a function isn't readily apparent, you should continue the search until you either find one or can verify that one does not yet exist. An application that's written with MFC but looks nothing like it on the inside is a waste of the technology that went into it. When you use MFC and the Developer Studio the way it was intended, you'll find yourself creating applications faster then you ever thought you could. Thank you for your support.

## CD Notes

There is no accompanying project on the CD for this example.

# Example 2   Creating an MFC Application Using the AppWizard

## Objective

You would like to create a Dialog, an SDI, or an MDI application using the Developer Studio's AppWizard.

## Steps

### Create an Application Using the AppWizard

1. Click on the Developer Studio's File/New menu commands to bring up the New dialog box. Select MFC AppWizard (exe), then enter the name and the directory in which you want to create your project (Figure 5.1). Be careful — this name is used externally and internally in almost all of your project files, so any typos will be difficult to fix later.

**Figure 5.1    Specify your application's filename and location.**

Example 2   Creating an MFC Application Using the AppWizard   **135**

2. Pick the application type in the first step of the AppWizard (Figure 5.2). If you don't already know, see the previous example for an idea of application type on which to base your project. The rest of this example assumes you have selected an SDI or MDI application.

## Figure 5.2    Pick your application type.

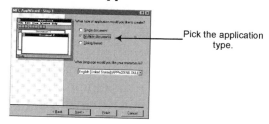

Pick the application type.

3. The next AppWizard step allows you to specify what type of database support you want in your application. Picking Header Files Only causes the AppWizard to simply add the MFC classes that support database access. (You can then use Example 72 or 72 to access an ODBC or DAO database.) Picking Database View without File Support or Database View with File Support causes the AppWizard to create a simple database editor, with a special View and Document class. If you select Database View without File Support, the AppWizard won't add the standard file open commands to your application's menu (i.e., File/New, File/Open, etc.). In theory, if you're only accessing a database, you won't need these commands anyway — the correct database will be opened automatically when the application starts. However, if your application will be accessing both flat files and database files, you should select Database View with File Support.

4. The next AppWizard step allows you to specify what type of COM support you want in your application. For the examples in this book, simply take the default options.

5. The next AppWizard step allows you to pick several general application options (Figure 5.3). You can choose whether or not your application will have a toolbar or status bar, will add print commands to your menu, and will include support for e-mail or network communications. The Recent File List is a list of the last n files your application opened, which can be automatically maintained by your application. You get to determine what n is here. Click on the Advanced button for more advanced options.

**Figure 5.3** **Pick your application's options.**

❶ Pick simple application options.

❷ Open advanced application options.

6. The first page of the advanced options allows you to pick the title that will appear in your application's caption bar. If you are creating an application that will be serializing its documents to disk, you can pick the file extension that your application appends to those documents. You can then edit the text that will appear in the filter field of the File Dialog that appears when opening or saving documents (Figure 5.4).

**Figure 5.4** **Specify your application's title, default file extension, and File Dialog text.**

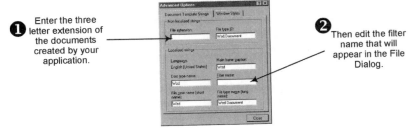

❶ Enter the three letter extension of the documents created by your application.

❷ Then edit the filter name that will appear in the File Dialog.

7. The second page of the advanced options allows you to add view splitting capabilities to your application, which will provide your user with a menu command that will allow them to dynamically split their view (Example 37). You can also determine whether your application's Main window or Child windows are initially maximized or minimized and whether your user can resize them (Figure 5.5 and Example 4).

Example 2    Creating an MFC Application Using the AppWizard    **137**

**Figure 5.5    Specify your application's Frame Window Options.**

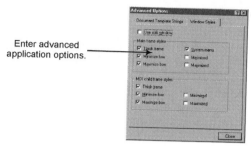

Enter advanced
application options.

8. In the next step of the AppWizard, you must decide whether to statically link with the MFC library or to link with a shared MFC DLL, instead (Figure 5.6). Statically linking to MFC makes your application much larger, but you would never have to make sure the correct version of the MFC DLL is currently installed on your user's system. If you plan to create your own DLL's that also use the MFC library, you must link to the MFC DLL.

**Figure 5.6    Choose how to link to MFC.**

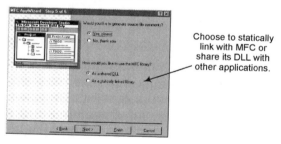

Choose to statically
link with MFC or
share its DLL with
other applications.

9. In the final AppWizard step, you are allowed to change your application's View Class. For a description of the choices, see the last example. For any View Classes not listed, pick the default `CView` class — you can edit the name later (Figure 5.7).

**Figure 5.7    Pick a View Class.**

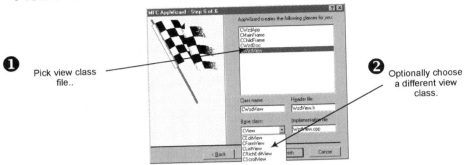

10. The AppWizard will now proceed to create all the classes you need to create a fully-executable, albeit feature-poor, application. Simply click on the Studio's Build/Build All menu commands to create your executable.

## Notes

- If you forget to add something to your application using the AppWizard, you can still add it manually later on, although it isn't nearly as easy. To find out what you have to add to your application, start by creating two new projects, one with the desired feature and one without. Then do a difference on the source code generated for these two projects. (A version control package would help.) Simply incorporate the differences into your application.

- Except where noted, most of the examples in this book are based on creating an MDI application using all of the defaults.

- If you do want to change the name of your application later, start by deleting all of the binary files in your directory. Then use a search and replace utility that can span multiple files to replace the old name with the new name. Make sure you replace all forms of the project name, including all-capitals, all-lowercase, and mixed case. (Search and replace utilities are available on the Internet.)

## CD Notes

- There is no accompanying project on the CD for this example.

Example 2   Creating an MFC Application Using the AppWizard   **139**

# Example 3   Creating a Class Using the ClassWizard

## Objective

You would like to add a class to your MFC application to either extend an MFC class or to exist on its own.

## Steps

### Extend an Existing MFC Class

1. Click on your Developer Studio's View/ClassWizard menu commands to open the MFC ClassWizard dialog. Then click on the Add Class button (Figure 5.8). A drop-down menu will appear from which you should select New... to open the New Class dialog box.

**Figure 5.8    Create a new Class with the ClassWizard.**

Click on "Add Class" button.

2. Enter the name of your new class. Add a "C" to the start of your class name. (The ClassWizard will omit this "C" when creating your class's .h and .cpp files.) Then select the base class from the list of available MFC classes (Figure 5.9). If you choose CRecordSet, the ClassWizard will also lead you through the steps necessary to bind your class with a database table. To derive from the CWnd class, choose "generic CWnd." To derive from CSplitterWnd, choose "splitter." If the MFC class from which you want to derive is not listed (as is the case with CToolBar), pick a similar name (such as CToolBarCtrl) and then edit the resulting files.

**Figure 5.9   Select a Base Class.**

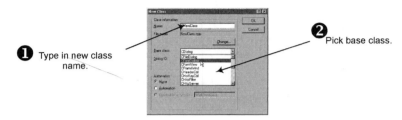

➊ Type in new class
      name.

➋ Pick base class.

## Create a NonMFC Class

1. If you don't want to use an MFC class as your base class, click on the Studio's Insert/New Class menu commands to open an alternative New Class dialog box. This alternative version includes an additional combo box in which you can specify a Class Type. Picking Generic Type allows you to specify your own base class, if any.

   Okay, you've just added a class, but... oops, you misspelled its name and now you want to delete it so you can add one with the correct spelling. You look for a Delete Class button, still can't find it after half an hour of looking, and suddenly the misspelling doesn't look half bad. So you swallow your pride and keep it. Or you continue with the next step and learn how to delete the class manually.

## Delete a Class From the ClassWizard

1. First, you must delete the .cpp and .h files that the ClassWizard created, from both the project list of files and your project's subdirectory.

2. You'd think that would be all you'd have to do, but noooo — the next time you bring up the ClassWizard, the misspelled class will still be there as a specter of your mistake. The ClassWizard keeps a record of every class you create in a separate .clw file. No problem — just delete the .clw file, too. The next time you invoke the ClassWizard, it will tell you it can't find the .clw file and ask if you want to create a new one, to which you answer Yes. The ClassWizard will then create a new .clw file using the .h files in your project's directory.

Example 4    Initializing the Application Screen    **141**

## Notes

- To incorporate a class from another project, copy the appropriate files from that project's directory. The ClassWizard won't recognize this new class until you do one more step: delete the .clw file in your project and reinvoke the ClassWizard. When the ClassWizard can't find it's .clw file, it will ask you if you want to recreate it. Answer Yes.
- The ClassWizard in version 6.0 of the Developer Studio automatically updates its .clw file.

## CD Notes

- There is no accompanying project on the CD for this example.

# Example 4    Initializing the Application Screen

## Objective

You would like to position and size your application's initial screen.

## Strategy

We have two options. The first is to make the appropriate choices in the AppWizard's advanced options when creating the application. If, however, you want to change your selection in an existing application, we will add code to the `CMainFrame`'s `PreCreateWindow()` to control the initial position and size of our application's main window.

## Steps

### Use the AppWizard

1. Refer to Example 2, the 7<sup>th</sup> step, on page 136. Click on the Advanced button and select the Window Styles tab. Picking a Thick Frame allows your user to resize your application's window by dragging the lower-right corner. Picking Minimized or Maximized forces your application's window to be initially minimized or maximized.

If you want to change your choices later, you will need to edit the CMain-Frame::PreCreateWindow() function directly. Direct editing also allows you to make some additional changes to your application's initial appearance.

### Edit CMainFrame::PreCreateWindow()

1. To center your application and make it fill only ninety percent of your screen, you can use the following code.

```
BOOL CMainFrame::PreCreateWindow(CREATESTRUCT& cs)
{
    // center window at 90% of full screen
    int xSize = ::GetSystemMetrics (SM_CXSCREEN);
    int ySize = ::GetSystemMetrics (SM_CYSCREEN);
    cs.cx = xSize*9/10;
    cs.cy = ySize*9/10;
    cs.x = (xSize-cs.cx)/2;
    cs.y = (ySize-cs.cy)/2;
    return CMDIFrameWnd::PreCreateWindow(cs);
}
```

2. If you would also like to eliminate the document's title from your application's title bar, add the following to PreCreateWindow().

```
cs.style &= ~ FWS_ADDTOTITLE;
```

3. If you would also like to remove the minimize and maximize buttons from your application's title bar, add

```
cs.style &= ~(WS_MAXIMIZEBOX|WS_MINIMIZEBOX);
```

4. If you would like to make the size of your application fixed, such that dragging the lower-right corner of the window would have no effect, then add

```
cs.style &= ~WS_THICKFRAME;
```

Example 4    Initializing the Application Screen    **143**

5. If you would like your application to be maximized when initially executed, then locate ShowWindow() in your application class and change it to use the SW_SHOWMAXIMIZED flag instead of m_nCmdShow.

```
pMainFrame->ShowWindow(SW_SHOWMAXIMIZED); //or SW_SHOWMINIMIZED
pMainFrame->UpdateWindow();
```

6. If you would like to initially maximize a child window in an MDI application, add PreCreateWindow() to your CChildFrm class and add the following to it.

```
BOOL CChildFrame::PreCreateWindow(CREATESTRUCT& cs)
{
    cs.style = WS_CHILD | WS_VISIBLE | WS_OVERLAPPED |
    WS_CAPTION | WS_SYSMENU | FWS_ADDTOTITLE |
    WS_THICKFRAME | WS_MINIMIZEBOX | WS_MAXIMIZEBOX |
    WS_MAXIMIZE;
    return CMDIChildWnd::PreCreateWindow(cs);
}
```

II    5

## Notes

- If you initially maximize your application's window, you should also set an initial size for it in CMainFrame::PreCreateWindow(). When your user clicks on your application's Restore button, your application's window will snap down to whatever size you set in PreCreateWindow().

- This example always sets the initialize size of your application's window to a fixed size and position. Any changes your user makes to the window size or position are not saved. To save the window size and position, see the next example. If you use the next example, however, you should still use this example. The first time your application runs on a system, it won't have any saved settings, so it will need to use these initial settings.

- If you don't set an initial size and position for your window, the Windows operating system will pick one itself based on a cascading algorithm. Each new application's window is created just to the right and below the last application as shown in Chapter 1.

## CD Notes

- When executing the project on the CD, the application will be initially maximized (fill the screen) and unsizable.

# Example 5   Saving the Application Screen

## Objective

You would like to save the size, position, and status of your application's screen, including the location and size of any toolbars or dialog bars so that they can be restored the next time your application runs.

## Strategy

When our application closes, we will save the size and position of our main window as well as the status of the toolbars and status bar to a location in the system registry. Then, when our application is executed again, we will retrieve this information and restore our window and toolbars, et al.

## Steps

### Save the Settings

1. First, we will define the location in the system registry where we will be saving this information in a global include file. "Company" would be your company's name.

```
#define COMPANY_KEY "Company"
#define SETTINGS_KEY "Settings"
#define WINDOWPLACEMENT_KEY "Window Placement"
```

2. In the InitInstance() member function of our application class, add COMPANY_KEY to SetRegistryKey().

```
SetRegistryKey(COMPANY_KEY);
```

3. Use the ClassWizard to add a WM_CLOSE message handler to your CMain-Frame class. There you can save your bar positions and sizes using Save-BarState(). To get your application's current size and position, use

Example 5    Saving the Application Screen    **145**

GetWindowPlacement() and save its results to the system registry using
WriteProfileBinary().

```
void CMainFrame::OnClose()
{
    // save state of control bars
    SaveBarState("Control Bar States");
    // save size of screen
    WINDOWPLACEMENT wp;
    GetWindowPlacement(&wp);
    AfxGetApp()->WriteProfileBinary(SETTINGS_KEY,
        WINDOWPLACEMENT_KEY, (BYTE*)&wp,
        sizeof(WINDOWPLACEMENT));
    CMDIFrameWnd::OnClose();
}
```

## Restore the Settings

1. To restore your toolbars to their original state after reexecuting your application, add the following to the start of the OnCreate() member function in your CMainFrame class.

```
LoadBarState("Control Bar States");
```

2. To restore your application's main window from the system registry, locate ShowWindow() in your Application Class and replace it with the following code. Note that we now use SetWindowPlacement() to restore the main window to its original size and position.

```
BYTE *p;
UINT size;
WINDOWPLACEMENT *pWP;
if (GetProfileBinary(SETTINGS_KEY, WINDOWPLACEMENT_KEY, pWP, &size))
{
    pMainFrame->SetWindowPlacement(pWP);
    delete []pWP;
}
else
```

```
{
    pMainFrame->ShowWindow(m_nCmdShow);
}
pMainFrame->UpdateWindow();
```

## Notes

- To save other options to the system registry, as well as for more on accessing the system registry from your application, see Example 9.
- This example will only position your application's main window after your user has executed your application once. To initialize your application window for the first time, please see the previous example.

## CD Notes

- When executing the project on the CD, you can reposition and resize the main window, then exit the application. When you reexecute the application, whatever size or position it had on terminating will be restored.

# Example 6   Processing Command Line Options

## Objective

You would like your application to process command line flags as seen here.

```
>myapp /c /d
```

## Strategy

An MFC application already processes several standard flags using the `ParseParam()` member function of the `CCommandLineInfo` class. To add our own flags while still supporting these other flags, we will derive our own class from `CCommandLineInfo` and then override `ParseParam()`.

Example 6    Processing Command Line Options    **147**

# Steps

## Create a New `CCommandLineInfo` Class

1. Use the ClassWizard to create a new class derived from `CCommandLineInfo`. Add a Boolean or string member variable to this new class for each new flag your application will process.

```
class CWzdCommandLineInfo : public CCommandLineInfo
{
public:
    BOOL m_bAFlag;
    BOOL m_bCFlag;
    BOOL m_bDAFlag;
    CString m_sArg;
```

2. Also add a `ParseParam()` function to override the base class's `ParseParam()` function.

```
// Operations
public:
    void ParseParam(const TCHAR* pszParam,BOOL bFlag,BOOL bLast);
};
```

3. Implement `ParseParam()` as follows.

```
void CWzdCommandLineInfo::ParseParam(const TCHAR* pszParam,
    BOOL bFlag,BOOL bLast)
{
    CString sArg(pszParam);
    if (bFlag)
    {
        m_bAFlag = !sArg.CompareNoCase("a");
        m_bCFlag = !sArg.CompareNoCase("c");
        m_bDAFlag = !sArg.CompareNoCase("da");
    }
```

II

```
    // m_strFileName gets the first nonflag name
    else if (m_strFileName.IsEmpty())
    {
        m_sArg=sArg;
    }
    CCommandLineInfo::ParseParam(pszParam,bFlag,bLast);
}
```

Note that the `pszParam` argument contains the next item on the command line. The `bFlag` argument is `TRUE` if `pszParam` was proceeded by a - (hyphen) or / (forward slash) character, which has since been removed. The `bLast` argument is `TRUE` if this is the last argument on the line. Make sure to call the base class's `ParseParam()` at the end or the standard flags won't be processed.

4. For a complete listing of the Command Line Info class, please see "Listings — Command Line Info Class" on page 150.

## Incorporate the New Command Line Info Class in Application Class

1. Locate the `ParseCommandLine()` function in your Application Class and substitute this new class for the `CCommandLineInfo` class.

```
// Parse command line for standard shell commands, DDE, file open
CWzdCommandLineInfo cmdInfo;
ParseCommandLine(cmdInfo);
```

2. Your command line options are now available as member variables of the `cmdInfo` variable.

```
if (cmdInfo.m_bAFlag)
{
      :    :    :
}
```

3. To make these options available throughout your application, embed the `cmdInfo` variable in your Application Class and access its member variables.

```
AfxGetApp()->m_cmdInfo.m_bAFlag;
```

Example 6   Processing Command Line Options   **149**

# Notes

- The standard MFC flags are as follows. The actual processing of these standard command line flags takes place in `ProcessShellCommand(cmdInfo)` just after the `ParseCommandLine()` function in your Application Class.

| nothing | Causes your application to try to open a new document. |
|---|---|
| filename | Causes your application to try open to open the filename as a document. |
| /p filename | Causes your application to open and print the given filename to the default printer. |
| /pt filename printer driver port | Same as above but to the specified printer |
| /dde | Causes your application to startup and wait for DDE commands. |
| /Automation /Embedding /Unregister /Unregserver | COM flags |

- Processing nonflags, such as names, can be a bit tricky. The first nonflag to come along is assumed to be a document filename. Once a filename has been found, however, you can grab any additional nonflags for your own purposes. That is, unless a /pt flag is encountered, in which case the next three nonflag arguments are used to initialize printing. To simplify things you may want to disable the /pt flag by not passing it on to `ParseParam()` in the base class.

- Of course, if you don't want to continue to support the standard MFC flags shown previously, you have a much freer hand. Any flag or nonflag option can be used, as long as you don't call the base class's `ParsePa-ram()`. However, don't give up the functionality these standard flags provide just because you can.

# CD Notes

- When executing the project on the CD, set a breakpoint on the `ParseCmdLine()` function in `Wzd.cpp` and watch as the new `CWzdCommandLineInfo` converts command line arguments into flags that can be used later in the application.

# Listings — Command Line Info Class

```
#if !defined WZDCOMMANDLINEINFO_H
#define WZDCOMMANDLINEINFO_H

// WzdCommandLineInfo.h : header file
//

///////////////////////////////////////////////////////////////////
// CWzdCommandLineInfo window

class CWzdCommandLineInfo : public CCommandLineInfo
{

// Construction
public:
    CWzdCommandLineInfo();

// Attributes
public:
    BOOL m_bAFlag;
    BOOL m_bCFlag;
    BOOL m_bDAFlag;
    CString m_sArg;

// Operations
public:
    void ParseParam(const TCHAR* pszParam,BOOL bFlag, BOOL bLast);

// Overrides

// Implementation
public:
    virtual ~CWzdCommandLineInfo();
};


///////////////////////////////////////////////////////////////////

#endif
```

Example 6    Processing Command Line Options    **151**

```cpp
// WzdCommandLineInfo.cpp : implementation file
//

#include "stdafx.h"
#include "wzd.h"
#include "WzdCommandLineInfo.h"

#ifdef _DEBUG
#define new DEBUG_NEW
#undef THIS_FILE
static char THIS_FILE[] = __FILE__:
#endif

/////////////////////////////////////////////////////////////////////
// CWzdCommandLineInfo

CWzdCommandLineInfo::CWzdCommandLineInfo()
{
    m_bAFlag = FALSE;
    m_bCFlag = FALSE;
    m_bDAFlag = FALSE;
    m_sArg=_T("");
}


CWzdCommandLineInfo::~CWzdCommandLineInfo()
{
}


/////////////////////////////////////////////////////////////////////
void CWzdCommandLineInfo::ParseParam(const TCHAR* pszParam, BOOL bFlag,
    BOOL bLast)
{
    CString sArg(pszParam);
    if (bFlag)
    {
    m_bAFlag = !sArg.CompareNoCase("a");
    m_bCFlag = !sArg.CompareNoCase("c");
    m_bDAFlag = !sArg.CompareNoCase("da");
    }
```

```
// m_strFileName gets the first nonflag name
else if (m_strFileName.IsEmpty())
{
    m_sArg=sArg;
}
CCommandLineInfo::ParseParam(pszParam,bFlag,bLast);
}
```

# Example 7   Dynamically Changing Application Icons

## Objective

You would like to change or even animate your application's icon (Figure 5.10). This icon also appears in your system's task bar and allows you to show your application's progress, even when it is minimized.

**Figure 5.10   Your application's Icon can be changed in your main window and the system's task bar.**

Your application's
icon can be
changed at any
time.

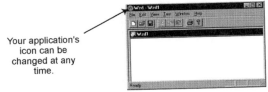

## Strategy

We will use MFC's `CWnd::SetIcon()` function to change the icon.

Example 7   Dynamically Changing Application Icons   **153**

# Steps

## Change the Application's Icon

1. Use the following to change your application's icon dynamically. As seen here, you can also use `CWinApp::LoadIcon()` to load the icon first from your application's resources.

```
AfxGetMainWnd()->SetIcon(
    AfxGetApp()->LoadIcon(IDI_STATUS_ICON), // icon handle
    TRUE);                      // FALSE=16x16 bit icon
```

Note that we are using `CWinApp::LoadIcon()` to load the icon from our application's resources

# Notes

- Not only will `SetIcon()` change an application's icon, but it will also change the icon of any application window that has a system menu, including Child Frame windows and dialog boxes. For example, to change the icon of a Child Frame window from a View Class, use the following.

```
GetParentFrame()->SetIcon(
    AfxGetApp()->LoadIcon(IDI_STATUS_ICON),
    TRUE);
```

- Use `SetIcon()` only when you need to change an icon at run time. Otherwise, you should change your icon using the Developer Studio's Icon Editor. Edit the `IDR_MAINFRAME` icon under the Icon folder in your application's resources.

---

**NOTE:** make sure you edit both the $32 \times 32$-bit icon and the $16 \times 16$-bit icon. Many a novice has edited only the $32 \times 32$-bit icon and spent hours trying to figure out why their icon hasn't changed. When creating an MDI application, make sure to also edit the other icon in the Icon folder to change the Child Frames icon, too.

---

II    5

- As mentioned previously, you can use this method to change your application's icon when minimized to indicate to your user that it's still processing a command. Just check `AfxGetMainWnd()->IsIconic()` and, if `TRUE`, update the icon using `SetIcon()`.
- The `CButton` class also has a `SetIcon()` function. However, in this case, it's used to set the icon that will appear on the face of a button. Make sure, however, you also use the `BS_ICON` button style when creating the button.

## CD Notes

- When executing the project on the CD, click on the Test/Wzd commands to change the icon of the application.

# Example 8    Prompting the User for Preferences

## Objective

You would like to maintain your user's preferences (Figure 5.11).

## Figure 5.11    Use Property Sheets and Pages to maintain your user's preferences.

## Strategy

We will use Property Sheets and Property Pages to prompt our user for their preferences. First, we will use the Menu Editor to add an Options menu to the main menu. Then, we will use the ClassWizard to handle this command by creating a Property Sheet. To this Property Sheet, we will add Property Pages for the options our application will support.

Example 7    Dynamically Changing Application Icons    **155**

# Steps

## Create the Property Pages

1. Use the Dialog Editor to create one or more dialog templates (Example 38). These templates should contain all the preferences your application will support. Each template style should be a Child with a Thin border and no system menu. Whatever title you choose for this template will become the name seen on the tab for this preference page. Try to keep the pages about the same size. However, the size of the entire Property Sheet will be based on the size of the largest dialog template added to it.

2. Use the ClassWizard to create a class for each of the dialog templates. Derive them from CPropertyPage. Create a member variable for each of the controls in your dialog box (Example 39), but also add a message handler for each control that indicates that the control has been modified. In that handler, call SetModified(TRUE). This will tell the Property Sheet to enable the Apply button.

```
void CFirstPage::OnChange()
{

    SetModified(TRUE);

}
```

3. In the class of the first Property Page, use the ClassWizard to override OnOK(). Also in that class, send our own user-defined windows message called WM_APPLY to the CMainFrame class. Also, call SetModified(FALSE) to turn off the Apply button. The OnOK() function of each page is called whenever the OK or Apply buttons are pressed, but we only need to process it in one page. We will see the effect of the WM_APPLY message later.

```
#define WM_APPLY WM_USER+1
 :     :     :
void CFirstPage::OnOK()
{

    AfxGetMainWnd()->SendMessage(WM_APPLY);
    SetModified(FALSE);

    CPropertyPage::OnOK();

}
```

II

4. For a complete listing of a Property Page class, see "Listings — Property Page Class" on page 158.

## Create the Property Sheet

1. Use the Menu Editor to create a new main menu item, Options, with a single submenu item, Preferences.
2. Use the ClassWizard to process the Preferences submenu item in your CMainFrame class.
3. Process this Preferences command by first creating a Property Sheet with the CPropertySheet class and then adding the classes you created previously as the pages to this Property Sheet using AddPage(). Initialize the member variables of the pages with any current settings, and then display the Property Sheet using DoModal().

```
void CMainFrame::OnOptionsPreferences()
{
    CPropertySheet sheet(_T("Preferences"),this);
    m_pFirstPage=new CFirstPage;
    m_pSecondPage=new CSecondPage;

    sheet.AddPage(m_pFirstPage);
    sheet.AddPage(m_pSecondPage);

    m_pFirstPage->m_bOption1 = m_bFirstOption1;
    m_pFirstPage->m_sOption2 = m_sFirstOption2;
    m_pSecondPage->m_nOption1 = m_nSecondOption1;
    m_pSecondPage->m_sOption2 = m_sSecondOption2;

    sheet.DoModal();
    delete m_pFirstPage;
    delete m_pSecondPage;
}
```

Notice that in the last step, the values from our Property Pages are never stored in the application from which they came. That's where that WM_APPLY message comes in.

Example 7    Dynamically Changing Application Icons    **157**

4. You will need to manually add a message handler for your WM_APPLY window message by adding the following item to the message map in Main-Frm.cpp. Make sure to put it below the //}}AFX_MSG_MAP notation or the ClassWizard may delete it.

```
ON_MESSAGE_VOID(WM_APPLY,OnApply)
```

Define OnApply() in MainFrm.h as

```
afx_msg void OnApply();
```

5. Now implement OnApply() in CMainFrame so that it stores the member variables from your Property Pages back into the application. As mentioned previously, this message will be sent anytime the Apply or OK buttons are clicked on the Property Sheet.

```
void CMainFrame::OnApply()
{
    m_bFirstOption1 = m_pFirstPage->m_bOption1;
    m_sFirstOption2 = m_pFirstPage->m_sOption2;
    m_nSecondOption1 = m_pSecondPage->m_nOption1;
    m_sSecondOption2 = m_pSecondPage->m_sOption2;
}
```

## Notes

- Application preferences have, by convention, been handled in the CMainFrame class, although where the actual option is stored varies depending on where it's used.

- If you would like to add your own buttons to the Property Sheet, derive your own class from CPropertySheet and use that derived class instead in the previous example. To add your own buttons, you must first make the Property Sheet big enough to handle them. Use the ClassWizard to add a message handler to your new class that handles WM_CREATE. Use MoveWindow() there to make the sheet the size you want. To create your own buttons, use the method shown in Example 46 on creating control windows anywhere.

- One of the first things you assume when you see the tab control tool in the Dialog Editor is that it somehow has the same functionality as the Property Sheet. In fact, it has almost none. The tab control is more akin to the list control, in that it simply keeps track of what page you're on and it's up to you to fill a page with a dialog box. If at all possible, try to avoid using a tab control alone.

## CD Notes

When executing the project on the CD, click on the Options/Preferences menu commands to open a Property Sheet.

## Listings — Property Page Class

```
#if !defined AFX_FIRSTPAGE_H
#define AFX_FIRSTPAGE_H


// FirstPage.h : header file
//
///////////////////////////////////////////////////////////////////
// CFirstPage dialog

class CFirstPage : public CPropertyPage
{
    DECLARE_DYNCREATE(CFirstPage)

// Construction
public:
    CFirstPage();
    ~CFirstPage();

// Dialog Data
    //{{AFX_DATA(CFirstPage)
    enum { IDD = IDD_FIRST_PAGE };
    BOOL m_bOption1;
    CString m_sOption2;
    //}}AFX_DATA
```

Example 7   Dynamically Changing Application Icons   **159**

```cpp
// Overrides
    // ClassWizard generate virtual function overrides
    //{{AFX_VIRTUAL(CFirstPage)
    public:
    virtual void OnOK();
    protected:
    virtual void DoDataExchange(CDataExchange* pDX);    // DDX/DDV support
    //}}AFX_VIRTUAL

// Implementation
protected:

    // Generated message map functions
    //{{AFX_MSG(CFirstPage)
    afx_msg void OnChange();
    //}}AFX_MSG
    DECLARE_MESSAGE_MAP()
};


/////////////////////////////////////////////////////////////////
// FirstPage.cpp : implementation file
//

#include "stdafx.h"
#include "wzd.h"
#include "FirstPage.h"
#include "WzdProject.h"

#ifdef _DEBUG
#define new DEBUG_NEW
#undef THIS_FILE
static char THIS_FILE[] = __FILE__;
#endif
```

```
/////////////////////////////////////////////////////////////
// CFirstPage property page

IMPLEMENT_DYNCREATE(CFirstPage, CPropertyPage)

CFirstPage::CFirstPage() : CPropertyPage(CFirstPage::IDD)
{
    //{{AFX_DATA_INIT(CFirstPage)
    m_bOption1 = FALSE;
    m_sOption2 = _T("");
    //}}AFX_DATA_INIT
}

CFirstPage::~CFirstPage()
{
}

void CFirstPage::DoDataExchange(CDataExchange* pDX)
{
    CPropertyPage::DoDataExchange(pDX);
    //{{AFX_DATA_MAP(CFirstPage)
    DDX_Check(pDX, IDC_CHECK, m_bOption1);
    DDX_Text(pDX, IDC_EDIT, m_sOption2);
    //}}AFX_DATA_MAP
}

BEGIN_MESSAGE_MAP(CFirstPage, CPropertyPage)
    //{{AFX_MSG_MAP(CFirstPage)
    ON_BN_CLICKED(IDC_CHECK, OnChange)
    ON_EN_CHANGE(IDC_EDIT, OnChange)
    //}}AFX_MSG_MAP
END_MESSAGE_MAP()

/////////////////////////////////////////////////////////////
// CFirstPage message handlers

void CFirstPage::OnChange()
{
    SetModified(TRUE);
}
```

Example 9    Saving and Restoring User Preferences    **161**

```
// only needed on one page!
void CFirstPage::OnOK()
{
    AfxGetMainWnd()->SendMessage(WM_APPLY);
    SetModified(FALSE);

    CPropertyPage::OnOK();
}
```

# Example 9    Saving and Restoring User Preferences

## Objective

You would like to save and restore the options your user picked in the previous example.

## Strategy

We will be using six CWinApp functions to load and save values to the system registry.

## Steps

### Configure Your Application

1. Add the following #defines to your Mainframe Class's definition file. Substitute "Company" for your company name and "Optionx" for descriptive names of your application's options.

```
#define COMPANY_KEY "Company"
#define SETTINGS_KEY "Settings"
#define OPTION1_KEY "Option1"
#define OPTION2_KEY "Option2"
#define OPTION3_KEY "Option3"
#define OPTION4_KEY "Option4"
```

2. In your `CMainFrame` class, modify the `SetRegistryKey()` call in the `Init-Instance()` function to use your company's system registry key.

```
SetRegistryKey(COMPANY_KEY);
```

## Save Application Options

1. Add a new member function to your `CMainFrame` class that will save your options. There, you can use up to three different `CWinApp` functions to save integer, string, or binary options.

```
void CMainFrame::SaveOptions()
{
// integer options
    AfxGetApp()->WriteProfileInt(SETTINGS_KEY,OPTION2_KEY,
        m_nOption2);
// string options
    AfxGetApp()->
        WritePro-
fileString(SETTINGS_KEY,OPTION1_KEY,m_sOption1);
// binary options
    AfxGetApp()->WriteProfileBinary(SETTINGS_KEY, OPTION3_KEY,
        (BYTE*)&m_dOption3, sizeof(m_dOption3));
}
```

2. Use the ClassWizard to add a `WM_CLOSE` message handler to your `CMain-Frame` class and call `SaveOptions()` from there.

## Restore Application Options

1. Add another member function to your `CMainFrame` class that will load your options. There you can use three other `CWinApp` function to load integer, string, and binary options from the system registry.

```
void CMainFrame::LoadOptions()
{
    // integer options
    m_nOption2=AfxGetApp()->
        GetProfileInt(SETTINGS_KEY,OPTION2_KEY, 3);
```

Example 9   Saving and Restoring User Preferences   **163**

```
    // string options
    m_sOption1=AfxGetApp()->
        GetProfileString(SETTINGS_KEY,OPTION1_KEY,"Default");
    // binary options
    BYTE *p;
    UINT size;
    m_dOption3=33.3;
    if (AfxGetApp()->GetProfileBinary(SETTINGS_KEY,OPTION3_KEY,
        &p, &size))
    {
        memcpy(&m_dOption3,p,size);
        delete []p;
    }
}
```

2. Call `LoadOptions()` at the start of the `OnCreate()` function in your `CMainFrame` class.

`GetProfileBinary()` and `WriteProfileBinary()` are undocumented and may not be available on all versions of MFC. If it isn't on yours, the following steps can be used to create alternative `LoadOptions()` and `SaveOptions()`. The following methods also allow you to save and load options from anywhere in the system registry.

## Alternative `LoadOptions()` **and** `SaveOptions()`

1. Add one more item to your project include file and substitute your application's name for "Wzd".

```
#define APPLICATION_KEY "Software\\Company\\Wzd\\Settings"
```

2. Then use the following for your `SaveOptions()` function. This function opens the system registry, writes your options to it, and then closes it using the Window API directly.

```
void CMainFrame::SaveOptions()
{
    // opens system registry for writing
    HKEY key;
    DWORD size, type, disposition;
```

```
if (RegOpenKeyEx(HKEY_CURRENT_USER,APPLICATION_KEY,0,
    KEY_WRITE,&key)!=ERROR_SUCCESS)
{
    RegCreateKeyEx(HKEY_CURRENT_USER,APPLICATION_KEY,0,"",
        REG_OPTION_NON_VOLATILE,KEY_ALL_ACCESS,NULL,
        &key,&disposition);
}

// writes an option that's a string
type = REG_SZ;
size = m_sOption1.GetLength();
RegSetValueEx(key,OPTION1_KEY,0,type,
    (LPBYTE)LPCSTR(m_sOption1),size);

// writes an option that's an integer
type = REG_DWORD;
size = 4;
RegSetValueEx(
    key,OPTION2_KEY,0,type,(LPBYTE)&m_nOption2,size);

// writes all other options
type = REG_BINARY;
size = sizeof(DOUBLE);
RegSetValueEx(
    key,OPTION3_KEY,0,type,(LPBYTE)&m_dOption3,size);

// closes system registry
RegCloseKey(key);

}
```

Example 9    Saving and Restoring User Preferences    **165**

3. Use the following for your LoadOptions(). It will open the system registry, read your options back in, and then close the system registry.

```
void CMainFrame::LoadOptions()
{
    HKEY key;
    DWORD size, type;
    if (RegOpenKeyEx(
        HKEY_CURRENT_USER,APPLICATION_KEY,0,KEY_READ,&key) ==
        ERROR_SUCCESS)
    {
        // read string options
        size=128;
        type = REG_SZ;
        LPSTR psz = m_sOption1.GetBuffer(size);
        RegQueryValueEx(
            key,OPTION1_KEY,0,&type,(LPBYTE)psz,&size);
        m_sOption1.ReleaseBuffer();

        // read integer options
        size=4;
        type = REG_DWORD;
        RegQueryValueEx(key,OPTION2_KEY,0,&type,
            (LPBYTE)&m_nOption2,&size);

    // read all other options as binary bytes
    size=sizeof(DOUBLE);
    type = REG_BINARY;
    RegQueryValueEx(key,OPTION3_KEY,0,&type,
        (LPBYTE)&m_dOption3,&size);

    RegCloseKey(key);
    }
}
```

II  5

## Notes

- You can edit the options you write to the system registry by using the REGEDIT.EXE application found in your Windows directory. You'll find your application's entries under HKEY_CURRENT_USER/Software/Company.

- You'll notice that you can save an option as an integer, a string, or a binary. In fact, all options can be saved as binary values. Whenever possible, however, use an integer or string because these values can be more easily edited using the REGEDIT.EXE utility. Strings will appear as strings and integers will appear as integers. Binary values simply appear as a string of bytes.

- If you comment out SetRegistryKey() in step 2 of "Configure Your Application" on page 161, your options will be saved to a file instead of the system registry. This file will be in your \Windows directory and be named app.ini, where app is the name of your application. Older windows applications used to save their options this way.

## CD Notes

- When executing the project on the CD, set a break point on LoadOptions1() and SaveOptions1() in Mainfrm.cpp. Then run the application and terminate it to watch options being loaded and saved to the system registry.

# Example 10   Terminating Your Application

## Objective

You would like to ask your user if they're sure they want to terminate your application.

## Strategy

We'll use the ClassWizard to add a message handler to the Mainframe Class to handle a WM_CLOSE message in which we will ask our question and then conditionally continue to terminate the application.

Example 10   Terminating Your Application   **167**

# Steps

## Ask the Question

1. Use the ClassWizard to add a WM_CLOSE message handler to the CMain-Frame class.
2. Add the following code to this message handler.

```
void CMainFrame::OnClose()
{
    if (AfxMessageBox("Do you really want to exit?",MB_YESNO)
        == IDYES)
    {
        CMDIFrameWnd::OnClose();
    }
}
```

II   5

# Notes

- When the user clicks the Close button in the upper-right corner of your application — or when they select Exit in the File menu — the system sends a WM_CLOSE message to your application. By intercepting the message here, we can prevent the default action of this command, which is to send WM_DESTROY messages to all of the Child windows of this application.

- In a standard MFC application, you would never need to use this example. Since MFC applications are document-centric, closing your application should be based solely on whether or not your document(s) has been modified. When you modify a document, you should use the Set-Modified(TRUE) member function of the CDocument class to mark your document as modified. Then, when your user goes to terminate your application, MFC will automatically check each document to see if they've been modified, and if so, will ask your user if they want to save their changes. If yes, MFC will automatically call your OnSaveModified() routine. If your user refuses to save a document, the close is canceled. If no document has been modified, your application will terminate without a peep.

- For an example of using the ClassWizard to add a message handler, see Example 59.

## CD Notes

- When executing the project on the CD, you can then click the Close button to cause a dialog prompt asking if you really want to terminate. Answering No will cause the application to continue to run.

# Example 11 Creating a Splash Screen

## Objective

You would like to create a splash screen to display a company logo and copyright (Figure 5.12).

**Figure 5.12 Your application Splash Screen can show your company logo and copyright.**

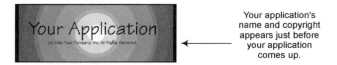

Your application's name and copyright appears just before your application comes up.

## Strategy

We will create our splash screen at the earliest point we can, which is in the InitInstance() function of our Application Class. Our splash screen will appear in a plain window using one of the bitmap classes we will create in Example 57 so that our window can have all the colors in the rainbow.

## Steps

### Create a Splash Screen Window Class

1. Use the ClassWizard to create a plain window class derived from generic CWnd.
2. Add your own Create() member function to this class. There you will load the bitmap that our splash screen will display and create the window

Example 11   Creating a Splash Screen   **169**

centered in the screen. Use the bitmap class in Example 57 to load the bitmap — it will allow your bitmap to retain its palette when it paints to the screen.

```
void CWzdSplash::Create(UINT nID)
{
    m_bitmap.LoadBitmapEx(nID,FALSE);
    int x = (::GetSystemMetrics (SM_CXSCREEN)-
        m_bitmap.m_Width)/2;
    int y = (::GetSystemMetrics (SM_CYSCREEN)-
        m_bitmap.m_Height)/2;
    CRect rect(x,y,x+m_bitmap.m_Width,y+m_bitmap.m_Height);
    CreateEx(0,AfxRegisterWndClass(0),"",
        WS_POPUP|WS_VISIBLE|WS_BORDER,rect,NULL,0);
}
```

3. Use the ClassWizard to add a `WM_PAINT` message handler to this window class. There you will draw the bitmap to the screen using `BitBlt()`.

```
void CWzdSplash::OnPaint()
{
    CPaintDC dc(this); // device context for painting
    // get bitmap colors
    CPalette *pOldPal =
        dc.SelectPalette(m_bitmap.GetPalette(),FALSE);
    dc.RealizePalette();
    // get device context to select bitmap into
    CDC dcComp;
    dcComp.CreateCompatibleDC(&dc);
    dcComp.SelectObject(&m_bitmap);
    // draw bitmap
    dc.BitBlt(0,0,m_bitmap.m_Width,m_bitmap.m_Height, &dcComp,
        0,0,SRCCOPY);
    // reselect old palette
    dc.SelectPalette(pOldPal,FALSE);
}
```

4. For a complete listing of this splash window class, please refer to "Listings—Splash Window Class" on page 170.

## Incorporate the Splash Screen Class into `InitInstance()`

1. Create an instance of this splash window class at the start of the `InitInstance()` function in your Application Class. Call its `Create()` and force it to paint.

```
CWzdSplash wndSplash;
wndSplash.Create(IDB_WZDSPLASH);
wndSplash.UpdateWindow();      //send WM_PAINT
```

2. Since we created the splash window class on the stack, this window will be automatically destroyed once `InitInstance()` returns. Therefore, if your application takes a lot of time to initialize, you won't have to put any delays into your application to ensure that the splash screen is visible long enough to read. On the other hand, if your application takes little time to initialize or if you're worried that faster machines will turn your splash screen into a blip, add the following lines somewhere in your `InitInstance()` to delay your application.

```
// add if splash screen too short
Sleep(2000);  <<<<<<<
```

## Notes

- To create an animated splash screen, derive your splash screen window from `CAnimateCtrl`. In `Create()`, load an `.avi` file instead of a bitmap file, and create the window centered. See Example 43 for more on the `CAnimateCtrl` class.

## CD Notes

- When executing the project on the CD, a splash screen will initially appear and then disappear before the application window appears.

## Listings—Splash Window Class

```
#if !defined WZDSPLASH_H
#define WZDSPLASH_H

// WzdSplash.h : header file
//
```

Example 11    Creating a Splash Screen    **171**

```
#include "WzdBitmap.h"

////////////////////////////////////////////////////////////////
// CWzdSplash window

class CWzdSplash : public CWnd
{

// Construction
public:
    CWzdSplash();

// Attributes
public:

// Operations
public:
    void Create(UINT nBitmapID);

// Overrides
    // ClassWizard generated virtual function overrides
    //{{AFX_VIRTUAL(CWzdSplash)
    //}}AFX_VIRTUAL

// Implementation
public:
    virtual ~CWzdSplash();

    // Generated message map functions
protected:
    //{{AFX_MSG(CWzdSplash)
    afx_msg void OnPaint();
    //}}AFX_MSG
    DECLARE_MESSAGE_MAP()
private:
    CWzdBitmap m_bitmap;
};

////////////////////////////////////////////////////////////////
```

II    5

```cpp
#include "WzdBitmap.h"

/////////////////////////////////////////////////////////////////
// CWzdSplash window

class CWzdSplash : public CWnd
{

// Construction
public:
    CWzdSplash();

// Attributes
public:

// Operations
public:
    void Create(UINT nBitmapID);

// Overrides
    // ClassWizard generated virtual function overrides
    //{{AFX_VIRTUAL(CWzdSplash)
    //}}AFX_VIRTUAL

// Implementation
public:
    virtual ~CWzdSplash();

    // Generated message map functions
protected:
    //{{AFX_MSG(CWzdSplash)
    afx_msg void OnPaint();
    //}}AFX_MSG
    DECLARE_MESSAGE_MAP()
private:
    CWzdBitmap m_bitmap;
};

/////////////////////////////////////////////////////////////////
```

Example 11    Creating a Splash Screen    **173**

```
#endif

// WzdSplash.cpp : implementation file
//

#include "stdafx.h"
#include "WzdSplash.h"

#ifdef _DEBUG
#define new DEBUG_NEW
#undef THIS_FILE
static char THIS_FILE[] = __FILE__;
#endif

/////////////////////////////////////////////////////////////////////
// CWzdSplash

CWzdSplash::CWzdSplash()
{
}

CWzdSplash::~CWzdSplash()
{
}
```

II  5

```
BEGIN_MESSAGE_MAP(CWzdSplash, CWnd)
    //{{AFX_MSG_MAP(CWzdSplash)
    ON_WM_PAINT()
    //}}AFX_MSG_MAP
END_MESSAGE_MAP()


/////////////////////////////////////////////////////////////
// CWzdSplash message handlers

void CWzdSplash::OnPaint()
{
    CPaintDC dc(this);      // device context for painting
    // get bitmap colors
    CPalette *pOldPal = dc.SelectPalette(m_bitmap.GetPalette(),FALSE);
    dc.RealizePalette();

    // get device context to select bitmap into
    CDC dcComp;
    dcComp.CreateCompatibleDC(&dc);
    dcComp.SelectObject(&m_bitmap);

    // draw bitmap
    dc.BitBlt(0,0,m_bitmap.m_Width,m_bitmap.m_Height, &dcComp, 0,0,SRCCOPY);

    // reselect old palette
    dc.SelectPalette(pOldPal,FALSE);
}

void CWzdSplash::Create(UINT nID)
{
    m_bitmap.LoadBitmapEx(nID,FALSE);

    int x = (::GetSystemMetrics (SM_CXSCREEN)- m_bitmap.m_Width)/2;
    int y = (::GetSystemMetrics (SM_CYSCREEN)- m_bitmap.m_Height)/2;
    CRect rect(x,y,x+m_bitmap.m_Width,y+m_bitmap.m_Height);
    CreateEx(0,AfxRegisterWndClass(0),"",
        WS_POPUP|WS_VISIBLE|WS_BORDER,rect,NULL,0);
}
```

# 6

# Menus

With all the new and exciting ways for your user to interact with your application, the plain old menu is sometimes overlooked. The Application Wizard automatically creates a generic menu for your Main window when you create an application, but with a little brute force you can make them more engaging.

**Example 12    Using the Menu Editor**   We will use the Menu Editor to add additional commands to the menu generated by the Application Wizard.

**Example 13    Adding a Menu Command Handler**   We will use the Class Wizard to automatically direct a menu click to the member function of one of your classes for processing.

**Example 14    Dynamically Changing Menus Based on Document Viewed**   We will look at how to automatically update the main menu with new commands whenever a particular view is opened.

**Example 15    Enabling and Disabling Menu Commands**   We will look at how to gray and ungray menu items.

**Example 16**     **Check-marking Menu Commands** We will look at how to put check marks next to menu items to indicate their current state.

**Example 17**     **Radio-marking Menu Commands** We will put a dot next to a group of menu items to indicate that one member of the group is currently active.

**Example 18**     **Dynamically Modifying Menus** We will look at adding and deleting items from our menu as the program is running.

**Example 19**     **Dynamically Modifying the System Menu** We will look at how to add commands to the system menu, which is the menu that appears when you click on an application's icon.

**Example 20**     **Triggering a Menu Command** We will look at how to cause a menu item to be "clicked" from within the program.

**Example 21**     **Creating Popup Menus** We will create a floating menu when our user right-clicks their mouse in our application's view.

# Example 12   Using the Menu Editor

## Objective

You would like to add, delete, or change a command in your application's menu or create a new menu.

---

NOTE: Any item you add to your menu here will appear grayed out and inactive until you add a menu command handler for it. See the next example for how to add a handler.

---

## Steps

### Add a Menu Item to Your Menu Using the Menu Editor

1. To create a new menu, click on the Developer Studio's Insert/Resource menu commands to open the Insert Resource dialog box. Select Menu from the list and click New.

Example 12    Using the Menu Editor    **177**

2. To edit an existing menu, locate its ID in the Menu folder of your applications resources and double-click on it.

3. To add a new item to a menu, use the mouse to drag the empty focus rectangle to the location of the new item (Figure 6.1).

**Figure 6.1    Drag the empty focus rectangle to a new location to add an item to the menu.**

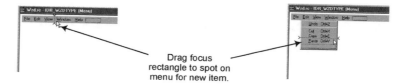

4. Then either double-click or right-click the rectangle and select Properties to open the Properties dialog box (Figure 6.2).

**Figure 6.2    Use the Properties dialog box to change a menu item's properties.**

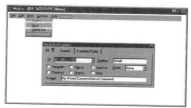

5. You can define your own menu ID in the ID field or, if left empty, an ID will be created for you. This ID becomes the Command ID sent to your application when this menu item is clicked.

6. Enter the text for this menu item in the Caption field. Put an ampersand (&) in front of the character with which you want your user to be able to access this menu item from the keyboard. That character will appear with an underline in the menu.

7. Fill the Prompt edit box with a help message that you would like to appear in the Status Bar when the mouse cursor is over this menu item. Append a new-line character to the end of this prompt and then add a smaller help message. The smaller message, also called a bubble help message, will appear over any toolbar button that also generates this

command. As seen here, the first message can be detailed, while the second should be brief.

```
Opens the file system.\nOpen File
```

8. You can usually ignore the Checked, Grayed, and Inactive check boxes, since you will be dynamically updating these properties yourself (Example 15). Check the Separator check box if this will simply be a separator in this menu (Figure 6.3).

**Figure 6.3     Check the Separator check box to add a Menu Separator.**

A separator with no menu command.

9. For menu bar items, check the Help check box if you would like this menu item and all the items to its right to be right justified in the menu bar (Figure 6.4).

**Figure 6.4     Check the Help check box items in the menu bar.**

"Help" property causes itself and all menu commands to its right to be right justified.

10. If you want to split a large menu up into more than one column, you can select a Break at the start of the new column. Column simply creates a new column, while Bar creates a column and then adds a separator between it and the last column. Figure 6.5 shows a menu without Bar.

Example 12   Using the Menu Editor   **179**

**Figure 6.5   Select Break/Column to split large menus into more than one column.**

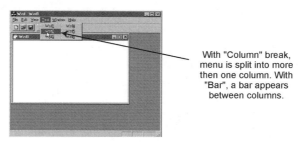

With "Column" break, menu is split into more then one column. With "Bar", a bar appears between columns.

---

**NOTE:** Several menu properties, including column breaks, will not be reflected in the image of the menu you are editing. These properties will only appear when your application is built and running.

---

11. To modify an existing menu item, simply select it with the mouse and open its Properties dialog box. To delete a menu item, select it with the mouse and then press the Delete key on your keyboard.

## Notes

- You can also create a menu dynamically (Example 21).

## CD Notes

- There is no accompanying project on the CD for this example.

# Example 13   Adding a Menu Command Handler

## Objective

You would like to use the ClassWizard to automatically add a menu command handler to one of your classes. This handler will actually be a member function of that class, called by MFC's message map system when a menu item is clicked. The function that the ClassWizard adds will initially be empty.

# Steps

## Add a Command Handler Using the ClassWizard

1. Click on the Developer Studio's View/ClassWizard menu commands to open the ClassWizard dialog box. Locate and select the class you want to process this menu command from the Class name combo box.

---

NOTE: Although a class name may appear in this combo box, it may not be automatically eligible to process menu commands. Please see the section "Notes" on page 181.

---

2. The Object ID's list box contains the IDs of all available menu and toolbar commands. Find the appropriate ID and select it. Then select COMMAND from the Messages list box. Now click the Add Function button to actually add this command handler to your class. Clicking on the Edit Function will cause the Studio's Text Editor to open this class file positioned at this new function. This function will initially be empty of code. The menu item it's associated with will now appear active when you run the application. However, clicking on it will obviously have no effect.

3. See Figure 6.6 to review these steps.

**Figure 6.6**  **Follow these steps to add a menu command with ClassWizard.**

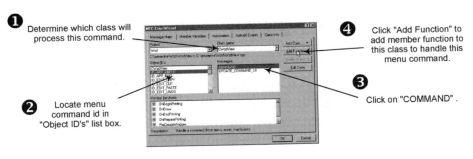

❶ Determine which class will process this command.

❷ Locate menu command id in "Object ID's" list box.

❸ Click on "COMMAND".

❹ Click "Add Function" to add member function to this class to handle this menu command.

Example 12   Using the Menu Editor   **181**

## Delete a Command Using the ClassWizard

1. Repeat the steps to add a command handler, but then click on the Delete Function button to allow the ClassWizard to delete this menu handler from this class. Actually, all the ClassWizard removes is this function's entry in the message map. It's up to you to delete the actual function from your .cpp file.

# Notes

- If you picked UPDATE_COMMAND_UI from the Messages list box instead of selecting COMMAND, the ClassWizard would add a user interface update handler to your class. This handler allows you to check mark, disable, or even change the text of a menu item (Examples 15 and 16).

- Until you add a command handler to your application to support a new menu command, that command will appear grayed out and disabled. If you prefer not to have to add a command handler to all of your menu items to make them appear active, you could instead add the following line to your CMainFrame Class's constructor.

```
CMainFrame::CMainFrame()
{

    m_bAutoMenuEnable =FALSE;
}
```

- Even though the ClassWizard will allow you to add a command handler to any class in your project, MFC's messaging system might not currently call that class to process command messages. Menu command messages follow a set path through your application's classes, as outlined in Chapter 3. In most cases, your class will probably be in this path, especially if the AppWizard created the class for you. However, if you are adding a new class to your application, you may need to manually insert the path into this message flow (Example 61).

# CD Notes

- There is no accompanying project on the CD for this example.

# Example 14 Dynamically Changing Menus Based on Document Viewed

## Objective

In an MDI application, you would like a different menu to appear depending on which document you are currently viewing (Figure 6.7).

**Figure 6.7** **Register more than one document template type to change menus based on the document viewed.**

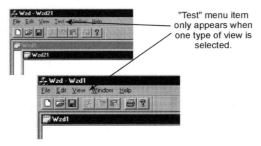

"Test" menu item only appears when one type of view is selected.

## Strategy

Actually, you get this functionality for free when you register more than one type of document template in your MDI application. An MDI application that edits more than one type of document can use this feature to add special commands to the menu bar depending on what it's currently editing. Case in point is Developer Studio's Image menu command, which only appears when you're editing an image (e.g., bitmap, icon, etc.).

## Steps

### Define a New Document Template in Your Application Class

1. Use the ID Editor to create the new ID for this document type. In this case, we are using an ID of IDR_WZDTYPE1.

2. Select the ResourceView of your WorkSpace window and open the Menu folder. Locate the ID for the menu of your application's current menu, select it, and then press Ctrl-c and Ctrl-v to make a copy of it. Give this

Example 14    Dynamically Changing Menus Based on Document Viewed    **183**

copied menu the ID of your new document type by right-clicking on the ID and clicking on Properties.

3. Repeat the last step to create a new icon for this document type. Icons are located in the Icon folder.

4. Repeat the last step to create a new document string type in the string table located in the String Table folder. Locate the string that looks like this

```
IDR_WZDTYPE        "\nWzd\nWzd\n\n\nWzd.Document\nWzd Document"
```

You should also change these values to reflect your new document type. For now, simply add a one (1) to the end of each mention of the application's name, which in this case is Wzd.

5. Use the ClassWizard to optionally create a new View Class and Document Class. See Example 1 to determine from which MFC classes to derive your classes.

6. You can now use the resources and classes you've created to create a new document template in your Application Class. Add the following code directly after the code that creates your current document template.

```
pDocTemplate = new CMultiDocTemplate(
    IDR_WZDTYPE1,                    <<new document id type
    RUNTIME_CLASS(CWzdDoc),          <<new document class
    RUNTIME_CLASS(CChildFrame),      <<MDI child frame
    RUNTIME_CLASS(CWzd2View));        <<new view class
    AddDocTemplate(pDocTemplate);
```

7. You can now use the Menu Editor to change your second, copied menu. When the user opens a document, they are given the choice of opening the original document type or this new type. If both types are open at the same time, and they both have different menus, clicking either view will cause your application's menu to change appropriately.

## Notes

- When using two document types, your application will automatically prompt you to ask which document type you would like to use when creating a new document. When opening an existing document, if your application can't figure out which type of view to use, it will again prompt you for a document type. The message that appears in this prompt is the second Wzd value that is shown in step 4.
- For more on Document/Views in general, please see Chapters 1 and 2.

## CD Notes

- When executing the project on the accompanying CD, you are given a choice of opening a Wzd document or a Wzd22 document. Pick Wzd, then create a new document with File/New and pick Wzd22 this time. Watch the menu while alternately clicking on one or the other view. Notice that the Test menu item appears only for the Wzd22 document.

# Example 15   Enabling and Disabling Menu Commands

## Objective

You would like to enable or disable a menu item (Figure 6.8).

**Figure 6.8    Add an interface message handler to selectively enable or disable a menu command.**

Example 15    Enabling and Disabling Menu Commands    **185**

# Strategy

If a new menu item has no command handler, your application will automatically cause it to appear disabled. To enable a menu item in this situation only requires you to add a command handler for this menu command using the ClassWizard (Example 13). However, if you want to selectively enable or disable a menu item based on some condition in your application, we will use the ClassWizard here to add an interface message handler.

# Steps

## Add a User Interface Command Handler Using the ClassWizard

1. Follow the steps in Example 13 as if you were going to add a menu command to one of your classes, but instead of selecting COMMAND, select UPDATE_COMMAND_UI. The ClassWizard will then create a user interface handler for your class.

2. Add the following code to this new handler. In this example, if m_bWzd is TRUE, your menu item will become active. FALSE will have the opposite effect.

```
void CWzdView::OnUpdateWzdButton(CCmdUI* pCmdUI)
{
    pCmdUI->Enable(m_bWzd);
}
```

# Notes

- When your application first opens a menu, it checks to see if there's a UPDATE_COMMAND_UI message handler for each menu item in that menu. If there isn't, your application then checks to see if you have any kind of COMMAND message handler for this menu item. If there is no such handler, your application will automatically disable this menu item, making it appear gray. You can disable this automatic feature by setting m_bAutoMenuEnable to FALSE in CMainFrame's constructor. However, this may be advisable only if you're slapping together a demo. In normal development, this feature helps to remind you what's left to do.

- Not only does your application enable and disable menu items, but it also does the same for toolbar buttons, dialog bar controls, and status bar panes. Since the controls in these bars are always visible, your application does this updating all the time when it's idle. Dialog boxes, both modal and modeless, are not affected.

- As a user, I have always hated disabled menu items and toolbar buttons. You know that a command is unavailable, but you don't know why or how to remedy the situation. I suggest that rather than disabling a menu item, you cause it to display a message box indicating that an item is unavailable and what can be done to make it available.

- For an example of using the ClassWizard, see Example 13. For more information on how MFC updates the user interface, see Chapter 3. Also, refer to the next example.

## CD Notes

- When executing the project on the accompanying CD, the Wzd menu command under Options appears enabled. No big whoop.

# Example 16  Check-marking Menu Commands

## Objective

You would like to add a check mark to a menu item (Figure 6.9).

## Figure 6.9  Add a user interface message handler to check-mark a menu command.

## Strategy

We will use the ClassWizard to add a user interface message handler that will allow us to check any menu item.

Example 16   Check-marking Menu Commands   **187**

## Steps

### Use the ClassWizard to Add a User Interface Handler

1. Follow the steps from the previous example to add a user interface handler to one of your application's classes, but now use another member function of CCmdUI to check-mark this menu item.

```
void CWzdView::OnUpdateWzdType(CCmdUI* pCmdUI)
{
    pCmdUI->SetCheck(m_bWzd);
}
```

## Notes

- Every time you open a menu, MFC allows you to update the status of every item in that menu. MFC constructs a CCmdUI class object containing a CMenu object that wraps the menu object and any other information you might need to update the items' status. MFC then passes this CCmdUI object to you using the same system it uses to send command messages to your classes, which allows you to update the status of a menu item in the same place you actually process the command.

- CCmdUI is overridden so that the same code you write to check a menu item will also depress a toolbar button. MFC just passes one or the other CCmdUI to your code, such that SetCheck() can be checking a menu item or depressing a button.

- Menu items are updated whenever one is opened. Toolbar buttons are updated whenever your application is idle.

- For an example of using the ClassWizard, see Example 13. For more information on how MFC updates the user interface, see Chapter 3. Also, refer to both the previous example and the next example.

## CD Notes

- When executing the project on the accompanying CD, the Options/Wzd command appears with a check mark.

II

6

# Example 17    Radio-marking Menu Commands

## Objective

You would like a dot to appear next to one of a grouping of menu items that reflect a changing mode (Figure 6.10).

## Figure 6.10    Add an interface message handler to every menu item in a group to radio-mark a menu command.

## Strategy

We will use the ClassWizard to add an interface message handler to every menu item in this group, which will allow us to put a dot next to the current mode.

## Steps

### Use the ClassWizard to Add Command Handlers to a Group of Menu Items

1. Use the ClassWizard to add a command handler to every member of a group of menu items (Example 13). In these handlers, we will set the appropriate mode to be used later.

```
// set mode based on what menu command was selected
void CWzdView::OnWzd1Type()
{
    m_nWzdMode=CWzdView::mode1;
}

void CWzdView::OnWzd2Type()
```

Example 17    Radio-marking Menu Commands    **189**

```
{
    m_nWzdMode=CWzdView::mode2;

}
void CWzdView::OnWzd3Type()
{
    m_nWzdMode=CWzdView::mode3;
}
```

## Use the ClassWizard to Add User Interface Handlers to this Same Group

1. Follow the steps from Example 15 to add a user interface handler to this group of menu items. Now use yet another member function of CCmdUI to radio-mark these menu items depending on what mode was set by the command handlers.

```
// put dot next to correct menu item
void CWzdView::OnUpdateWzd1Type(CCmdUI* pCmdUI)
{
    pCmdUI->SetRadio(m_nWzdMode==CWzdView::mode1);
}

void CWzdView::OnUpdateWzd2Type(CCmdUI* pCmdUI)
{
    pCmdUI->SetRadio(m_nWzdMode==CWzdView::mode2);
}

void CWzdView::OnUpdateWzd3Type(CCmdUI* pCmdUI)
{
    pCmdUI->SetRadio(m_nWzdMode==CWzdView::mode3);
}
```

## Notes

- Actually, the only difference between SetRadio() and SetCheck() is that the first puts a dot next to each menu item. There is no other behind-the-scenes processing, as there is with radio buttons in a dialog box — it's all up to you to determine what items are in a group.

- For more information on how MFC updates the user interface, see Chapter 3. Also, see the examples that come before this one.

### CD Notes

- When executing the project on the accompanying CD, picking one of the Wzd commands under Options will cause the radio dot to move to it.

# Example 18    Dynamically Modifying Menus

## Objective

You would like to add, delete, or change a command in your menu while your application is running.

## Strategy

We will use `CWnd::GetMenu()` to access the main menu. `GetMenu()` returns a pointer to a `CMenu` object that has several member functions that allow us to modify a menu.

## Steps

### Wrap the Menu Object in an MFC Class

1. First, we will find and wrap your application's main menu object in a `CMenu` object.

```
// Get the Main Menu
CMenu* pMainMenu = AfxGetMainWnd()->GetMenu();
```

2. Then, we will locate the submenu we want to edit by searching for it. Since menus are identified by position rather than any sort of ID, we must look for the desired menu by locating any menu command ID contained

Example 18   Dynamically Modifying Menus   **191**

in that menu. In this case, we will look for a menu that contains a menu item with an ID equal to IDC_WZD_TYPE.

```
CMenu* pSubMenu = NULL;
for (int i=0; i<(int)pMainMenu->GetMenuItemCount(); i++)
{
    pSubMenu = pMainMenu->GetSubMenu(i);
    if (pSubMenu && pSubMenu->GetMenuItemID(0) == IDC_WZD_TYPE)
    {
        break;
    }
}
ASSERT(pSubMenu);
```

Now that we have an MFC class that wraps the menu we want, we can use its member functions to dynamically edit that menu.

## Edit the Menu Dynamically

1. To append this menu with a menu command called Wzd2 with a command ID of IDC_WZD2_TYPE, we would use

```
pSubMenu->AppendMenu(0,IDC_WZD2_TYPE,"Wzd&2");
```

2. To insert Wzd3 before Wzd2, we would use

```
pSubMenu->InsertMenu(IDC_WZD2_TYPE,MF_BYCOMMAND,
    IDC_WZD3_TYPE,"Wzd&3");
```

3. To convert Wzd2 to Wzd4, we would use

```
pSubMenu->ModifyMenu(IDC_WZD2_TYPE,MF_BYCOMMAND,
    IDC_WZD4_TYPE,"Wzd&4");
```

4. To remove the second item in this menu, we would use

```
pSubMenu->RemoveMenu(1,MF_BYPOSITION);
```

## Notes

- Notice that submenus are maintained by position, rather than ID. To find the submenu you want to change, you either have to know its exact position or you have to look for a menu command ID that it contains.

## CD Notes

- When executing the project on the accompanying CD, set a break point in `OnWzdType()` in `WzdView.cpp`. Then click on the Options/Wzd command and watch as `OnWzdType()` modifies the Options menu.

# Example 19    Dynamically Modifying the System Menu

## Objective

You would like to add or delete a command from the system menu (Figure 6.11).

## Figure 6.11    You can add menu items to the system menu.

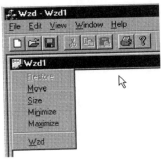

## Strategy

We will use `CWnd::GetSystemMenu()` to get a `CMenu` class pointer to a system menu. We will then use the member functions of `CMenu` to modify the system menu.

Example 19   Dynamically Modifying the System Menu   **193**

# Steps

## Wrap the System Menu with a CMenu **Class**

1. You can wrap the system menu of any window by using the following code in the window class that controls that window.

```
CMenu* pSysMenu = GetSystemMenu(FALSE);
```

## Edit the System Menu

1. To remove the Close menu item from this system menu, you can use

```
pSysMenu->RemoveMenu(SC_CLOSE, MF_BYCOMMAND);
```

2. To add a new menu item, in this case Wzd with a command ID of IDC_WZD_TYPE, you can use

```
pSysMenu->AppendMenu(0,IDC_WZD_TYPE,"&Wzd");
```

## Process Messages from New System Menu Items

1. To process the Wzd command we added in the last step, use the Class-Wizard to add a WM_SYSCOMMAND message handler to this class. There, you can look for this command ID.

```
void CChildFrame::OnSysCommand(UINT nID, LPARAM lParam)
{
    if (nID==IDC_WZD_TYPE)
    {
        AfxMessageBox("Hello.");
    }

    CMDIChildWnd::OnSysCommand(nID, lParam);
}
```

## Notes

- The IDs for the other standard system menu commands are as follows.

```
SC_SIZE
SC_MOVE
SC_MINIMIZE
SC_MAXIMIZE
SC_NEXTWINDOW
SC_PREVWINDOW
SC_RESTORE
```

- Removing the Close menu item from the system menu also causes the Close button in the upper-right corner of a window to become disabled.
- The system menu is most often used when your application is minimized and sitting in the task bar. However, considering its low profile, you may not want to implement much functionality in the system menu.

## CD Notes

- When executing the project on the accompanying CD, set a break point in OnCreate() in Childfrm.cpp and watch as the child frame window's system menu is modified. You can also set a break point on OnSysCommand() and then click on the new Wzd command in the system menu to watch it being processed.

# Example 20    Triggering a Menu Command

## Objective

You would like to send a command to your application as if the user clicked on it from the menu or toolbar.

## Strategy

We will emulate the generation of a menu or toolbar command and send a WM_COMMAND message to the Mainframe Class's window.

Example 21   Creating Popup Menus   **195**

## Steps

### Trigger a Menu Command

1. Put the following code where you want to send the command. Substitute ID_FILE_OPEN with the menu command ID you want to trigger.

```
AfxGetMainWnd()->SendMessage(WM_COMMAND,ID_FILE_OPEN);
```

## Notes

- We could have sent this command message to any of our application's windows. However, by sending it to the Mainframe Class's window, this command will be routed throughout our application, as described in Chapter 3.

- You can use CWnd::SendMessage() to send any type of command to any window. See Appendix C for how to get a pointer to another application's class. If you want to emulate a key press or a mouse command, use CWnd::PostMessage() instead.

## CD Notes

- When executing the project on the accompanying CD, you can click on the Options/Wzd menu command and watch as it performs the same function as having clicked on the File/Open command.

# Example 21   Creating Popup Menus

## Objective

You would like to create a popup menu when your user right-clicks the view (Figure 6.12).

**Figure 6.12** **You can create a popup menu to display when your user right-clicks.**

## Strategy

We will use `CMenu::CreatePopupMenu()` and `CMenu::TrackPopupMenu()` to create our popup menu. We will also use the other member function of `CMenu` to dynamically populate our menu with commands.

## Steps

### Create and Populate a Popup Menu

1. Use the ClassWizard to add a `WM_RBUTTONDOWN` message handler to your class.
2. In this new handler, create a `CMenu` class object and a popup menu object.

```
CMenu menu;
menu.CreatePopupMenu();
```

3. To append a menu item to this menu, you can use

```
menu.AppendMenu(0, IDC_WZD1_TYPE, "Wzd&1");
```

4. To append a command with a check mark next to it, use

```
menu.AppendMenu(MF_CHECKED, IDC_WZD2_TYPE, "Wzd&2");
```

5. To put a separator line into this popup menu, use

```
menu.AppendMenu(MF_SEPARATOR, 0, "");
```

6. To add a command that's grayed out and unavailable to the user, use

```
menu.AppendMenu(MF_GRAYED, IDC_WZD3_TYPE, "Wzd&3");
```

Example 21   Creating Popup Menus   **197**

7. To add radio button dots to a group of menu items, you can use

```
menu.CheckMenuRadioItem(IDC_WZD3_TYPE, IDC_WZD4_TYPE,
    IDC_WZD4_TYPE, MF_BYCOMMAND);
```

The first two arguments of `CheckMenuRadioItem()` are the beginning and end of a range of command IDs. The third argument is the menu item next to which you want to put the dot — all other items in the group will be cleared of any dot. If you prefer to use menu positions, you can use `MF_BYPOSITION` for the fourth argument.

8. To cause one menu item to be the default menu item, such that it appears in boldface and is executed when the user presses Enter, you can use

```
::SetMenuDefaultItem(menu.m_hMenu, IDC_WZD4_TYPE,
    MF_BYCOMMAND);
```

## Display the Popup Menu

1. To display this popup menu where the right mouse button was clicked, use

```
CPoint pt;
GetCursorPos(&pt);
menu.TrackPopupMenu(TPM_RIGHTBUTTON, pt.x, pt.y, this);
```

Your application will now pause in `TrackPopupMenu()` until the user has clicked either on the menu or outside of it. When the menu is clicked, a `WM_COMMAND` message is sent to the window specified in the fourth argument of `TrackPopupMenu()`, along with the command ID.

2. After `TrackPopupMenu()` is finished, you need to destroy the popup menu object with `DestroyMenu()`. Since the CMenu object was allocated on the stack in this example, returning will also destroy it.

```
menu.DestroyMenu();
```

Rather than create your popup menu on the fly, you can also load a predefined menu directly from your application's resources.

### Create a Popup Menu with an Application Resource

1. Use the Menu Editor to create a new menu. Add the necessary menu items to this menu.

2. Create the Menu class object on the stack as before, but now use `CMenu::LoadMenu()` to load your new menu from your resources.

```
CMenu menu;
menu.LoadMenu(IDR_WZD_MENU);          // get menu resource
CMenu* pPopup = menu.GetSubMenu(0); // get pointer to popup menu
```

3. To disable items in this popup menu, you can use

```
pPopup->EnableMenuItem(ID_POPUP_WZD1, MF_BYCOMMAND|MF_GRAYED);
```

4. To add a check mark to an item, use

```
pPopup->CheckMenuItem(2,MF_BYPOSITION|MF_CHECKED);
```

5. You can display and destroy this popup menu as before.

## Notes

- Notice that we are dealing with two objects here: the `CMenu` class object and a Windows popup menu object. You can think of the `CMenu` object as your C++ class wrapper around a Windows resource. The popup menu object is actually a popup window that uses #32768 as its Window Class. For more on MFC objects and windows objects, see Chapter 1.

- Notice that just as with a regular menu, the messages sent from a popup menu are a `WM_COMMAND` command message with a command ID. For more on messaging, see Chapter 3.

## CD Notes

- When executing the project on the accompanying CD, right-click the mouse anywhere on the view to invoke a custom popup menu.

# Toolbars and Status Bars

The Application Wizard will automatically create a generic toolbar for your application. The status bar at the bottom of your application is also standard issue, with only the status of the Shift and Num keys updated there. The examples in this chapter will show you how to customize these bars.

**Example 22     Using the Toolbar Editor**   We will use the Toolbar Editor to modify our toolbar and create new toolbars.

**Example 23     Enabling and Disabling Toolbar Buttons**   We will cause toolbar buttons to gray and ungray, depending on the state of our application.

**Example 24     Adding Words to Your Toolbar Buttons**   We will add a simple word to the bottom of each toolbar button to indicate its purpose.

**Example 25     Nonstandard Toolbar Sizes**   We will look at ways to change the size of the toolbar buttons with which our application was created.

**Example 26**     **Keeping Toolbar Buttons Depressed**    We will look at a way to keep a toolbar button depressed to indicate this function is engaged.

**Example 27**     **Keeping Only One Toolbar Button in a Group Depressed**    Similar to the previous example, we will look at allowing only one in a group of buttons to be depressed at any given time.

**Example 28**     **Adding NonButton Controls to a Toolbar**    We will add a combo box to a toolbar.

**Example 29**     **Modifying Your Application's Status Bar**    We will use the String Table Editor to modify our application's status bar.

**Example 30**     **Updating Status Bar Panes**    We will use the Class Wizard to allow one of our classes to update the text message appearing in a status pane.

**Example 31**     **Adding Other Controls to Your Status Bar**    We will add a button and a progress control to a status bar.

# Example 22    Using the Toolbar Editor

## Objective

You would like to edit your application's toolbar or add additional toolbars to your application using the Toolbar Editor.

---

**NOTE:** Any new toolbar buttons you add to your application will initially appear grayed and disabled until you have added a command handler to process the button. Example 13 shows you how to add a menu command handler. The method for adding a toolbar command handler is the same.

---

Example 22    Using the Toolbar Editor    **201**

# Steps

## Create a New Toolbar Using the Developer Studio

1. To create a new toolbar, click on the Developer Studio's Insert/Resource menu commands to open the Insert Resource dialog box. Then select Toolbar from the list and click New.

## Edit a Toolbar with the Toolbar Editor

1. To edit an existing toolbar, locate its ID in the Toolbar folder of your application's resources and double-click on it. This will open the Toolbar Editor.

2. To add a new button to a toolbar using the Toolbar Editor, use the mouse to drag the blank toolbar button at the end of the toolbar to the location of the new button. You can then use the Bitmap Editor to add an image to this button. To open the Properties dialog box for this button, double-click on its image in the upper window (Figure 7.1).

**Figure 7.1    Add buttons with the Toolbar Editor.**

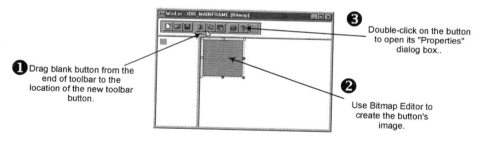

❸ Double-click on the button to open its "Properties" dialog box..

❶ Drag blank button from the end of toolbar to the location of the new toolbar button.

❷ Use Bitmap Editor to create the button's image.

3. To add a spacer to this toolbar (a blank spot between buttons), determine which button is currently over the location at which you want to put the spacer and grab it with the mouse. Then drag this button slightly to the right and let go — a blank space should now appear. To delete a separator, drag the button that's currently to its right slightly over it. To delete a button, simply drag it off the toolbar. Use your Studio's Undo command if you make a mistake.

4. The Properties dialog box for each button displays ID and Prompt edit boxes identical to that of a menu item's Properties box. See Example 12

for their meanings. The size edit boxes allow you to change the size not only of this button, but every button. See Example 25 for more on creating nonstandard toolbar buttons.

5. Adding a command handler for a toolbar button is identical to that of adding a menu command handler (Example 13). In fact, toolbar buttons are typically used as a pictorial representation of a menu command and rarely exist without a menu equivalent, although this is strictly by convention. Technically, you can add a toolbar button without a menu equivalent.

## Add a New Toolbar to Your Application

1. After creating a new toolbar using the Developer Studio and the Toolbar Editor, you must add the following code to your CMainFrame class's Create() function, just after the original toolbar is added.

```
if (!m_wndToolBar1.Create(this) ||
    !m_wndToolBar1.LoadToolBar(IDR_TOOLBAR1))
{
    TRACE0("Failed to create toolbar 1\n");
    return -1;         // fail to create
}
m_wndToolBar1.SetBarStyle(m_wndToolBar1.GetBarStyle() |
    CBRS_TOOLTIPS | CBRS_FLYBY | CBRS_SIZE_DYNAMIC);
m_wndToolBar1.EnableDocking(CBRS_ALIGN_ANY);
```

2. New toolbars are added to your application's window vertically, one row after the other. If you would prefer that your toolbars are added horizontally next to each other, you can use the following code.

```
DockControlBarLeftOf(&m_wndToolBar1, &m_wndToolBar);
```

You can find the listing for DockControlBarLeftOf() on page 204.

3. Floating toolbars can be dragged from your application's window and "float" in its own window (actually a Mini Frame Window controlled by MFC's CMiniFrameWnd class). If your user presses the close button on a floating toolbar, the toolbar is not destroyed like other windows, but is simply hidden. To allow your user to bring this toolbar back, you will need to add an additional menu command to your application's View command, one per new toolbar. Each of these commands will simply

Example 22   Using the Toolbar Editor   **203**

redisplay a hidden toolbar, as seen here. See Example 13 for how to add a menu command.

```
void CMainFrame::OnToolDecision()
{
    ShowControlBar(&m_wndToolBar1,m_wndToolBar1.GetStyle() &
        WS_VISIBLE) == 0, FALSE);
}
```

4. You should also put a check mark next to the menu item of a toolbar that is currently visible. See Example 16 for how to add a message handler that will add a check mark to a menu command. You can then fill in this handler with this code.

```
void CMainFrame::OnUpdateToolBar1(CCmdUI* pCmdUI)
{
    pCmdUI->SetCheck((m_wndToolBar1.GetStyle() & WS_VISIBLE) != 0);
}
```

## Notes

- Besides using `DockControlBarLeftOf()`, you can also specify exactly where a toolbar should go in the calling arguments of `CToolBar`'s `Create()`. For example, the following code would put a toolbar initially at an `x,y` location of `100,0`. CMainFrame, however, may then move it somewhere else.

```
if (!m_wndToolBar1.Create(this, 100, 0) ||
    :    :    :
```

- The last step shows how to initially place a toolbar to the left of the others, but to save the exact position your user places a toolbar between sessions with your application. Please see Example 5.
- You add a toolbar command handler to your application the same way you add a menu command handler (Example 13). Although you can use `m_bAutoMenuEnable` to enable menu commands that don't currently have a command handler, it will not enable toolbar buttons without a handler.

## CD Notes

- There is no accompanying project on the CD for this example.

### Listings — DockControlBarLeftOf()

```
void CMainFrame::DockControlBarLeftOf(CToolBar* Bar1,
    CToolBar* Bar2)
{

    CRect rect;
    RecalcLayout();
    Bar2->GetWindowRect(&rect);
    rect.OffsetRect(1,0);
    DockControlBar(Bar1,AFX_IDW_DOCKBAR_LEFT,&rect);
}
```

# Example 23   Enabling and Disabling Toolbar Buttons

## Objective

You would like to disable or enable a toolbar button (disabled buttons appear grayed out).

## Strategy

As with menu commands, a toolbar button will appear disabled until you use the ClassWizard to add a command handler for it (Example 13). Otherwise, we can conditionally enable and disable a toolbar button based on some condition in our application by adding a user interface handler to our class.

## Steps

### Add a User Interface Handler

1. Follow the steps shown in Example 13 for adding a menu command handler, except choose UPDATE_COMMAND_UI instead of COMMAND.

Example 24   Adding Words to Your Toolbar Buttons   **205**

2. Add the following code to this new handler, where m_bWzd is TRUE if the menu item should be enabled.

```
void CWzdView::OnUpdateWzdType(CCmdUI* pCmdUI)
{
    pCmdUI->Enable(m_bWzd);
}
```

## Notes

- A menu item that corresponds to a toolbar button by ID is simultaneously enabled and disabled with that button.
- The status of a toolbar button isn't updated and this routine won't be called until your application is idle. If you can't wait that long, get a pointer to this toolbar class (probably in CMainFrame) and call its UpdateWindow() member function.
- For more information on how MFC updates the user interface, see Chapter 3. Also, refer to the examples that come after this one.

## CD Notes

- When executing the project on the accompanying CD, you will notice a new toolbar button, which is also enabled.

# Example 24   Adding Words to Your Toolbar Buttons

## Objective

You would like to create a toolbar that includes one or two words on each button (Figure 7.2).

**Figure 7.2**    **Add a set of text strings to create toolbar buttons with captions.**

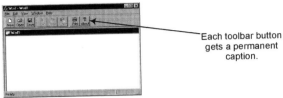

Each toolbar button gets a permanent caption.

## Strategy

We will add a set of text strings to our toolbar using CToolBarC-trl::AddStrings(). We will then associate each button with one of those strings by using CToolBarCtrl::InsertButton(). To continue to use the Toolbar Editor to create and edit our application's toolbars, we will dynamically associate a toolbar control with its string. We will use the ClassWizard to package this functionality in our own toolbar class.

## Steps

### Create an Array of Toolbar Captions

1. Use the String Editor to add a group of new strings to your application. These strings should be the short captions that will appear under your toolbar buttons, one per button.

2. Define an array in your CMainFrame class that associates a toolbar command ID with an ID of the string caption you created in the last step. We will create our own structure, called TOOLTEXT, to make this association. This structure simply contains two variables, one for the command ID and one for the string ID.

```
static TOOLTEXT tooltext[] =
{
    ID_FILE_NEW , IDS_FILE_NEW,
    ID_FILE_OPEN, IDS_FILE_OPEN,
    ID_FILE_SAVE, IDS_FILE_SAVE,
    ID_EDIT_CUT , IDS_FILE_CUT,
    ID_EDIT_COPY, IDS_FILE_COPY,
```

Example 24  Adding Words to Your Toolbar Buttons  **207**

```
    ID_EDIT_PASTE, IDS_FILE_PASTE,
    ID_FILE_PRINT, IDS_FILE_PRINT,
    ID_APP_ABOUT , IDS_APP_ABOUT,
};
#define TOOLTEXT_NUM (sizeof(tooltext)/sizeof(TOOLTEXT))
```

We will now create a new toolbar class. We will add a function to this class called `LoadToolBarEx()` that will load a toolbar resource and then dynamically add a caption to each button.

## Create a New Toolbar Class

1. Use the ClassWizard to create a new toolbar class derived from `CToolB-arCtrl`. Since we really want to derive this class from `CToolBar`, but the ClassWizard doesn't support `CToolBar`, we must manually change all references from `CToolBarCtrl` to `CToolBar` in our new class.

2. Add one new member function to this class called `LoadToolBarEx()`, to which you will pass the previous table. There you will start by using `CToolBar::LoadToolBar()` to load a toolbar resource.

```
BOOL CWzdToolBar::LoadToolBarEx(UINT nID,
    TOOLTEXT *pToolText,int nCnt)
{
    BOOL bRet;
    if (bRet=LoadToolBar(nID))
    {
```

3. Once loaded, you will loop through the caption table created in an earlier step ("Create an Array of Toolbar Captions" on page 206). There you will be locating the toolbar command IDs to which it refers and has located, we load that toolbar button's string from the string resources.

```
for (int i=0;i<nCnt;i++)
{
    // find button
    TBBUTTON tb;
    int inx=CommandToIndex(pToolText[i].idCommand);
    GetToolBarCtrl().GetButton(inx,&tb);
```

```
// get text for button
CString str;
str.LoadString(pToolText[i].idString);
```

4. Next, add this loaded string to the toolbar control.

```
// add a second NULL to string for AddStrings()
int nLen = str.GetLength() + 1;
TCHAR * pStr = str.GetBufferSetLength(nLen);
pStr[nLen] = 0;

// add new button using AddStrings
tb.iString=GetToolBarCtrl().AddStrings(pStr);
str.ReleaseBuffer();

// (no ModifyButton() function)
GetToolBarCtrl().DeleteButton(inx);
GetToolBarCtrl().InsertButton(inx,&tb);
}
```

The toolbar control returns with an index to that string, which we can then store with the button. Notice that AddStrings() requires two null characters after the last string, causing us to add an extra null character. Also, note that since there is no ModifyButton(), we have to delete and reinsert each button.

5. Use CToolBar::SetSizes() to make the buttons in this toolbar larger so that they can accommodate the new captions. The first argument to Set-Sizes() refers to how much of the bitmap image is required for each button. The second argument refers to how big to actually make the button.

```
// make buttons larger to handle added text
CSize sizeImage(16,15);
CSize sizeButton(35,35);
SetSizes(sizeButton, sizeImage);
```

6. To see the complete listing of this toolbar class, see "Listings — Toolbar Class" on page 209.

Example 24    Adding Words to Your Toolbar Buttons    **209**

## Implement the New Toolbar Class

1. Substitute this new toolbar class for your application's original toolbar class in `CMainFrame`.

```
// call the new CWzdToolBar::LoadToolBarEx() function
if (!m_wndToolBar.Create(this) ||
    !m_wndToolBar.LoadToolBarEx(IDR_MAINFRAME,
    (TOOLTEXT*)&tooltext,TOOLTEXT_NUM))
{
    TRACE0("Failed to create toolbar\n");
    return -1;                                    // fail to create
}
```

**II**

**7**

## Notes

- The captions added in this example are in addition to the bubble help captions that appear when you hold the mouse cursor over a toolbar button for longer than a half-second.

## CD Notes

- When executing the project on the accompanying CD, you will notice that the toolbar buttons have been enlarged to include a descriptive word for each.

## Listings — Toolbar Class

```
#if
!defined(AFX_WZDTOOLBAR_H__3A5CD903_E412_11D1_9B7D_00AA003D8695__INCLUDED_)
#define AFX_WZDTOOLBAR_H__3A5CD903_E412_11D1_9B7D_00AA003D8695__INCLUDED_

#if _MSC_VER >= 1000
#pragma once
#endif // _MSC_VER >= 1000
```

```
// WzdToolBar.h : header file
//


/////////////////////////////////////////////////////////////////////////
// CWzdToolBar window

typedef struct t_TOOLTEXT {
    UINT idCommand;
    UINT idString;
} TOOLTEXT;

class CWzdToolBar : public CToolBar
{

// Construction
public:
    CWzdToolBar();

// Attributes
public:

// Operations
public:
    BOOL LoadToolBarEx(UINT nID,TOOLTEXT *pToolText,int nCnt);

// Overrides
    // ClassWizard generated virtual function overrides
    //{{AFX_VIRTUAL(CWzdToolBar)
    //}}AFX_VIRTUAL

// Implementation
public:
    virtual ~CWzdToolBar();
```

Example 24    Adding Words to Your Toolbar Buttons    **211**

```
        // Generated message map functions
protected:
    //{{AFX_MSG(CWzdToolBar)
        // NOTE - the ClassWizard will add and remove member functions here.
    //}}AFX_MSG

    DECLARE_MESSAGE_MAP()
};


/////////////////////////////////////////////////////////////////////////////

//{{AFX_INSERT_LOCATION}}
// Microsoft Developer Studio will insert additional declarations immediately
// before the previous line.

#endif
// !defined(AFX_WZDTOOLBAR_H__3A5CD903_E412_11D1_9B7D_00AA003D8695__INCLUDED_)

// WzdToolBar.cpp : implementation file
//

#include "stdafx.h"
#include "wzd.h"
#include "WzdToolBar.h"

#ifdef _DEBUG
#define new DEBUG_NEW
#undef THIS_FILE
static char THIS_FILE[] = __FILE__;
#endif


/////////////////////////////////////////////////////////////////////////////
// CWzdToolBar

CWzdToolBar::CWzdToolBar()
{
}
```

```
CWzdToolBar::~CWzdToolBar()
{
}



BEGIN_MESSAGE_MAP(CWzdToolBar, CToolBar)
    //{{AFX_MSG_MAP(CWzdToolBar)
        // NOTE - the ClassWizard will add and remove mapping macros here.
    //}}AFX_MSG_MAP
END_MESSAGE_MAP()

/////////////////////////////////////////////////////////////////////////////
// CWzdToolBar message handlers

BOOL CWzdToolBar::LoadToolBarEx(UINT nID,TOOLTEXT *pToolText,int nCnt)
{
    BOOL bRet;
    if (bRet=LoadToolBar(nID))
    {
        // loop through tooltext adding text to buttons
        for (int i=0;i<nCnt;i++)
        {
            // find button
            TBBUTTON tb;
            int inx=CommandToIndex(pToolText[i].idCommand);
            GetToolBarCtrl().GetButton(inx,&tb);

            // get text for button
            CString str;
            str.LoadString(pToolText[i].idString);

            // add a second NULL to string for AddStrings()
            int nLen = str.GetLength() + 1;
            TCHAR * pStr = str.GetBufferSetLength(nLen);
            pStr[nLen] = 0;

            // add new button using AddStrings
            tb.iString=GetToolBarCtrl().AddStrings(pStr);
            str.ReleaseBuffer();
```

Example 25　Nonstandard Toolbar Sizes　**213**

```
            // (no ModifyButton() function)
        GetToolBarCtrl().DeleteButton(inx);
        GetToolBarCtrl().InsertButton(inx,&tb);
    }


    // make buttons larger to handle added text
    CSize sizeImage(16,15);
    CSize sizeButton(35,35);
    SetSizes(sizeButton, sizeImage);
    }
    return bRet;
}
```

# Example 25　**Nonstandard Toolbar Sizes**

## Objective

You would like to give your toolbar larger buttons (Figure 7.3).

## Figure 7.3　**Enlarge toolbar buttons with the Toolbar Editor.**

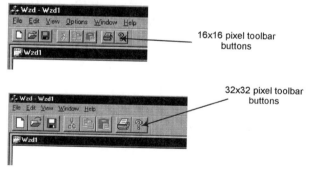

16x16 pixel toolbar buttons

32x32 pixel toolbar buttons

## Strategy

We will use the Toolbar Editor to increase the size of our toolbar buttons. We will then redraw or cut and paste a new image onto each button face. We will also look at CToolBar::SetSizes(), which allows us to programmatically change the size of a toolbar's buttons.

# Steps

## Change Toolbar Button Sizes with the Toolbar Editor

1. Find your toolbar ID in the Toolbar folder of your application's resources. Double-click on it to open the Toolbar Editor. Select your Developer Studio's Image/Grid Settings... menu commands to open the Grid Settings dialog box. There you can set the size of the bitmap portion of your new toolbar buttons. In this example, we made them $25 \times 25$ pixels.

2. Now grab the grabber box in the bottom-right corner of your toolbar button until it becomes the correct size.

3. Use the Toolbar Editor to enlarge any existing button images to this new size. You can start by using the cut tool to select the image and then dragging the grabber box until the image is large enough. Then touch up the resulting image. You can't get by with just enlarging the image alone, because there really aren't enough pixels to automatically create a smooth image. You can also cut and paste a larger image into this button from another source.

4. And that's it. Your toolbar resource automatically tells your application to make your toolbar buttons larger to accommodate this new image size.

## Change Toolbar Button Sizes with `CToolbar::SetSizes()`

1. Occasionally, you will need to enlarge buttons at run time. An example of this can be found in Example 24, where we needed to enlarge our toolbar buttons to accommodate a caption. To enlarge buttons at run time, you can use the following.

```
// set toolbar size to 32 by 32 pixels
SIZE sizeButton, sizeImage;
sizeImage.cx = 25;
sizeImage.cy = 25;
sizeButton.cx = sizeImage.cx + 7; //allow for spacing around image
sizeButton.cy = sizeImage.cy + 7;
m_wndToolBar.SetSizes(sizeButton, sizeImage);
```

Example 26    Keeping Toolbar Buttons Depressed    **215**

## Notes

- For your convenience, an enlarged version of the standard toolbar is included on the CD. To use this image in your application, open the project for this example on the CD in another Developer Studio and use cut and paste to transfer those images to your new application.

## CD Notes

- When executing the project on the accompanying CD, you will notice that the toolbar buttons and icons have been enlarged.

# Example 26    Keeping Toolbar Buttons Depressed

## Objective

You would like a toolbar button to remain pressed after it has been clicked (Figure 7.4).

**Figure 7.4    Add a user interface handler to keep buttons depressed.**

## Strategy

We will use the ClassWizard to add a user interface handler that will allow us to keep a button depressed.

## Steps

### Add a User Interface Handler

1. Follow the steps in Example 24 to add a user interface handler to your application. Then add the following code to this new handler.

```
void CWzdView::OnUpdateWzdButton(CCmdUI* pCmdUI)
{
    // this same command checks any menu items too
    pCmdUI->SetCheck(m_bWzd);
}
```

## Notes

- A depressed button is actually the toolbar equivalent of a checked menu item. When you check a menu item, you get a depressed toolbar button for free. The CCmdUI class object is overridden, such that when MFC is updating a toolbar button rather than a menu item, different SetCheck() functions are called for each menu item and each toolbar button.
- The status of a toolbar button isn't updated (and therefore this routine isn't called) until your application is idle. If you can't wait that long, get a pointer to this toolbar class (probably in CMainFrame) and call its UpdateWindow() member function.
- For more information on how MFC updates the user interface, see Chapter 3.

## CD Notes

- When executing the project on the accompanying CD, you will notice a new button that is permanently depressed.

# Example 27   Keeping Only One Toolbar Button in a Group Depressed

## Objective

You would like to keep only one button in a group of buttons depressed to signify that your application is in a particular mode (Figure 7.5).

Example 26   Keeping Toolbar Buttons Depressed   **217**

**Figure 7.5** **Add an interface message handler to every toolbar in a group to keep one button in that group depressed.**

## Strategy

We will use the ClassWizard to add an interface message handler to every toolbar button in this group. We will then call `CCmdUI::SetRadio()` to depress the appropriate button.

## Steps

## Add a Menu Command Handler for Each Member of the Group

1. Use the ClassWizard to add a command handler for each button in the group (Example 13). Use these handlers to set the appropriate mode for your application.

```
void CWzdView::OnWzd1Button()
{
    m_nWzdMode=CWzdView::mode1;
}


void CWzdView::OnWzd2Button()
{
    m_nWzdMode=CWzdView::mode2;
}
```

```
void CWzdView::OnWzd3Button()
{
    m_nWzdMode=CWzdView::mode3;
}
```

## Add a User Interface Handler for Each Member of the Group

1. Use the ClassWizard to add a user interface handler for each button in the group (Example 15). In each of these handlers, use SetRadio() to tell a button to be depressed, based on the current mode.

```
void CWzdView::OnUpdateWzd1Button(CCmdUI* pCmdUI)
{
    pCmdUI->SetRadio(m_nWzdMode==CWzdView::mode1);
}

void CWzdView::OnUpdateWzd2Button(CCmdUI* pCmdUI)
{
    pCmdUI->SetRadio(m_nWzdMode==CWzdView::mode2);
}

void CWzdView::OnUpdateWzd3Button(CCmdUI* pCmdUI)
{
    pCmdUI->SetRadio(m_nWzdMode==CWzdView::mode3);
}
```

## Notes

- There is no difference between using SetCheck() or SetRadio() when it comes to toolbar buttons. Both SetCheck() and SetRadio() make a depressed toolbar button. However, if these buttons are being used in conjunction with a group of menu items, you should use SetRadio() so that dots appear next to the menu items.
- For more information on how MFC updates the user interface, see Chapter 3.

Example 28    Adding NonButton Controls to a Toolbar    **219**

## CD Notes

- When executing the project on the accompanying CD, you can click on each of the Wzd buttons in the toolbar to depress it and undepress all other buttons.

# Example 28    Adding NonButton Controls to a Toolbar

## Objective

You would like to add a combo box or other control window to a toolbar (Figure 7.6).

**Figure 7.6    Add a Combo Box to Your Toolbar.**

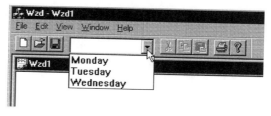

## Strategy

To put another control or any child window into a toolbar, you must first tell the toolbar to put a big spacer where you want to put your window. Then, you manually create the control window yourself in that spot. To retain the ability to use the Toolbar Editor to edit this toolbar, we will create a special toolbar button ID that, when used by a toolbar button, will be changed by our software into a combo box. We will be encapsulating this functionality in our own toolbar class.

# Steps

## Create a New Toolbar Class

1. Use the Class Wizard to create your own toolbar class derived from `CToolBarCtrl`. Then, use your text editor to substitute all references to `CToolBarCtrl` to `CToolBar` in the created `.cpp` and `.h` files.

2. Add a new function to this class called `LoadToolBarEx()`.

```
BOOL CWzdToolBar::LoadToolBarEx(UINT id)
{
```

3. In `LoadToolBarEx()`, start out by using `CToolBar::LoadToolBar()` to load a toolbar resource.

```
// load toolbar info
BOOL bRet;
bRet=CToolBar::LoadToolBar(id);
```

4. Next, we will look for a special button command ID. In this example, we are calling it `IDC_WZD_COMBO`. A toolbar button created with this ID using the Toolbar Editor will be converted to a combo box by this new toolbar class. To look for this ID in this toolbar, we use the following.

```
// find where our combo box will go
int pos=CommandToIndex(IDC_WZD_COMBO);
```

5. We then convert this toolbar button into an extra wide spacer with

```
// covert button in toolbar into a spacer for our combo box
SetButtonInfo(pos,IDC_WZD_COMBO,TBBS_SEPARATOR,COMBOLEN);
```

where `COMBOLEN` is the size of that spacer in pixels.

6. Our final step in `LoadToolBarEx()` is to create a control window where we created this spacer.

```
// create combo box
CRect rect;
GetItemRect(pos,&rect);
rect.bottom+=COMBODROP;     //how far will combo drop down?
m_ctrlWzdCombo.Create(WS_CHILD|WS_VISIBLE|CBS_DROPDOWNLIST,
    rect, this, IDC_WZD_COMBO);
```

Example 28    Adding NonButton Controls to a Toolbar    **221**

7. You should also handle all of the control notifications from this combo box right here in this toolbar class, but you must add them manually. The complete listing, "Listings — Toolbar Class" on page 221, shows how to do this.

## Implement the New Toolbar Class

1. Substitute this new toolbar class for any presently used toolbar class in CMainFrame. In this example, we changed all CToolBar references to CWzdToolBar. Then, we changed the LoadToolBar() in CMainFrame's OnCreate() to LoadToolBarEx().

2. Use the Toolbar Editor to add a new button to your toolbar at the point you would like to insert a combo box. Give this button the special command ID you specified previously, which in this example was IDC_WZD_COMBO. Multiple toolbars in your application can have this same special button ID.

## Notes

- You can also use this method to add other controls to a toolbar as long as they're thin. That includes buttons (push, check, and radio), progress controls, edit boxes, static controls, and the new date/time control.

- A toolbar appears at first to be one Parent window with several Child window buttons. In fact, a toolbar is one big Control window that draws several "buttons" itself and then handles all of the mouse action for those buttons.

## CD Notes

- When executing the project on the accompanying CD, you will notice a new combo box control in the toolbar.

## Listings — Toolbar Class

```
#if
!defined(AFX_WZDTOOLBAR_H__27649E31_C807_11D1_9B5D_00AA003D8695__INCLUDED_)
#define AFX_WZDTOOLBAR_H__27649E31_C807_11D1_9B5D_00AA003D8695__INCLUDED_

#if _MSC_VER >= 1000
#pragma once
```

```
#endif // _MSC_VER >= 1000

// WzdToolBar.h : header file
//

/////////////////////////////////////////////////////////////////////////////
// CWzdToolBar window

class CWzdToolBar : public CToolBar
{

// Construction
public:
    CWzdToolBar();

    BOOL LoadToolBarEx(UINT id);

// Attributes
public:

// Operations
public:

// Overrides
    // ClassWizard generated virtual function overrides
    //{{AFX_VIRTUAL(CWzdToolBar)
    //}}AFX_VIRTUAL

// Implementation
public:
    virtual ~CWzdToolBar();

    // Generated message map functions
protected:
    //{{AFX_MSG(CWzdToolBar)
    //}}AFX_MSG
    afx_msg void OnDropdownCombo();
    afx_msg void OnCloseupCombo();

    DECLARE_MESSAGE_MAP()
```

Example 28    Adding NonButton Controls to a Toolbar    **223**

```
private:
    CString        m_sSelection;
    CComboBox       m_ctrlWzdCombo;
};


/////////////////////////////////////////////////////////////////////////////

//{{AFX_INSERT_LOCATION}}
// Microsoft Developer Studio will insert additional declarations immediately
// before the previous line.

#endif
// !defined(AFX_WZDTOOLBAR_H__27649E31_C807_11D1_9B5D_00AA003D8695__INCLUDED_)

// WzdToolBar.cpp : implementation file
//

#include "stdafx.h"
#include "wzd.h"
#include "WzdToolBar.h"

#ifdef _DEBUG
#define new DEBUG_NEW
#undef THIS_FILE
static char THIS_FILE[] = __FILE__;
#endif
```

II

7

```
#define COMBOPOS 3           // position of combo box in toolbar
#define COMBOLEN 120         // length of combo box in pixels
#define COMBODROP 100        // length of drop of combo box in pixels


///////////////////////////////////////////////////////////////////////
// CWzdToolBar

CWzdToolBar::CWzdToolBar()
{
    m_sSelection=_T("");
}

CWzdToolBar::~CWzdToolBar()
{
}



BEGIN_MESSAGE_MAP(CWzdToolBar, CToolBar)
    //{{AFX_MSG_MAP(CWzdToolBar)
    //}}AFX_MSG_MAP
    ON_CBN_CLOSEUP(IDC_WZD_COMBO, OnCloseupCombo)
    ON_CBN_DROPDOWN(IDC_WZD_COMBO, OnDropdownCombo)
END_MESSAGE_MAP()

///////////////////////////////////////////////////////////////////////
// CWzdToolBar message handlers

BOOL CWzdToolBar::LoadToolBarEx(UINT id)
{
    // load toolbar info
    BOOL bRet;
    bRet=CToolBar::LoadToolBar(id);

    // find where our combo box will go
    int pos=CommandToIndex(IDC_WZD_COMBO);

    // covert button in toolbar into a spacer for our combo box
    SetButtonInfo(pos,IDC_WZD_COMBO,TBBS_SEPARATOR,COMBOLEN);
```

Example 28    Adding NonButton Controls to a Toolbar    **225**

```
        // create combo box
        CRect rect;
        GetItemRect(pos,&rect);
        rect.bottom+=COMBODROP; //how far will combo drop down?
        m_ctrlWzdCombo.Create(WS_CHILD|WS_VISIBLE|CBS_DROPDOWNLIST, rect, this,
            IDC_WZD_COMBO);

        return bRet;
    }


void CWzdToolBar::OnDropdownCombo()
{
    m_ctrlWzdCombo.ResetContent();

    m_ctrlWzdCombo.AddString("Monday");
    m_ctrlWzdCombo.AddString("Tuesday");
    m_ctrlWzdCombo.AddString("Wednesday");
    m_ctrlWzdCombo.SelectString(-1,m_sSelection);
}


void CWzdToolBar::OnCloseupCombo()
{
    int i;
    if ((i=m_ctrlWzdCombo.GetCurSel())!=CB_ERR)
    {
        m_ctrlWzdCombo.GetLBText(i, m_sSelection);
    }
    else
    {
        m_ctrlWzdCombo.AddString(m_sSelection);
        m_ctrlWzdCombo.SelectString(-1,m_sSelection);
    }
}
```

II

# Example 29   Modifying Your Application's Status Bar

## Objective

You would like to add additional indicators to your application's status bar (Figure 7.7).

**Figure 7.7     Add panes to the Status Bar using the String Table Editor and the Text Editor.**

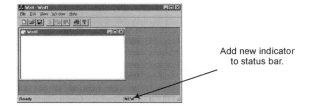

Add new indicator to status bar.

> NOTE: Any status indicator you add using this example will remain blank until you add a user interface handler to update it. The next example shows you how to add such a handler.

## Strategy

There currently is no Status Bar Editor. To add indicators to a status bar, we will first use the String Table Editor to store the text that will conditionally appear in the status bar. Then, we will use the Text Editor to add a line to the CMainFrame class to tell the CStatusBar class to create another indicator "pane".

## Steps

### Add a Status String Using the String Table Editor

1. Click on your application's ResourceView tab and locate the string table in the String Table folder. Double-click on the string table's ID to bring up the String Table Editor.

2. To add a new indicator string to this table, locate the ID_INDICATOR_REC string and click the Studio's Insert/New String menu commands. This

Example 29   Modifying Your Application's Status Bar   **227**

will act to both insert a new string and open its properties dialog box (Figure 7.8).

## Figure 7.8   Use the String Editor to add a status string.

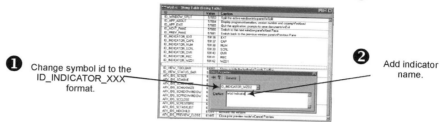

① Change symbol id to the ID_INDICATOR_XXX format.

② Add indicator name.

3. By convention, you should give your indicator ID a name similar to ID_INDICATOR_XXX where XXX describes the indicator. Then, enter a text string in Caption. This string will appear in the status bar when its pane is enabled. The length of this string will determine the length of the pane in which this indicator will appear in the status bar. If you pad the front and back of this caption with spaces, the pane will appear wider than the text string. Click the close button on this Properties dialog box and the change will be made to the table.

## Implement this New String in the Mainframe Class

1. Add this indicator's ID to the indicators[] array in CMainFrame. Locate this array in Mainfrm.cpp and add your new ID as seen below. The location at which you add the ID determines the location at which this indicator will appear in the status bar.

```
static UINT indicators[] =
{
    ID_SEPARATOR,             // status line indicator
    ID_INDICATOR_WZD1,        //<<<<<< new >>>>>>
    ID_INDICATOR_CAPS,
    ID_INDICATOR_NUM,
    ID_INDICATOR_SCRL,
};
```

2. Continue with the next example to find out how to make the text in this indicator appear conditionally.

## Notes

- The horizontal lines in the String Table Editor indicate a gap in the numbering of the strings. For example, ID_INDICATOR_REC has a value of 59141 and ID_VIEW_TOOLBAR then jumps to 59392, so there is a horizontal line between them. You can only insert a new string at one of these horizontal lines, for obvious reasons. Even if you had picked ID_INDICATOR_NUM as the point at which you wanted to add your new string, the String Table Editor would have defaulted instead to the point after ID_INDICATOR_REC.

## CD Notes

- There is no accompanying project on the CD for this example.

# Example 30    Updating Status Bar Panes

## Objective

You would like to enable and possibly change a pane in your application's status bar (Figure 7.9).

**Figure 7.9    Enable the panes in the status bar to indicate the status of your application through the icons in the System Tray.**

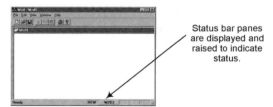

Status bar panes are displayed and raised to indicate status.

## Strategy

The same mechanism that allows our application to update the status of its toolbar buttons can also be used to update the status of the panes in the status bar, although with different effects. A disabled status bar pane, rather than showing grayed text, will show nothing at all. A pane that has been "checked" will appear raised from the screen.

Example 30   Updating Status Bar Panes   **229**

# Steps

## Manually Add User Interface Handlers for Each Status Bar Pane

1. Because the ClassWizard doesn't currently handle status bar indicators, you must manually add a user interface handler for each of the status bar panes you want to enable or disable.

```
BEGIN_MESSAGE_MAP(CWzdView, CView)
    //{{AFX_MSG_MAP(CWzdView)
    //}}AFX_MSG_MAP
    ON_UPDATE_COMMAND_UI(ID_INDICATOR_WZD1,
        OnUpdateIndicatorWzd1)
    ON_UPDATE_COMMAND_UI(ID_INDICATOR_WZD2,
        OnUpdateIndicatorWzd2)
END_MESSAGE_MAP()
```

Make sure to put these macros outside of the brackets ({{}}) that the ClassWizard uses. The ID you use is the pane's ID.

2. Define your indicator handlers in your .h file.

```
// Generated message map functions
protected:
    //{{AFX_MSG(CWzdView)
    //}}AFX_MSG
    afx_msg void OnUpdateIndicatorWzd1(CCmdUI *pCmdUI);
    afx_msg void OnUpdateIndicatorWzd2(CCmdUI *pCmdUI);
    DECLARE_MESSAGE_MAP()
```

## Update the Status Bar Pane

1. To turn a status indicator on, you can use

```
void CWzdView::OnUpdateIndicatorWzd1(CCmdUI *pCCmdUI)
{
    pCCmdUI->Enable(TRUE);
}
```

II

2. To turn a status indicator on and change it's name to NEW, or anything else, you can use

```
void CWzdView::OnUpdateIndicatorWzd1(CCmdUI *pCCmdUI)
{

    pCCmdUI->Enable(TRUE);
    pCCmdUI->SetText("NEW");

}
```

3. To turn an indicator on and check it (thus causing it to appear raised), you can use

```
void CWzdView::OnUpdateIndicatorWzd2(CCmdUI *pCCmdUI)
{

    pCCmdUI->Enable(TRUE);
    pCCmdUI->SetCheck();

}
```

## Notes

- The status bar's panes, like toolbar buttons, are updated when your application is idle. To force it to update earlier, you can use `UpdateWindow()`.

```
m_statusbar.UpdateWindow();
```

- You might have noticed that you call the same `SetCheck()` function here that you called for updating the menu and the toolbar interface. In fact, those instances of `SetCheck()` (and the other member functions of `CCmdUI`) are not the same, other than in name. They are overridden members of a `CCmdUI` base class from which four other classes have been derived to handle each type of bar or the menu. When a bar or the menu is about to be updated, the appropriate `CCmdUI` derivative is created for each pane, button, or control and is sent to your handler using the `OnCmdMsg()` mechanism described in Chapter 3. For more on the `CCmdUI` class and updating the status of the user interface, see Chapter 3.
- To add other controls to your status bar, see the next example.

Example 31   Adding Other Controls to Your Status Bar   **231**

## CD Notes

- When executing the project on the accompanying CD, you will notice that there are two new panes in the status bar, one depressed and one raised.

# Example 31   Adding Other Controls to Your Status Bar

## Objective

You would like to add a progress bar and a button control to your status bar (Figure 7.10).

## Figure 7.10   Add controls to the status bar.

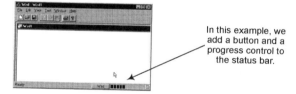

In this example, we add a button and a progress control to the status bar.

## Strategy

We will add two new items to the status bar using the method we demonstrated in the previous example. This time, however, the text items we add will be blank text strings that will merely act as place holders for control windows we will be creating dynamically ourselves. We will then use the CStatusBar's GetItemRect() member function to get the dimensions of these placeholders and manually create our controls over them.

## Steps

### Create Placeholders in the Status Bar for Our Control Windows

1. Create two new strings in the string table called ID_INDICATOR_WZDBUTTON and ID_INDICATOR_WZDPROGRESS and define them as strings of blanks. The

number of blanks will determine the size of your controls in the status bar.

2. Add these string IDs to your CMainFrame's indicator array in the location at which you would like them to appear in the status bar.

```
// add new id's to indicators in Mainfrm.cpp
static UINT indicators[] =
{
    ID_SEPARATOR,                   // status line indicator
    ID_INDICATOR_WZDBUTTON,         <<<<<<<<<<<
    ID_INDICATOR_WZDPROGRESS,       <<<<<<<<<<<<<<<
    ID_INDICATOR_CAPS,
    ID_INDICATOR_NUM,
    ID_INDICATOR_SCRL,
};
```

3. If your project were compiled now, you should see two empty spots in the status bar.

## Create a New Status Bar Class

1. Use the ClassWizard to create your own CWzdStatusBar class, derived from CStatusBar.

2. Embed your new controls in this class.

```
CButton        m_WzdButton;
CProgressCtrl m_WzdProgressCtrl;
```

3. Use the ClassWizard to add a WM_CREATE message handler to this class. There, you will create these embedded controls.

```
int CWzdStatusBar::OnCreate(LPCREATESTRUCT lpCreateStruct)
{
    if (CStatusBar::OnCreate(lpCreateStruct) == -1)
        return -1;

    CRect rect(0,0,0,0);
    m_WzdButton.Create("Wzd",WS_CHILD,rect,this,
        IDC_WZD_BUTTON);
```

Example 31 Adding Other Controls to Your Status Bar **233**

```
    CFont *pFont=CFont::FromHandle((HFONT)::
        GetStockObject(ANSI_VAR_FONT));
    m_WzdButton.SetFont(pFont);


    m_WzdProgressCtrl.Create(WS_CHILD|WS_VISIBLE,rect,this,
        IDC_WZD_PROGRESS);


    return 0;
}
```

Notice that we're creating these control windows with an initial size of $0 \times 0$ because we will be fixing them up immediately, anyway. We also are using a different font for the button text because the default font is boldfaced.

4. Use the ClassWizard to add a WM_SIZE message handler to this class. There, you will position and size your new controls.

```
void CWzdStatusBar::OnSize(UINT nType, int cx, int cy)
{
    CStatusBar::OnSize(nType, cx, cy);

    UINT inx;
    CRect rect;

    inx=CommandToIndex(ID_INDICATOR_WZDBUTTON);
    GetItemRect(inx,&rect);
    m_WzdButton.MoveWindow(rect);

    inx=CommandToIndex(ID_INDICATOR_WZDPROGRESS);
    GetItemRect(inx,&rect);
    m_WzdProgressCtrl.MoveWindow(rect);

}
```

Notice that we first we use CStatusBar::CommandToIndex() to determine the index of the spacer in which our control will sit. Then we get the dimensions of that spacer using CStatusBar::GetItemRect() and move our control window into place with CWnd::MoveWindow().

5. To see a complete listing of the new status bar class, see "Listings — Status Bar Class" on page 235.

## Implement this New Status Bar Class

1. Substitute the new status bar class for your current status bar class in `CMainFrame`.

```
protected:  // control bar embedded members
    CWzdStatusBar   m_wndStatusBar;      <<<
    CToolBar        m_wndToolBar;
```

2. When your application is created, it will contain these two new controls.

3. You can access these controls through your status bar member variable.

```
m_wndStatusBar.m_wndButton().ShowWindow(SW_SHOW);
m_wndStatusBar.m_wndProgressCtrl().SetPos(m_nInc);
```

4. The button control will send its command message to `CMainFrame` just like a menu command and, therefore, can be processed by any class in the command message chain.

5. There's one last problem related to the button control: when clicked, it receives input focus and becomes the default button. A thin rectangle is drawn on the button face, and the button itself becomes bold around the border. To undo these effects, add the following to the message handler that processes this button.

```
void CMainFrame::OnWzdButton()
{
    // these commands keep the button from being highlighted
    SetFocus();
    m_wndStatusBar.GetButton().SetButtonStyle(BS_PUSHBUTTON);
        :    :    :
}
```

## Notes

- Using placeholders rather than calculating the dimensions of the controls ourselves allows us more control over the positioning of our button and process control.

Example 31 Adding Other Controls to Your Status Bar **235**

## CD Notes

- When executing the project on the accompanying CD, click on the Text/Progress menu command. A button and progress control will appear in the status bar. Continuing to press the Progress menu item will cause the progress control to increase. Pressing the Wzd button in the status bar will cause a window to appear to indicate the button was pressed.

## Listings — Status Bar Class

```
#if !defined(AFX_WZDSTATUSBAR_H__DDA34AC3_D491_11D1_9B68_00AA003D8695__INCLUDED_)
#define AFX_WZDSTATUSBAR_H__DDA34AC3_D491_11D1_9B68_00AA003D8695__INCLUDED_

#if _MSC_VER >= 1000
#pragma once
#endif // _MSC_VER >= 1000

// WzdStatusBar.h : header file
//
```

II

7

```cpp
/////////////////////////////////////////////////////////////////////////////
// CWzdStatusBar window

class CWzdStatusBar : public CStatusBar
{
// Construction
public:
    CWzdStatusBar();

// Attributes
public:
    CButton &GetButton(){return m_WzdButton;};
    CProgressCtrl &GetProgressCtrl(){return m_WzdProgressCtrl;};

// Operations
public:

// Overrides
    // ClassWizard generated virtual function overrides
    //{{AFX_VIRTUAL(CWzdStatusBar)
    //}}AFX_VIRTUAL

// Implementation
public:
    virtual ~CWzdStatusBar();

    // Generated message map functions
protected:
    //{{AFX_MSG(CWzdStatusBar)
    afx_msg int OnCreate(LPCREATESTRUCT lpCreateStruct);
    afx_msg void OnSize(UINT nType, int cx, int cy);
    //}}AFX_MSG

    DECLARE_MESSAGE_MAP()
private:
    CButton       m_WzdButton;
    CProgressCtrl m_WzdProgressCtrl;
};
```

Example 31    Adding Other Controls to Your Status Bar    **237**

```
/////////////////////////////////////////////////////////////////////////////

//{{AFX_INSERT_LOCATION}}
// Microsoft Developer Studio will insert additional declarations immediately
// before the previous line.

#endif // !defined(
    AFX_WZDSTATUSBAR_H__DDA34AC3_D491_11D1_9B68_00AA003D8695__INCLUDED_)


// WzdStatusBar.cpp : implementation file
//

#include "stdafx.h"
#include "wzd.h"
#include "WzdStatusBar.h"

#ifdef _DEBUG
#define new DEBUG_NEW
#undef THIS_FILE
static char THIS_FILE[] = __FILE__;
#endif


/////////////////////////////////////////////////////////////////////////////
// CWzdStatusBar

CWzdStatusBar::CWzdStatusBar()
{
}


CWzdStatusBar::~CWzdStatusBar()
{
}



BEGIN_MESSAGE_MAP(CWzdStatusBar, CStatusBar)
    //{{AFX_MSG_MAP(CWzdStatusBar)
    ON_WM_CREATE()
    ON_WM_SIZE()
    //}}AFX_MSG_MAP
END_MESSAGE_MAP()
```

II

7

```
///////////////////////////////////////////////////////////////////////////
// CWzdStatusBar message handlers

int CWzdStatusBar::OnCreate(LPCREATESTRUCT lpCreateStruct)
{
    if (CStatusBar::OnCreate(lpCreateStruct) == -1)
        return -1;

    CRect rect(0,0,0,0);
    m_WzdButton.Create("Wzd",WS_CHILD,rect,this,IDC_WZD_BUTTON);
    CFont *pFont=CFont::FromHandle((HFONT)::GetStockObject(ANSI_VAR_FONT));
    m_WzdButton.SetFont(pFont);

    m_WzdProgressCtrl.Create(WS_CHILD|WS_VISIBLE,rect,this,IDC_WZD_PROGRESS);

    return 0;
}

void CWzdStatusBar::OnSize(UINT nType, int cx, int cy)
{
    CStatusBar::OnSize(nType, cx, cy);

    UINT inx;
    CRect rect;

    inx=CommandToIndex(ID_INDICATOR_WZDBUTTON);
    GetItemRect(inx,&rect);
    m_WzdButton.MoveWindow(rect);

    inx=CommandToIndex(ID_INDICATOR_WZDPROGRESS);
    GetItemRect(inx,&rect);
    m_WzdProgressCtrl.MoveWindow(rect);

}
```

8

# Views

The View in an SDI or MDI application is the primary mechanism for your user to interact with your application and, in particular, the document that your application is editing. All of the examples in this chapter relate to the view, from creating a view out of a Dialog Box to splitting up a view into multiple views.

**Example 32    Scroll Views**   We will add a scroll view to our application. Scroll views are ideal for graphic design applications.

**Example 33    Changing the Mouse Cursor**   We will look at how to conditionally change the shape of the cursor, which is also usually a requirement of a graphic design application.

**Example 34    Wait Cursors**   We will look at how to turn the mouse cursor into an hour glass shape for a lengthy operation.

**Example 35    Form Views**   We will create a view out of a dialog box. Dialog boxes define the size and position of a set of control windows and are reviewed in the next two chapters.

**Example 36     List View**   We will create a view containing a list of textual information with the ability to graphically highlight individual lines.

**Example 37     Dynamically Splitting a View**   We will look at how to manually add view splitting capabilities to your application. The AppWizard also gives you an opportunity to automatically add this feature when you first create your application.

# Example 32   Scroll Views

## Objective

You would like your view to be able to automatically scroll around images that are too large to fit in your view (Figure 8.1).

## Figure 8.1     Add a scroll view to enable scroll bars.

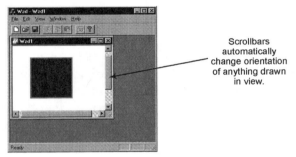

Scrollbars automatically change orientation of anything drawn in view.

## Strategy

We will use the AppWizard to create our application with a scroll view. We will also look at how to add a scroll view to our application after it's already been through the AppWizard. Scroll views are derived from `CScrollView`, which itself is derived from the standard MFC `CView` class. We will use `CScrollView::SetScrollSizes()` to set the size of our view in pixels. If the user resizes a view below this size, horizontal and vertical scroll bars will appear so the user can continue to see the entire view.

Example 32   Scroll Views   **241**

# Steps

## Create a Scroll View with the AppWizard

1. In the last step of creating your application using the AppWizard, you are shown an inventory of the classes that will be created for your application. Select the `CXxxView` class where `Xxx` is your project's name. Then, in the base class combo box, select `CScrollView` and click on Finish.

## Create a Scroll View with the ClassWizard

1. Use the ClassWizard to create a new view class derived from `CScroll-View`. Substitute this new class for the one currently used in your Application Class's `InitInstance()` to define your application's document template.

```
// add new view class to document template
CMultiDocTemplate* pDocTemplate;
pDocTemplate = new CMultiDocTemplate(
    IDR_WZDTYPE,
    RUNTIME_CLASS(CWzdDoc),
    RUNTIME_CLASS(CChildFrame),
    RUNTIME_CLASS(CWzdScrollView));       <<<<<<<
AddDocTemplate(pDocTemplate);
```

You can then delete the old View Class from your project. Or, if this is going to be the View Class of a new document template, simply add it to that template.

## Setup Your Scroll View

1. Use the ClassWizard to override `CScrollView`'s `OnInitialUpdate()`. There you can use `CScrollView::SetScrollSizes()` to set the minimum pixel size of your view. In other words, if your user shrinks their view

below this size, scrollbars appear so that they can scroll to see the entire viewing area again.

```
void CWzdScrollView::OnInitialUpdate()
{
    CScrollView::OnInitialUpdate();

    CSize sizeTotal;
    sizeTotal.cx = 250; // size required to display image
    sizeTotal.cy = 250; // size required to display image
    SetScrollSizes(MM_TEXT, sizeTotal);
}
```

## Notes

- To determine the scrolling area size, start by determining the size of your image. For example, if you plan to view an 11" × 8 1/2" sheet of paper and you are allowing 100 pixels per inch, then your vertical size should be 11 × 100 = 1100 pixels high.

- The `CScrollView` class provides two services. The first is its automatic creation of scrollbars. The second is its processing of those scrollbars by adjusting your view's Viewport. Viewports and window ports, which are discussed in Chapter 4, are a mechanism that allows you to draw your images without having to worry about what part will appear in the view or where they will appear in the view. Since a Scroll View automatically adjusts this viewport, all you have to worry about is drawing your figures using a Device Context. Unfortunately, this only applies to images you draw using a Device Context. If your view will contain other Child windows, you will have to move these manually whenever the view is scrolled. Use the ClassWizard to add `WM_VSCROLL` and `WM_HSCROLL` message handlers. Then use `CWnd::MoveWindow()` to move the Child windows.

- Scroll Views are mostly used in graphic and CAD applications. However, Form Views are also derived from Scroll Views to allow you to scroll around a large form. Since this is somewhat uncosmetic when the form is not that large, we will see how to turn off the scroll bars in a Form View in an upcoming example.

Example 33   Changing the Mouse Cursor   **243**

## CD Notes

- When executing the project on the accompanying CD, a drawn box will appear in the view without scrollbars. However, if you grab the side of the view using the mouse and shrink it to hide half of the box, scrollbars will appear, which will allow you to scroll to see the entire box.

# Example 33   Changing the Mouse Cursor

## Objective

You would like to change the mouse cursor shape depending on what drawing tool your user selects. Or you would like the default mouse cursor in a window to be something other than an arrow, such as the cross seen in Figure 8.2.

**Figure 8.2**   **Define your own window class to change the mouse cursor shape.**

## Strategy

We will be using two strategies to change the cursor shape. In the first, we will change the default window cursor by defining our own Window Class that uses our new cursor. In the second strategy we will use the SetCursor() function of CWnd.

## Steps

### Define a New Class Cursor

1. To change the default cursor of a window, such that any time the mouse cursor moves across the client area of that window it becomes that

shape, you must define your own Window Class for that window. Start by using the ClassWizard to override the `PreCreateWindow()` function of the window in question. In this example, we will be using our View Class's window. Then, define the Window Class using `AfxRegisterWnd-Class`.

```
BOOL CWzdView::PreCreateWindow(CREATESTRUCT& cs)
{
    cs.lpszClass = AfxRegisterWndClass(
        CS_DBLCLKS,                                     // double clicks are
                                                        // passed through
        AfxGetApp()->LoadStandardCursor(IDC_CROSS),     // stock cursor, but you
                                                        // can also load your own
        (HBRUSH)(COLOR_WINDOW+1),                        // normal background color
        AfxGetApp()->LoadIcon(IDR_MAINFRAME));          // normal icon
    return CView::PreCreateWindow(cs);
}
```

## Change Cursors with `CWnd::SetCursor()`

1. To change the cursor based on your application's mode, use the Class-Wizard to add a message handler for the `WM_SETCURSOR` message. In that handler, determine if you are in a particular mode; if so, use the `SetCursor()` function of `CWnd` to change the cursor shape.

```
BOOL CWzdView::OnSetCursor(CWnd* pWnd, UINT nHitTest,
    UINT message)
{
    if (m_bDrawMode)
    {
        SetCursor(AfxGetApp()->LoadCursor(IDC_DRAW_CURSOR));
        return TRUE;
    }

    return CView::OnSetCursor(pWnd, nHitTest, message);
}
```

Example 34    Wait Cursors    **245**

2. You can also use `SetCursor()` at other times to temporarily change the shape of the cursor.

```
void CWzdView::OnLButtonDown(UINT nFlags, CPoint point)
{
    ::SetCursor(AfxGetApp()->LoadCursor(IDC_DRAW_CURSOR1));

    CView::OnLButtonDown(nFlags, point);
}
```

However, as in this example, the next mouse move message that comes through will revert the cursor to the shape it was before you used `SetCursor()`, making this method very fragile.

## Notes

- You can only change the shape of the mouse cursor when it is over the client area of your window.
- For an example of using the ClassWizard to add a message handler, see Example 13.

## CD Notes

- When executing the project on the accompanying CD, you will notice that the default mouse cursor in the view is a cross. Selecting the new Pencil button in the toolbar will cause the default mouse cursor to be a pencil. And pressing the left mouse button in the view will cause the mouse cursor to become a broken pencil.

# Example 34    Wait Cursors

## Objective

You would like to temporarily turn the mouse cursor into an hourglass to indicate that a lengthy process is occurring and the user should wait.

## Strategy

You could simply create an hourglass cursor using the method described in the last example. However, MFC provides a helper class called `CWaitCursor` that automates this functionality.

## Steps

### Use `CWaitCursor()` to Create an Hourglass Cursor

1. To turn the wait cursor on, simply create an instance of the `CWaitCursor` class on the stack just before the location at which the lengthy process is to occur.

```
void CClass::Foo()
{
    CWaitCursor wc;
    :    :    :
```

The constructor of this class puts up the hourglass cursor, the destructor restores the original cursor.

2. To return the normal cursor without destroying the instance of this class, you can use

```
SetCursor(AfxGetApp()->LoadStandardCursor(IDC_ARROW));
```

3. Then, if you want to restore the hourglass cursor, you can use

```
wc.Restore();
```

## Notes

- The problem with this approach is that your function will occasionally call some other function that will restore the cursor when it shouldn't. To solve this problem, simply call `CWaitCursor::Restore()` after the offending function. If you need to call `Restore()` from the lower reaches of your application, you can call

```
AfxGetApp()->RestoreWaitCursor();
```

- A wait cursor is more appropriate for short waits. For longer waits, you might consider using a modeless dialog box with some sort of message describing what process is occurring. This resolves the problem of a

Example 35   Form Views   **247**

lower function turning off the wait cursor. You can also include an Abort button on a dialog that allows your user to terminate the function if they no longer want to wait. See Example 41 for how to create a modeless dialog box.

- Another advantage of a modeless dialog box is that you can indicate your application's progress using a progress control. If progress is difficult or impossible to measure, you can use an animation control to display a continually looping animation. See Example 43 for how to put an animation on a dialog box.

## CD Notes

- When executing the project on the accompanying CD, clicking on the Test/Wzd menu commands will cause the mouse cursor to become an hour glass for two seconds, an arrow for one second, an hour glass for one second, and final revert to the default cursor.

# Example 35   Form Views

## Objective

You would like to incorporate a dialog template into your view (Figure 8.3).

**Figure 8.3    Create a Form View that turns a dialog template into a view.**

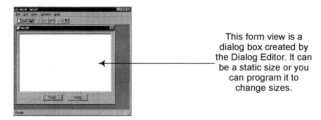

This form view is a dialog box created by the Dialog Editor. It can be a static size or you can program it to change sizes.

## Strategy

We will use the last step of the AppWizard to create a Form View, which turns a dialog template into a view. We will also look at using the Class-Wizard and Dialog Editor to add a Form View to an existing application. Our Form View is derived from CFormView, which is itself derived from

CScrollView. We will do some additional tweaking so that our Form View cannot be resized by our user. We will also look at how to turn off the scrollbars that CScrollView insists on drawing. Finally, we will look at how to resize the controls in a Form View if you plan to allow your user to resize their view.

# Steps

## Create a Form View with the AppWizard

1. In the last step of creating your application using the AppWizard, you are shown an inventory of the classes that will be created for your application. Select the CXxxView class, where Xxx is your project's name. Then, in the base class combo box, select CFormView and click on Finish.

2. You will notice that the AppWizard creates an additional dialog template in your resources when you create a Form View application. Use the Dialog Editor to add controls to this template and the ClassWizard to add handlers for these controls.

## Create a Form View with the ClassWizard

1. Create a new dialog template and add controls to it using the Dialog Editor. It doesn't have to be pretty yet — you can modify it again later. However, make sure its styles are as follows: child, no border, invisible, and no title.

2. Right-click on this form and select the ClassWizard from the popup menu. Enter a new class name, but instead of CDialog, make the base class CFormView. Use the ClassWizard to create a new class derived from CFormView.

3. Substitute this new class for the one currently used in your Application Class's InitInstance() to define your application's document template.

```
// add new view class to document template
CMultiDocTemplate* pDocTemplate;
pDocTemplate = new CMultiDocTemplate(
    IDR_WZDTYPE,
```

Example 35   Form Views   **249**

```
        RUNTIME_CLASS(CWzdDoc),
        RUNTIME_CLASS(CChildFrame),
        RUNTIME_CLASS(CWzdFormView));          <<<<<<<
    AddDocTemplate(pDocTemplate);
```

You can then delete the old View Class from your project. Or, if this is going to be the View Class of a new document template, simply add it to that template.

## Update the Form View

1. If you will be creating member variables using the ClassWizard for this Form View, you will need to call UpdateData() directly to exchange values between the form and those member variables. Retrieving information from the form can be accomplished on demand by using

```
UpdateData(TRUE);
```

2. Typically the OnUpdate() member function of a View Class is used to update the view with new data from the Document Class. Here, you should use UpdateData() to refresh your form with values from the document.

```
void CWzdView::OnUpdate(CView* pSender, LPARAM lHint,
    CObject* pHint)
{

    // store any data from document into member variables of
    // Form View class here

    UpdateData(FALSE);
}
```

## Fixed Form Views

1. If you don't want your user to be able to change your Form View's size, add the following lines to your PreCreateWindow() in your CChildFrame

class for an MDI application or in your `CMainFrame` class for an SDI application.

```
// removes min/max boxes
cs.style &= ~(WS_MAXIMIZEBOX|WS_MINIMIZEBOX);


// makes dialog box unsizable
cs.style &= ~WS_THICKFRAME;
```

2. By default, Form Views create a view that can be much larger than the dialog template from which you created them, giving them a somewhat awkward look. You'll have all of your controls concentrated in the left corner of the view and a great deal of empty space to the right. To shrink the view to the template size, use the ClassWizard to override the `OnInitialUpdate()` function of your Form View class. Then use `CScrollView`'s `ResizeParentToFit()` to shrink wrap the form.

```
void CWzdView::OnInitialUpdate()
{

    CFormView::OnInitialUpdate();

    // make frame the size of the original dialog box
    ResizeParentToFit();

    // get rid of those pesky scroll bars by making the point
    // at which they appear very small
    SetScrollSizes(MM_TEXT, CSize(20,20));

}
```

We added the last call to `SetScrollSizes()` to keep the `CScrollView` class from turning on the scrollbars. The `CScrollView` class turns on the scrollbars if it thinks the view is too small to contain the form. Since the view is being shrunk to just barely fit the form, on some platforms, `CScrollView` thinks it's time to turn on the scrollbars. To prevent this, we tell `CScrollView` that the size of the view is just $20 \times 20$ pixels. Since our form is larger than that, the bars stay off.

In some applications you may want to allow your user to resize their Form View, especially if the majority of your view will be a list control or an edit box with maybe a few buttons at the bottom. To allow your user to

Example 35 Form Views **251**

resize their Form View, we need to constantly resize and move the controls in the form, as seen next.

## Resizable Form Views

1. Don't make the changes seen previously to create a fixed Form View. In particular, don't add the following line to the `PreCreateWindow()` function of your frame window.

```
cs.style &= ~WS_THICKFRAME;
```

2. Use the ClassWizard to create control member variables for each of the controls in your dialog template. This simply makes it easier to call the `MoveWindow()` function of each of the control windows in the Form View.

3. Use the ClassWizard to add a `WM_SIZE` message handler to your Form View. Then, at the same location, use the `MoveWindow()` function of each of your controls to size and position them. You must resize and/or move your controls as the view window changes size.

```
void CWzdView::OnSize(UINT nType, int cx, int cy)
{
    CFormView::OnSize(nType, cx, cy);

    if (m_ctrlWzdList.m_hWnd)
    {

        CRect rect;
        m_ctrlWzdButton1.GetClientRect(&rect);

        // list control is always 10 pixels from corners and
        // above buttons (last two arguments are width & height)
        m_ctrlWzdList.MoveWindow(
            10,10,cx-20,cy-rect.Height()-30);

        // buttons are always the size they started 10 pixels
        // from bottom, 20 from each other and centered
        int strt = (cx - (rect.Width()*2+20))/2;
```

II

8

```
        rect.OffsetRect(strt,cy-rect.Height()-10);
        m_ctrlWzdButton1.MoveWindow(rect);


        rect.OffsetRect(rect.Width()+20,0);
        m_ctrlWzdButton2.MoveWindow(rect);
    }
```

We check to ensure that our control window classes have a window handle before we try to move them because a WM_SIZE message might be sent to our Form View before the control windows have even been created.

## Notes

- For an example of adding message handlers using the ClassWizard, see Example 13.

## CD Notes

- When executing the project on the accompanying CD, you will notice the view contains the controls of a dialog template. Resizing the view with the mouse will cause these controls to change size, too.

# Example 36   List View

## Objective

You would like to create a view that is simply a list of data with columns (Figure 8.4).

## Figure 8.4   Create a List View that is a list of data with columns.

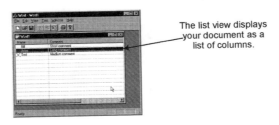

The list view displays your document as a list of columns.

Example 36   List View   **253**

## Strategy

We will use the AppWizard to create an application with a List View. We will also look at how to add a List View to an existing application using the ClassWizard. A List View uses the CListView, class which has an embedded List Control window. This Control window is accessible to us in the view. We will be manipulating it directly to change its style and add data to it.

## Steps

### Create a List View with the AppWizard

1. In the last step of creating your application using the AppWizard, you are shown an inventory of the classes that will be created for your application. Select the CXxxView class, where Xxx is your project's name. Then in the base class combo box, select CListView and click on Finish.

### Create a List View with the ClassWizard

1. Use the ClassWizard to create a new View Class derived from CListView. Substitute this new class for the one currently used in your Application Class's InitInstance() to define your application's document template.

```
// add new view class to document template
CMultiDocTemplate* pDocTemplate;
pDocTemplate = new CMultiDocTemplate(
    IDR_WZDTYPE,
    RUNTIME_CLASS(CWzdDoc),
    RUNTIME_CLASS(CChildFrame),
    RUNTIME_CLASS(CWzdListView));          <<<<<<<
AddDocTemplate(pDocTemplate);
```

You can then delete the old View Class from your project. Or, if this is going to be the View Class of a new document template, simply add it to that template.

## Setup Your List View in `OnInitialUpdate()`

1. Use the ClassWizard to override the `OnInitialUpdate()` function of the List View.

```
void CWzdView::OnInitialUpdate()
{
```

2. Set the list control's style. In this example, we are giving this list control a "report" style that sorts the first column alphabetically. We also are telling the control to always show its selection, even when our application doesn't have the focus.

```
GetListCtrl().ModifyStyle(0,LVS_REPORT| LVS_SHOWSELALWAYS|
    LVS_SORTASCENDING);
```

Notice that `GetListCtrl()` lets you access the actual list control class that's powering this view. Refer to your MFC documentation for other list control styles.

3. Next, we will be setting two list control extended styles. The first extended style causes lines to be drawn to separate the rows and columns of our list. The second allows our user to select an entire row. (By default, a list control allows only the first column to be selected):

```
GetListCtrl().SendMessage(LVM_SETEXTENDEDLISTVIEWSTYLE,0,
    LVS_EX_GRIDLINES|LVS_EX_FULLROWSELECT);
```

4. Since the report style we are using in this example has column headers, we will also need to define and name these columns with `CListCtrl::InsertColumn()`. The column width argument to `InsertColumn()` is in pixels, so we will use the average width of a character in pixels to help set this argument.

```
CDC* pDC = GetDC();
TEXTMETRIC tm;
pDC->GetTextMetrics(&tm);
GetListCtrl().InsertColumn(0,"Name",LVCFMT_LEFT,
    30 * tm.tmAveCharWidth, 0);
GetListCtrl().InsertColumn(1,"Comment",LVCFMT_LEFT,
    70 * tm.tmAveCharWidth, 1);
```

Example 36   List View   **255**

5. You can optionally place a bitmap image at the start of every row in a list control. However, you must first define a list of images available to that control using CListCtrl()::SetImageList().

```
m_ImageList.Create(IDB_STATUS_BITMAP, 15, 1, RGB(0,0,0));
GetListCtrl().SetImageList(&m_ImageList, LVSIL_STATE);
```

## Update Your List View

1. Use the ClassWizard to override the OnUpdate() function of CListView. There, we will copy data from our document into our list. In this example, our document consists of a list of data class objects that can be found in "Listings — Data Class" on page 266.

```
void CWzdView::OnUpdate(CView* pSender, LPARAM lHint,
    CObject* pHint)
{
    CList<CWzdInfo*,CWzdInfo*> *pList=GetDocument()->
        GetInfoList();
    GetListCtrl().DeleteAllItems();
    for (POSITION pos = pList->GetHeadPosition(); pos;)
    {
        CWzdInfo *pInfo = pList->GetNext(pos);
        AddItem(-1,pInfo);
    }
}
```

2. In OnUpdate, we call another function, AddItem, which will do the actual stuffing of data into the list control.

```
void CWzdView::AddItem(int i,CWzdInfo *pInfo)
{
    if (i==-1)
    {
        i=GetListCtrl().InsertItem(0, pInfo->m_sName);
    }
    else
    {
        GetListCtrl().SetItemText(i, 0,  pInfo->m_sName);
    }
```

```
GetListCtrl().SetItemText(i, 1,  pInfo->m_sComment);
GetListCtrl().SetItemData(i,(DWORD)pInfo);
// tells list control which bitmap to display at start of line
GetListCtrl().SetItemState(i, INDEXTOSTATEIMAGEMASK(
    pInfo->m_nState), LVIS_STATEIMAGEMASK);
}
```

## Work with a List View

1. To determine whether anything has been selected by your user, you can use

```
if (GetListCtrl().GetSelectedCount())
{
    // yes
}
```

2. To scan through only the selected items in your list control, you can use

```
int i=-1;
  while ((i = GetListCtrl().GetNextItem(i, LVIS_SELECTED)) != -1)
  {
      CWzdInfo *pInfo=(CWzdInfo *)GetListCtrl().GetItemData(i);

  }
```

3. To add a new item to your document, as well as to your list control, you can use

```
CList<CWzdInfo*,CWzdInfo*> *pList=GetDocument()->GetInfoList();
CWzdInfo *pInfo=new CWzdInfo("new","comment",CWzdInfo::NEW);
pList->AddHead(pInfo);
AddItem(-1,pInfo);
```

4. To make sure a particular line in the list control is visible, you can use `CListCtrl::EnsureVisible()`, which simply scrolls the listing until the selected item is revealed.

```
GetListCtrl().EnsureVisible(inx,FALSE);
```

Example 36    List View    **257**

The `inx` argument is the line number to reveal.

5. To cause an item in your list view to appear selected without the user actually selecting it, you can use the following.

```
LV_ITEM lvi;
lvi.mask = LVIF_STATE;
lvi.iItem = inx;
lvi.stateMask = 0x000f;
lvi.state = LVIS_SELECTED|LVIS_FOCUSED;
GetListCtrl().SetItemState(inx, &lvi);
```

The `inx` argument is the line number to select.

6. To delete a line from your list control, use

```
GetListCtrl().DeleteItem(inx);
```

7. To cause your control to redraw itself, you can use

```
GetListCtrl().Invalidate();
```

8. For more things you can do with a list control, refer to `CListCtrl` in your MFC documentation. To see this List View class in its entirety, see "Listings — List View Class" on page 258. You will find a listing of the data class we used in this example in "Listings — Data Class" on page 266.

## Notes

- Any control window can be turned into a view. Simply create an application with the `CView` class. Then, embed the desired MFC common control class in that `CView` class — as an example, use the `CButton` control class. In `CView`'s `OnCreate()` message handler, call `CButton`'s `Create()`. In `CView`'s `OnSize()` message handler, call `CButton`'s `MoveWindow()` to cause it to fill the screen.

## CD Notes

- When executing the project on the accompanying CD, you will notice that the view is filled with a list control. Set a breakpoint on `OnTestWzd()` in `WzdView.cpp`, then click on the Test/Wzd menu commands and step through to watch an item in this list being modified.

## Listings — List View Class

```
// WzdView.h : interface of the CWzdView class
//
/////////////////////////////////////////////////////////////////////

#if !defined(AFX_WZDVIEW_H__CA9038F0_B0DF_11D1_A18C_DCB3C85EBD34__INCLUDED_)
#define AFX_WZDVIEW_H__CA9038F0_B0DF_11D1_A18C_DCB3C85EBD34__INCLUDED_

#if _MSC_VER >= 1000
#pragma once
#endif // _MSC_VER >= 1000
#include <afxcview.h>
```

Example 36    List View    **259**

```
class CWzdView : public CListView
{
protected:      // create from serialization only
        CWzdView();
        DECLARE_DYNCREATE(CWzdView)

// Attributes
public:
    CWzdDoc* GetDocument();

// Operations
public:

// Overrides
    // ClassWizard generated virtual function overrides
    //{{AFX_VIRTUAL(CWzdView)
    public:
    virtual void OnDraw(CDC* pDC);  // overridden to draw this view
    virtual BOOL PreCreateWindow(CREATESTRUCT& cs);
    virtual void OnInitialUpdate();
    protected:
    virtual BOOL OnPreparePrinting(CPrintInfo* pInfo);
    virtual void OnBeginPrinting(CDC* pDC, CPrintInfo* pInfo);
    virtual void OnEndPrinting(CDC* pDC, CPrintInfo* pInfo);
    virtual void OnUpdate(CView* pSender, LPARAM lHint, CObject* pHint);
    //}}AFX_VIRTUAL

// Implementation
public:
    virtual ~CWzdView();
#ifdef _DEBUG
    virtual void AssertValid() const;
    virtual void Dump(CDumpContext& dc) const;
#endif
```

```
protected:

// Generated message map functions
protected:
    //{{AFX_MSG(CWzdView)
    afx_msg void OnTestWzd();
    //}}AFX_MSG
    DECLARE_MESSAGE_MAP()
private:
    int         m_AveCharWidth;
    CImageList  m_ImageList;
    int GetTextExtent(int len);
    void AddItem(int ndx,CWzdInfo *pInfo);

};

#ifndef _DEBUG  // debug version in WzdView.cpp
inline CWzdDoc* CWzdView::GetDocument()
    {return (CWzdDoc*)m_pDocument;}
#endif

/////////////////////////////////////////////////////////////////////////////

//{{AFX_INSERT_LOCATION}}
// Microsoft Developer Studio will insert additional declarations immediately
// before the previous line.

#endif // !defined(
    AFX_WZDVIEW_H__CA9038F0_B0DF_11D1_A18C_DCB3C85EBD34__INCLUDED_)

// WzdView.cpp : implementation of the CWzdView class
//

#include "stdafx.h"
#include "Wzd.h"

#include "WzdDoc.h"
#include "WzdView.h"
```

Example 36   List View   **261**

```
#ifdef _DEBUG
#define new DEBUG_NEW
#undef THIS_FILE
static char THIS_FILE[] = __FILE__;
#endif


/////////////////////////////////////////////////////////////////////////////
// CWzdView

IMPLEMENT_DYNCREATE(CWzdView, CListView)

BEGIN_MESSAGE_MAP(CWzdView, CListView)
    //{{AFX_MSG_MAP(CWzdView)
    ON_COMMAND(ID_TEST_WZD, OnTestWzd)
    //}}AFX_MSG_MAP
    // Standard printing commands
    ON_COMMAND(ID_FILE_PRINT, CView::OnFilePrint)
    ON_COMMAND(ID_FILE_PRINT_DIRECT, CView::OnFilePrint)
    ON_COMMAND(ID_FILE_PRINT_PREVIEW, CView::OnFilePrintPreview)
END_MESSAGE_MAP()


/////////////////////////////////////////////////////////////////////////////
// CWzdView construction/destruction

CWzdView::CWzdView()
{
    m_AveCharWidth=0;
}


CWzdView::~CWzdView()
{
}


BOOL CWzdView::PreCreateWindow(CREATESTRUCT& cs)
{
        // TODO: Modify the Window class or styles here by modifying
        // the CREATESTRUCT cs

        return CListView::PreCreateWindow(cs);
}
```

II

8

```
/////////////////////////////////////////////////////////////////////////
// CWzdView drawing

void CWzdView::OnDraw(CDC* pDC)
{
    CWzdDoc* pDoc = GetDocument();
    ASSERT_VALID(pDoc);

    // TODO: add draw code for native data here
}


/////////////////////////////////////////////////////////////////////////
// CWzdView printing

BOOL CWzdView::OnPreparePrinting(CPrintInfo* pInfo)
{
    // default preparation
    return DoPreparePrinting(pInfo);
}

void CWzdView::OnBeginPrinting(CDC* /*pDC*/, CPrintInfo* /*pInfo*/)
{
    // TODO: add extra initialization before printing
}

void CWzdView::OnEndPrinting(CDC* /*pDC*/, CPrintInfo* /*pInfo*/)
{
    // TODO: add cleanup after printing
}


/////////////////////////////////////////////////////////////////////////
// CWzdView diagnostics

#ifdef _DEBUG
void CWzdView::AssertValid() const
{
    CListView::AssertValid();
}
```

Example 36    List View    **263**

```
void CWzdView::Dump(CDumpContext& dc) const
{
    CListView::Dump(dc);
}


CWzdDoc* CWzdView::GetDocument() // non-debug version is inline
{
    ASSERT(m_pDocument->IsKindOf(RUNTIME_CLASS(CWzdDoc)));
    return (CWzdDoc*)m_pDocument;
}
#endif //_DEBUG


/////////////////////////////////////////////////////////////////////////////
// CWzdView message handlers


void CWzdView::OnInitialUpdate()
{

    m_ImageList.Create(IDB_STATUS_BITMAP, 15, 1, RGB(0,0,0));
    GetListCtrl().SetImageList(&m_ImageList, LVSIL_STATE);
    GetListCtrl().ModifyStyle(
        0,LVS_REPORT|LVS_SHOWSELALWAYS|LVS_SORTASCENDING);
    GetListCtrl().SendMessage(
        LVM_SETEXTENDEDLISTVIEWSTYLE,0,LVS_EX_GRIDLINES|LVS_EX_FULLROWSELECT);
    GetListCtrl().InsertColumn(0,"Name",LVCFMT_LEFT,GetTextExtent(30),0);
    GetListCtrl().InsertColumn(1,"Comment",LVCFMT_LEFT,GetTextExtent(70),1);

    CListView::OnInitialUpdate();

}
```

II

8

```
int CWzdView::GetTextExtent(int len)
{
    CDC* dc = GetDC();

    if (! m_AveCharWidth)
    {
        TEXTMETRIC tm;
        dc->GetTextMetrics(&tm);
        m_AveCharWidth = tm.tmAveCharWidth;
    }
    CSize size(m_AveCharWidth * len, 0);
    dc->LPtoDP(&size);
    ReleaseDC(dc);
    return size.cx;
}

void CWzdView::OnUpdate(CView* pSender, LPARAM lHint, CObject* pHint)
{
    CList<CWzdInfo*,CWzdInfo*> *pList=GetDocument()->GetInfoList();
    GetListCtrl().DeleteAllItems();
    for (POSITION pos = pList->GetHeadPosition(); pos;)
    {
        CWzdInfo *pInfo = pList->GetNext(pos);
        AddItem(-1,pInfo);
    }
}

void CWzdView::AddItem(int i,CWzdInfo *pInfo)
{
    if (i==-1)
    {
        i=GetListCtrl().InsertItem(0, pInfo->m_sName);
    }
    else
    {
        GetListCtrl().SetItemText(i, 0, pInfo->m_sName);
    }
```

Example 36    List View    **265**

```
      GetListCtrl().SetItemText(i, 1,  pInfo->m_sComment);
      GetListCtrl().SetItemData(i,(DWORD)pInfo);
      GetListCtrl().SetItemState(i, INDEXTOSTATEIMAGEMASK(pInfo->m_nState),
          LVIS_STATEIMAGEMASK);
}


void CWzdView::OnTestWzd()
{
    // determine if anything was selected
    if (GetListCtrl().GetSelectedCount())
    {
        // yes
    }

    // loop through selections
    int i=-1;
    while ((i = GetListCtrl().GetNextItem(i, LVIS_SELECTED)) != -1)
    {
        CWzdInfo *pInfo=(CWzdInfo *)GetListCtrl().GetItemData(i);

    }

    // add item to list
    CList<CWzdInfo*,CWzdInfo*> *pList=GetDocument()->GetInfoList();
    CWzdInfo *pInfo=new CWzdInfo("new","comment",CWzdInfo::NEW);
    pList->AddHead(pInfo);
    AddItem(-1,pInfo);

    // to modify
    i = 1;
    pInfo=(CWzdInfo *)GetListCtrl().GetItemData(i);
    //
    AddItem(i,pInfo);
```

```
// to ensure a line is visible
GetListCtrl().EnsureVisible(i,FALSE);

// to select a line
LV_ITEM lvi;
lvi.mask = LVIF_STATE;
lvi.iItem = i;
lvi.stateMask = 0x000f;
lvi.state = LVIS_SELECTED|LVIS_FOCUSED;
GetListCtrl().SetItemState(i, &lvi);

// to delete a line
GetListCtrl().DeleteItem(i);

// to redraw view
GetListCtrl().Invalidate();

}
```

## Listings — Data Class

```
#ifndef WZDINFO_H
#define WZDINFO_H


class CWzdInfo : public CObject
{
public:

enum STATES {
    OLD,
    NEW,
    MODIFIED,
    DELETED
};
```

Example 36    List View    **267**

```
        DECLARE_SERIAL(CWzdInfo)

    CWzdInfo();
    CWzdInfo(CString sName,CString sComment,int nState);

    void Set(CString sName,CString sComment,int nVersion, int nState);

    //misc info
    CString m_sName;
    CString         m_sComment;
    int m_nVersion;
    int m_nState;

    CWzdInfo& operator=(CWzdInfo& src);

};
#endif


// WzdInfo.cpp : implementation of the CWzdInfo class
//

#include "stdafx.h"
#include "WzdInfo.h"


/////////////////////////////////////////////////////////////////////////////
// CWzdInfo

IMPLEMENT_SERIAL(CWzdInfo, CObject, 1)


CWzdInfo::CWzdInfo()
{
    m_sName=_T("");
    m_sComment=_T("");
    m_nVersion=1;
    m_nState=CWzdInfo::NEW;
}
```

```
CWzdInfo::CWzdInfo(CString sName,CString sComment,int nState) :
    m_sName(sName),m_sComment(sComment),m_nState(nState)
{

}

void CWzdInfo::Set(CString sName,CString sComment,int nVersion, int nState)
{
    m_sName=sName;
    m_sComment=sComment;
    m_nVersion=nVersion;
    m_nState=nState;
}

CWzdInfo& CWzdInfo::operator=(CWzdInfo& src)
{
    if(this != &src)
    {
        m_sName = src.m_sName;
        m_sComment = src.m_sComment;
        m_nVersion = src.m_nVersion;
        m_nState = src.m_nState;
    }
    return *this;
}
```

# Example 37   Dynamically Splitting a View

## Objective

You would like to allow your user to split a view into two parts so that they can view different parts of the same document (Figure 8.5).

Example 37   Dynamically Splitting a View   **269**

**Figure 8.5**   **Split a View so your user can view different parts of the same document.**

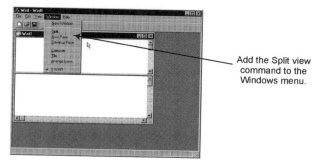

Add the Split view command to the Windows menu.

## Strategy

You can ask the AppWizard to add a split view command to your application. However, if you aren't that foresightful, or if you want to modify the default settings that the AppWizard creates for you, we will embed a CSplitterWnd class variable in our Mainframe or Child Frame Class and then override that class's OnCreateClient() to initialize this variable. We will also be adding a Split menu command to View.

## Steps

### Add a Split Command with the AppWizard

1. When creating your application, the 4th step of the AppWizard contains an Advanced button. Click on it, select the Window Styles tab, and click on Use split window. A new Split command will be added to your View menu command.

### Manually Add a Split Command

1. Embed a CSplitterWnd member variable in your application's CChild-Frame class for an MDI application or in your application's CMainFrame class for an SDI application.

```
CSplitterWnd m_wndSplitter;
```

2. Use the ClassWizard to override your Mainframe or Child Frame Class's `OnCreateClient()`. There, you will initialize your `CSplitterWnd` variable to create up to two vertical panes and up to two horizontal panes.

```
BOOL CChildFrame::OnCreateClient(LPCREATESTRUCT lpcs,
    CCreateContext* pContext)
{

    if (!m_wndSplitter.Create(this,
        2,              // maximum rows (maximum == 2)
        1,              // maximum columns (maximum == 2)
        CSize(5,5),     // minimum pane size in pixels
        pContext))
    {

        TRACE0("Failed to Create Splitter Window\n");
        return FALSE;

    }


    return TRUE; //CMDIChildWnd::OnCreateClient(lpcs,
        pContext);

}
```

Return `TRUE` here and don't call the base class's `OnCreateClient()`. Otherwise, the base class will simply override your splitter class.

3. The rest of the infrastructure for splitting a view is already available in your application, whether you originally asked for split windows or not. To activate this functionality, you simply have to add the menu items to your Window menu. For the items to add and the menu IDs you should use, use the AppWizard to create a new application with splitter windows and then cut and paste those menu items to your application.

## Notes

- Splitter windows can be created either dynamically or statically. When dynamic, they're limited to four panes. Statically created splitter windows are limited only by how small you can make a pane before it doesn't make sense. Please see your MFC documentation for the `CSplitterWnd` member function call to create static splitter windows.

Example 37    Dynamically Splitting a View    **271**

- The minimum pane size you set previously is actually the point at which the split window disappears. In other words, if your user moves the splitter bar to within this distance from the side of the view, the bar disappears and the view becomes whole again.

- The AppWizard gives your user a possible $2 \times 2$ split for up to four possible views. By convention, however, most applications only allow a view to be split into two views. To limit the choice to two views, set either number in the `CSplitterWnd::Create()` function to one (1).

- When your view is split, a second identical instance of your View Class is created, which refers to the exact same Document Class. Thus, when accessing other classes from within your application, you can only access one document from your view. However, you can access a list of views from your document. If you want your application to support split views, you should never store document information in your view. In fact, you should follow a policy that your View Class simply views and edits the document class and nothing more.

## CD Notes

- When executing the project on the accompanying CD, you can use the commands in the Window menu to split the view.

II

8

# 9

# Dialog Boxes and Bars

The Dialog Box is the next most important mechanism your application has, after the view, to retrieve information from your user. The control windows and conventions found in a Dialog Box can also be found in Dialog Bars, which are a hybrid of Dialog Box and Toolbar, and Form Views, which are a hybrid of Dialog Boxes and Views.

**Example 38    Using the Dialog Editor**   We will use the Dialog Editor to create a dialog template, which simply records the size and position of a set of control windows that will be created when a dialog box is created with this template.

**Example 39    Creating a Dialog Box Class**   We will use the Class Wizard to encapsulate the dialog template we created in the last example in our own C++ class and automatically allow data to be exchanged with the controls and member variables of that class.

**Example 40    Modal Dialog Boxes**   We will create our dialog box modally, which is to say our application waits until the termination of the dialog box before it can proceed.

**Example 41**     **Modeless Dialog Boxes**   We will create a modeless dialog box, which will allow our application to continue even though our user has yet to dismiss the dialog box.

**Example 42**     **Tabbing Between Controls in a Modeless Dialog Box**   We will look at how to restore the ability to tab between controls in our dialog box. This ability is free in a modal dialog box, but is disabled in a modeless dialog box.

**Example 43**     **Animation on a Dialog Box**   We will look at how to spruce up a dialog box that's displayed during a lengthy operation with an animation depicting the progress of that operation.

**Example 44**     **Message Boxes**   We will look at the capabilities of the specialized Message Box dialog box, which can prompt our user for Yes or No answers.

**Example 45**     **Dialog Bars**   We will create a standard dialog bar, which is a cross between a dialog box and a toolbar. A dialog bar's interior can be designed in the Dialog Editor, but it can be docked to the side of our application's window, like a toolbar.

# Example 38   Using the Dialog Editor

## Objective

You would like to add or modify a dialog box template in your application's resources. This template can then be used to create a dialog box or Property Page.

## Steps

### Create a New Dialog Template Using the Developer Studio

1. To create a new dialog box template, click on the Insert/Resource... menu commands of the Developer Studio to open the Insert Resource dialog box. Select Dialog and then click the New button.

Example 38  Using the Dialog Editor  **275**

### Edit an Existing Dialog Template Using the Dialog Editor

1. To edit an existing template, click the ResourceView tab of your Workspace window. Then, locate the ID of that template in the Dialog folder and double-click on it.

2. The Dialog Editor works by dragging controls from the Controls Toolbar into the dialog box. The properties for that control are opened when initially created, but you can also open them by right-clicking on the control and then selecting the Properties command in the popup menu. Properties vary between control types and roughly reflect the window styles available for that control. Not all styles that are available to a control are shown in the Properties dialog. For a complete list, refer to your MFC documentation. For a list of the most significant styles available to any particular control, see Appendix A. If a style is available to a control but not listed in its Properties dialog box, you will need to create the control yourself (Example 46).

> NOTE: Opening the Dialog Editor should also automatically display the Controls Toolbar (Figure 9.1). If not, you can open the Controls Toolbar by clicking on the Tools/Customize menu commands to open the Studio's Customize dialog box. Then, click on the Toolbars tab and locate the Controls entry in the list box to make sure it's checked.

**Figure 9.1   The Dialog Editor should automatically display the Controls Toolbar.**

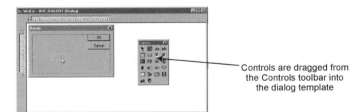

Controls are dragged from the Controls toolbar into the dialog template

3. To straighten out the controls you place in your template, use the menu commands available under the Studio's Layout command. See your MFC documentation for more about these commands.

4. The Layout/Tab Order menu commands allow you to specify the order in which these controls will receive input focus when your user presses the tab key. Actually, all this command does is reorder the control declarations

in the resource file. The Tab Order also determines the order in which the Control windows in the dialog will be drawn. This can be used to your advantage if you want two windows to overlap. The window with the higher tab order will appear on top when two windows share some of the same space.

5. The Studio's Layout/Test menu commands allow you to see what the template would look like as a dialog box. To escape this mode, either click on any button or hit the Escape key.

6. Specific dialog template styles and uses are discussed throughout this book.

## Notes

- When a dialog box is created, a dialog template determines which controls to create and with what styles. See Examples 40 and 41 for creating a dialog box from a dialog template.

- The size of a dialog template as well as the size and position of the controls in that template are all based on dialog units — not pixels. When a dialog template is used to create a dialog box, all dialog units are converted to pixels based on the font used by the dialog box. The larger the font, the larger the resulting dialog box. This normally isn't a problem — if you want to dynamically add your own controls to a dialog box, simply put a dummy placeholder control, such as a static control, where you want your control to appear and then use `GetWindowRect()` and `SetWindowRect()` to transfer this size to your new control. However, if you need to work with the template directly, the conversion between dialog units and pixels is discussed in Chapter 4.

## CD Notes

- There is no accompanying project on the CD for this example.

# Example 39   Creating a Dialog Box Class

## Objective

Having created a dialog box template as seen in the previous example, you would like to create a dialog box class to facilitate the creation of dialog boxes from this template.

Example 39   Creating a Dialog Box Class   **277**

## Strategy

We will use the ClassWizard to create a dialog class, which will help create a dialog box from a dialog template. We will also use the ClassWizard to add member variables to this dialog class that will correspond with the controls created in the dialog box.

## Steps

### Create a Dialog Class Using the ClassWizard

1. With the appropriate dialog box template open in the Dialog Editor, click on the Developer Studio's View/ClassWizard menu commands to open the ClassWizard dialog. The ClassWizard will start by asking if you want to create a new dialog class to support this template. Answer Yes. Then enter an appropriate class name in the New Class dialog box, making sure that it's derived from CDialog. Then click OK.

---

**NOTE:** To remove a class from the ClassWizard you must not only delete that class's .cpp and .h files from your project, but you must also delete the .clw file in your project's directory. This file is created by the ClassWizard to keep track of classes. If the ClassWizard can't find the .clw file, it will create a new one.

---

### Add Control Message Handlers to the Dialog Class

1. Once your dialog class is opened in the ClassWizard, you will find a list of the control IDs in your dialog template listed in the Message Maps tab of the ClassWizard. You can use the ClassWizard to add a message handler for each of these controls by first selecting the appropriate control ID, then locating and selecting a message ID in the Messages list box. The messages displayed depends on the type of control to which you are referring. Then click on the Add Function button to actually add the message handler (Figure 9.2).

## Figure 9.2  Use the ClassWizard to add a message handler.

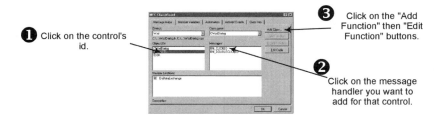

## Add Member Variables to the Dialog Class

1. You can also use the ClassWizard to add member variables to your dialog class that will exchange their data with the controls in the dialog box. To add these member variables, click on the ClassWizard's Member Variables tab. Then, click on the control ID for which you want to add a member variable. Then, click on the Add Variable button to open the Add Member Variable dialog box (Figure 9.3).

## Figure 9.3  Use the ClassWizard to add a member variable.

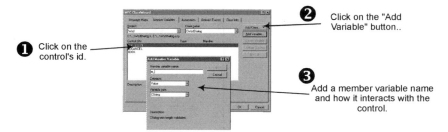

2. After entering a variable name, you can then pick how this variable will interact with its control. The interaction available depends on the control. An edit box, for instance, will exchange its contents with a string variable or an integer variable. To exchange contents with an integer variable, an edit box must internally convert its contents to an integer. All controls allow you to interact with them as a control variable. In effect, you can effortlessly subclass a control in your dialog box this way. However, the controls in the dialog box won't exchange data with your member variables unless `CWnd::UpdateData()` is called. `Update-Data(FALSE)` will take the values in your member variables and store them in their corresponding control window. `UpdateData(TRUE)` will

Example 39   Creating a Dialog Box Class   **279**

take the values currently in the controls and store them in their corresponding member variable.

3. You cannot modify a member variable once entered. Instead, you must delete a member variable and then reenter it.

4. You can have multiple member variables accessing the same control. For example, a control can exchange its data with a string, number, and control member variable at the same time. However, you can't add these variables automatically without a little manual intervention. First, add one type of member variable using the ClassWizard. Then, move the definitions that the ClassWizard created out of the {{}} brackets maintained in your source files (both the .cpp and .h files) for the ClassWizard. Once outside of these {{}} brackets, you are free to add yet another type of member variable for this control.

5. You can also have a message handler in your dialog class for the same control that has one or more member variables. A good use of this might be a button control. You might have a message handler that processes the button when clicked, but you might also have a member variable that subclasses the button so that you can easily hide the button with

```
m_wndButton.ShowWindow(SW_HIDE);
```

6. For some controls, you can also add validation. For example, you can specify a maximum length allowed for a string variable control. For an edit control that accepts a number, you can specify a numeric range.

## Notes

- The member variables you add to your dialog class can be easily accessed by the class creating the dialog box. See the next example for specifics.

- The control messages that the ClassWizard displays for you may not always encompass all of the possible messages generated by that control. In those rare cases, you will need to add the control message map macros yourself. See Appendix B for the appropriate macro.

- Data exchange is performed in the DoDataExchange() member function of your dialog class using MFC-supplied functions for all standard variable types.

## CD Notes

- There is no accompanying project on the CD for this example.

# Example 40   Modal Dialog Boxes

## Objective

You would like to use a dialog box to prompt your user for input while suspending your application (Figure 9.4).

**Figure 9.4**   **Use a modal dialog box to suspend your application until the user responds.**

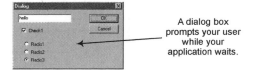

A dialog box prompts your user while your application waits.

## Strategy

We will call the DoModal() member function of the dialog class we created in the last example to create a modal dialog box. We will first put values into the member variables of this dialog class and then retrieve them when the dialog box closes.

## Steps

### Create a Modal Dialog Box

1. Create a dialog box resource and class, as shown in the last two examples. Add member variables for each of the controls in your dialog template to your dialog class.

2. Create an instance of your dialog class on the stack.

```
CWzdDialog dlg;
```

3. Initialize the member variables in your dialog class with values from your calling class. Then, call the DoModal() function of your dialog class. If

Example 40   Modal Dialog Boxes   **281**

DoModal() returns a status of IDOK, stuff the values in those member variables back into your class.

```
dlg.m_sName=sName;
dlg.m_nRadio=nRadio;
dlg.m_bCheck=bCheck;
if (dlg.DoModal()==IDOK)
{
    sName=dlg.m_sName;
    nRadio=dlg.m_nRadio;
    bCheck=dlg.m_bCheck;
}
```

## Notes

- For true C++ encapsulation, you shouldn't be able to access the member variables of your dialog class directly. However, by convention, the member variables of a dialog class and a Property Page are usually accessed directly. Since the life span of each is so short, this exposure to corruption isn't too deleterious.

- When creating your dialog template using your Dialog Editor, you may have noticed that you can give your dialog a System Modal style. In the old days of Windows 3.1, this meant that your dialog box would not only suspend all activity in your application until the user responded, but also suspend all activity on the system. In the current world of true multitasking applications, all System Modal will do for you is make your dialog a top-most window, meaning that no other window can cover it (except, of course, for another top-most window). Your user should be sufficiently annoyed to answer your question. System Modal dialogs are usually used in critical situations when the whole system is affected. This can include anything from system resources to printer failures.

## CD Notes

- When executing the project on the accompanying CD, clicking on the Test/Wzd menu commands causes a modal dialog box to appear, which freezes all input to the application until you close the dialog box.

# Example 41   Modeless Dialog Boxes

## Objective

You would like to create a dialog box that allows your application to continue operation, even if the user doesn't close the dialog box. A modeless dialog box can be used in lieu of a wait cursor while your application continues to work (Figure 9.5).

**Figure 9.5**   **Use a modeless dialog box to allow your application to run, whether the user responds or not.**

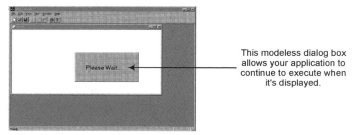

This modeless dialog box allows your application to continue to execute when it's displayed.

## Strategy

We will create a dialog template and class, as shown in Examples 38 and 39, but then we will use our new dialog class's `Create()` member function to create a modeless dialog box. To close our modeless dialog box, we will use `CWnd::DestroyWindow()`.

Example 41   Modeless Dialog Boxes   **283**

## Steps

### Create a Modeless Dialog Box

1. Use the Dialog Editor to create a dialog box resource in your application. Then, use the ClassWizard to create a dialog class. You can create a modeless dialog box using this dialog class.

```
CWzdDialog *pDlg;
pDlg = new CWzdDialog;
pDlg->Create(IDD_WZD_DIALOG);    // id of dialog box resource
pDlg->ShowWindow(SW_SHOW);       // dialog is initially hidden
```

Notice that we create our dialog class instance in the heap. If we had created it on the stack, as we did with our modal dialog box in the last example, our dialog class would have deconstructed when our function returned, automatically destroying our dialog box. Rather than create your dialog box in the heap, you can also embed it as a member variable of the class from which you are creating it.

```
CWzdDialog m_dlg;
```

### Destroy a Modeless Dialog Box

1. You can destroy a modeless dialog box with

```
pDlg->DestroyWindow();
```

2. Your user can also destroy a modeless dialog box by clicking on the close button in the upper-right corner of the window. However, when the user closes a dialog window this way, the operating system has no idea it should also delete your dialog class object. Therefore, you must do it yourself in your dialog class. Use the ClassWizard to override the Post-NcDestroy() function of your dialog class and delete its own instance.

```
void CWzdDialog::PostNcDestroy()
{
    CDialog::PostNcDestroy();

    delete this;
}
```

II

9

## Use a Wait Dialog Box

1. Example 34 showed you how to create a wait cursor to indicate to your user that your application is busy. For longer waits, it is advisable to use a modeless dialog box, instead. Simply populate it with a static "Please wait..." message, a progress control, or an animation control. You can even put a button on it to stop execution. Then create it just before your application becomes busy.

```
CWzdWaitDialog dlg;;
dlg.Create(IDD_WAIT_DIALOG);
dlg.ShowWindow(SW_SHOW);
dlg.UpdateWindow();
// processing.............
```

What's new here is the call to `CWnd::UpdateWindow()`. Our dialog window might not be getting its normal `WM_PAINT` message, so we need to force one with `UpdateWindow()`.

2. Once your function is finished, the wait dialog class will destroy itself. In its destructor, it will destroy the dialog window.

## Notes

- The ability to tab between controls in a dialog box is assumed to be off with a modeless dialog box. See Example 42 for how to reactivate this functionality.

## CD Notes

- When executing the project on the accompanying CD, you will see three items in the Test menu. Test/Create will create a modeless dialog that doesn't freeze your application like a modal dialog. Test/Destroy will destroy this dialog. Test/Wait will cause a dialog to appear that will disappear within two seconds when the function that created it finishes incrementing a simple loop.

Example 42   Tabbing Between Controls in a Modeless Dialog Box   **285**

# Example 42   Tabbing Between Controls in a Modeless Dialog Box

## Objective

You would like to tab between control in a modeless dialog box. You get this functionality automatically when the dialog box is modal, but not in a modeless dialog box.

## Strategy

When creating this dialog box template in the Dialog Editor, we will select one more style from the Properties for this dialog box.

## Steps

### Enable Tabbing in a Dialog Box

1. When in the Dialog Editor for this dialog box, right-click on the dialog box to bring up its properties. Then, select the Extended Styles tab and click on Control Parent. That's it.

## Notes

- Clicking on Control Parent causes the dialog box to be created with the WS_EX_CONTROLPARENT window style. In fact, it doesn't even need to be a dialog box to get this functionality. You can create any window with the WS_EX_CONTROLPARENT style. When you create control windows in this parent window, include the WS_TABSTOP style with each control. Once created, pressing the tab key will cause keyboard input focus to move from control window to control window in the order in which you created them.

- If you want to be able to move from window to window with the up and down arrow keys, see Example 47, in which we convert up and down arrow key strokes to Tab and Shift-Tab key strokes.

II

9

## CD Notes

- When executing the project on the accompanying CD, click on the Test/Wzd menu commands to open a modeless dialog. Pressing the tab key still causes the input focus to move from control to control.

# Example 43    Animation on a Dialog Box

## Objective

You would like to display an animation to indicate the progress of a function in your dialog box (Figure 9.6).

## Figure 9.6    Use an animation in a dialog box to indicate progress.

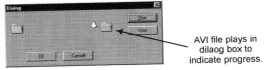

AVI file plays in dilaog box to indicate progress.

## Strategy

We will use the Dialog Editor to add an animation control to our dialog template. We will then load this control with an `.avi` file in the `OnInitDialog()` function of our dialog. We will then start and stop playing this `.avi` using member functions of the animation control class.

## Steps

### Import an `.avi` File into Your Application's Resources

1. Click on the Developer Studio's Insert/Resource menu items to open the Insert Resource dialog box. Then, click on the Import button and location the `.avi` file. For resource type, enter "AVI" (without the double quotes). Give this resource an appropriate ID. For this example, we will use `IDR_FILECOPY`.

Example 43   Animation on a Dialog Box   **287**

## Add an Animation Control to the Dialog Box

1. Use the Dialog Editor to add an animation control to your dialog template. (The animation control is the button in the Controls Toolbar that looks like a strip of film). For its Properties, make it centered and transparent.

2. If you haven't already done so, use the ClassWizard to create a dialog class for this dialog template. To load your `.avi` file into this control, you can use the following (assuming that `IDC_ANIMATE_CTRL` is the control's ID).

```
CAnimateCtrl *pCtrl=(CAninmate *)GetDlgItem(IDC_ANIMATE_CTRL);
pCtrl->Open(IDR_FILECOPY);
```

**II**

## Play the `.AVI` File

1. To tell the control to play the `.AVI` file, you can use

```
pCtrl->Play(0,      // first frame
    -1,             // last frame (-1=play every frame)
    -1);            // number of times to play avi (-1=
                    // play until manually stopped)
```

2. To stop the control, you can use

```
pCtrl->Stop();
```

**9**

## Notes

- `.avi` files are made up a sequence of bitmap files. Each bitmap is called a frame. There are several utilities that will create an `.avi` file available as shareware from the Internet, with an average price of $40.

- Animation in a dialog box is frequently used as a more ornate way of showing the progression of a lengthy operation. For the complete effect, you should hide the control until the operation is under way by using `CWnd::ShowWindow(SW_HIDE)`. Then, when running, you should simultaneously show the control and start playing it with

```
pCtrl->ShowWindow(SW_SHOW);
pCtrl->Play(0,-1,-1);
```

- To stop the control and hide it again when the process is over, use

```
pCtrl->ShowWindow(SW_HIDE);
pCtrl->Stop();
```

## CD Notes

- When executing the project on the accompanying CD, click on the Test/Wzd menu commands to open a dialog box. Click on Start to start the animation, Stop to stop it.

# Example 44   Message Boxes

## Objective

You would like to prompt your user with a simple yes or no question, without creating a new dialog template and class (Figure 9.7).

**Figure 9.7    Use a simple message box for yes or no questions.**

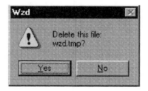

## Strategy

We will use two framework functions to create a message box: AfxMessage-Box() and AfxFormatString().

## Steps

### Work with Message Boxes

1. To create a simple message box, you can use

```
AfxMessageBox("Hello");
```

A modal dialog box is created with a single OK button, which waits for your user to click it.

Example 44   Message Boxes   **289**

2. To create a simple message box that uses a string from the string table in your application's resource file, use

```
AfxMessageBox(IDS_HELLO);
```

3. To dynamically edit the string table string, use AfxFormatString1().

```
AfxFormatString1(msg, IDS_HELLO_1, "Goodbye");
AfxMessageBox(msg);
```

In this example, IDS_HELLO_1 refers to the string "Hello %1" in the String Table. AfxFormatString1() substitutes the word "Goodbye" for %1 and in this way allows you to keep the majority of your message string in the String Table. Keeping the majority of the strings in the String Table enables your application to be easily translated to another language, like Spanish or French. You would only need to substitute one String Table for another. You can also put this String Table in a resource library (Example 85).

4. AfxFormatSting2() substitutes two strings, instead of one. In the following example, the message string is "Hello %1 %2".

```
AfxFormatString2(msg, IDS_HELLO_2, "and", "Goodbye");
AfxMessageBox(msg);
```

5. To ask a yes or no question, you would add a style flag to AfxMessage-Box().

```
if (AfxMessageBox(msg, MB_YESNO)==IDYES)
{
}
```

Other possible styles include MB_ABORTRETRYIGNORE, MB_OKCANCEL, MB_RETRYCANCEL, MB_YESNO, and MB_YESNOCANCEL. Other possible tests to check for include IDNO, IDABORT, IDCANCEL, IDIGNORE, IDOK, and IDRETRY.

6. To make a button other than the first button the default (clicked when the user presses Enter), OR the style flag with MB_DEFBUTTON2 or MB_DEFBUTTON3.

```
if (AfxMessageBox(msg,MB_YESNO|MB_DEFBUTTON2)==IDYES)
{
}
```

7. To make the icon in a message box something other than an exclamation point, you can OR the style flag with MB_ICONSTOP, MB_ICONINFORMATION, or MB_ICONQUESTION.

```
AfxMessageBox(msg,MB_ICONSTOP);
```

## Notes

- The %1 and %2 markers used with AfxFormatString1() and AfxFormatString2() can appear in any order and as often as you would like in the string.

- AfxMessageBox() is implemented with CWinApp::DoMessageBox(). To intercept the creation of all message boxes by your application, simply use the ClassWizard to override your Application Class's DoMessage-Box() function. There, you can append the message with a standard title, time, or even save the message to a log.

## CD Notes

- When executing the project on the accompanying CD, set a breakpoint on OnTestWzd() in WzdView.cpp. Then click the Test/Wzd menu commands and step through OnTestWzd() as it creates various message boxes.

## Porting Note

- AfxMessageBox() is a great way for your newly ported application to start displaying messages right away. Normally in a windows application, you must own or specify a window to create another window. All application windows belong to the application's main window, which itself belongs to the desktop window. In this way, all windows can be refreshed by the windows manager. But with AfxMessageBox(), the class member function, or even the static function that calls the class member function, doesn't have to own a window. AfxMessageBox() automatically locates and uses your application's main window as its owner window. Therefore, even a lowly data class can use AfxMessageBox() to display a message (although data classes don't normally display messages). The only requirement is that you include stdafx.h in the source that uses

Example 45   Dialog Bars   **291**

`AfxMessageBox()`. If that still causes a problem, you can use the Windows API call

```
::MessageBox();
```

but then you have to include "`windows.h`."

# Example 45   Dialog Bars

## Objective

You would like to create a dialog bar that has controls like a dialog box but floats and docks to the side of your application's main window like a toolbar (Figure 9.8).

## Figure 9.8   A dialog bar is a hybrid of a dialog box and a toolbar.

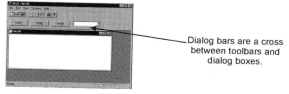

Dialog bars are a cross between toolbars and dialog boxes.

## Strategy

First, we will create a dialog box template using the Dialog Editor. Then, we will use the ClassWizard to create a dialog class for our template. To this dialog class, we will add functions to handle our dialog's controls and, because a dialog bar doesn't receive a `WM_INITDIALOG` message, we will also add our own `InitDialog()` function to call when the bar is first created. We will add a menu command to our application's View menu to hide and show our dialog bar. To allow the controls in our dialog bar to be updated by our application, we will override the `CMainFrame`'s `OnCmdMsg()`.

## Steps

### Create the Dialog Template and Dialog Class

1. Use the Dialog Editor to create a dialog template (Example 38). Try to keep the design thin, either vertically or horizontally. The intention of a

dialog bar is to remain on the screen as much as a toolbar, so it shouldn't obscure the view any more than necessary. If you can't make your template thin, try a modeless dialog box, instead.

2. Use the ClassWizard to create a class for this dialog resource (Example 39). Unfortunately, the base class necessary for creating a dialog bar is not one of the selections when creating a class with the ClassWizard, so pick CDialog and then change all of the references in the generated source files from CDialog class to CDialogBar.

## Fill in the Dialog Class

1. Add an InitDialog() function to this dialog class. There, you can initialize your controls. You can also give your dialog bar a name with SetWindowText().

```
SetWindowText("Wzd Dialog Bar");
```

This name will only appear when your dialog bar is floating.

2. At this point, you can use the ClassWizard to add message handlers and member variables to your dialog class. Any messages for which you don't create a message handler will be automatically passed on to the dialog bar's Parent window, which in this case is CMainFrame. In this way, button controls can act just like toolbar buttons because clicking them will cause their resulting command message to be processed in any class in the command message path (i.e., the Main Frame Class, the Document Class, etc.).

3. For an example listing of a dialog bar class, see "Listings — Dialog Bar Class" on page 295.

## Add a Dialog Bar to the Main Frame Class

1. Embed this dialog bar class in the CMainFrame class.

```
CWzdDialogBar    m_WzdDialogBar;
```

**Example 45   Dialog Bars   293**

2. Create the dialog bar after creating all of your application's toolbars in your CMainFrame class's OnCreate() function.

```
if (!m_WzdDialogBar.Create(this, IDD_WZD_DIALOG,
    CBRS_TOP,         // control bar is initially at top of frame
//  CBRS_BOTTOM,      // control bar is initially at bottom of frame
//  CBRS_LEFT,        // control bar is initially on left side of frame
//  CBRS_RIGHT,       // control bar is initially on right side of frame
    -1) ||
    !m_WzdDialogBar.InitDialog())
    {
        TRACE0("Failed to create dialog bar\n");
        return -1;        // fail to create
    }


    m_WzdDialogBar.EnableDocking(
    CBRS_ORIENT_HORZ    // can only dock to top or bottom of frame
//  CBRS_ORIENT_VERT    // can only dock to left or right of frame
//  CBRS_ORIENT_ANY     // can dock to any side of frame
    );
    DockControlBar(&m_WzdDialogBar);
```

Notice that we call our dialog bar class's InitDialog() function immediately after creating the dialog bar.

3. Buttons in your dialog bar will remain disabled unless you connect it to the command message path of your application. To do this, use the ClassWizard to override the OnCmdMsg() function of your CMainFrame class. There, you should call the OnCmdMsg() member function of your dialog bar class.

```
BOOL CMainFrame::OnCmdMsg(UINT nID, int nCode, void* pExtra,
    AFX_CMDHANDLERINFO* pHandlerInfo)
{
    if (m_WzdDialogBar.OnCmdMsg(nID,nCode,pExtra,pHandlerInfo))
    return TRUE;

    return CMDIFrameWnd::OnCmdMsg(nID, nCode, pExtra,
        pHandlerInfo);
}
```

The close button of a dialog bar doesn't actually destroy the window. Instead, it simply hides it. We will now add a menu command that will hide or show our dialog bar. This menu command will appear under our application's View menu.

### Add a Dialog Bar View Command

1. Use the Menu Editor to add a Dialog Bar command to your application's View menu.
2. Use the ClassWizard to add a command handler and a user interface handler for this new command to your CMainFrame class. Fill these handlers in as follows. In this example, m_WzdDialogBar is the name of the dialog bar's variable.

```
void CMainFrame::OnViewDialogbar()
{
    ShowControlBar(&m_WzdDialogBar, (m_WzdDialogBar.GetStyle()
        & WS_VISIBLE) == 0, FALSE);
}


void CMainFrame::OnUpdateViewDialogbar(CCmdUI* pCmdUI)
{
    pCmdUI->SetCheck((m_WzdDialogBar.GetStyle() &
        WS_VISIBLE) != 0);
}
```

## Notes

- As mentioned previously, you might also consider simply creating a modeless dialog box if you can't squeeze the controls you need into a window of toolbar dimensions (thin). You might also consider a modeless dialog box if the functionality you're looking for is only for a particular operation in your application. To create a modeless dialog box, see Example 41.
- The SaveControlBars() and LoadControlBars() functions featured in Example 5 will save the exact position and size of your dialog bars, too, when you exit and reenter your application.

Example 45 · Dialog Bars **295**

## CD Notes

- When executing the project on the accompanying CD, you will notice a dialog bar at the top with four controls. You will also notice that this bar can be docked and moved just like a toolbar.

## Listings — Dialog Bar Class

```
#if !defined(AFX_WZDDIALOGBAR_H__1EC84998_C589_11D1_9B5C_00AA003D8695__INCLUDED_)
#define AFX_WZDDIALOGBAR_H__1EC84998_C589_11D1_9B5C_00AA003D8695__INCLUDED_

#if _MSC_VER >= 1000
#pragma once
#endif // _MSC_VER >= 1000

// WzdDialogBar.h : header file
//
```

II

```
/////////////////////////////////////////////////////////////////////////
// CWzdDialogBar dialog

class CWzdDialogBar : public CDialogBar
{

// Construction
public:
    CWzdDialogBar(CWnd* pParent = NULL);   // standard constructor

    BOOL InitDialog();

// Dialog Data
    //{{AFX_DATA(CWzdDialogBar)
    enum {IDD = IDD_WZD_DIALOG};
        // NOTE: the ClassWizard will add data members here
    //}}AFX_DATA


// Overrides
    // ClassWizard generated virtual function overrides
    //{{AFX_VIRTUAL(CWzdDialogBar)
    protected:
    virtual void DoDataExchange(CDataExchange* pDX);    // DDX/DDV support
    //}}AFX_VIRTUAL

// Implementation
protected:

    // Generated message map functions
    //{{AFX_MSG(CWzdDialogBar)
    afx_msg void OnWzdButton1();
    afx_msg void OnWzdButton3();
    //}}AFX_MSG
    DECLARE_MESSAGE_MAP()
};
```

Example 45   Dialog Bars   **297**

```
//{{AFX_INSERT_LOCATION}}
// Microsoft Developer Studio will insert additional declarations immediately
// before the previous line.

#endif // !defined(
    AFX_WZDDIALOGBAR_H__1EC84998_C589_11D1_9B5C_00AA003D8695__INCLUDED_)


// WzdDialogBar.cpp : implementation file
//

#include "stdafx.h"
#include "wzd.h"
#include "WzdDialogBar.h"

#ifdef _DEBUG
#define new DEBUG_NEW
#undef THIS_FILE
static char THIS_FILE[] = __FILE__;
#endif

/////////////////////////////////////////////////////////////////////////////
// CWzdDialogBar dialog


CWzdDialogBar::CWzdDialogBar(CWnd* pParent /*=NULL*/)
{
    //{{AFX_DATA_INIT(CWzdDialogBar)
        // NOTE: the ClassWizard will add member initialization here
    //}}AFX_DATA_INIT
}


void CWzdDialogBar::DoDataExchange(CDataExchange* pDX)
{
    CDialogBar::DoDataExchange(pDX);
    //{{AFX_DATA_MAP(CWzdDialogBar)
        // NOTE: the ClassWizard will add DDX and DDV calls here
    //}}AFX_DATA_MAP
}
```

```
BEGIN_MESSAGE_MAP(CWzdDialogBar, CDialogBar)
    //{{AFX_MSG_MAP(CWzdDialogBar)
    ON_BN_CLICKED(IDC_WZD_BUTTON1, OnWzdButton1)
    ON_BN_CLICKED(IDC_WZD_BUTTON3, OnWzdButton3)
    //}}AFX_MSG_MAP
END_MESSAGE_MAP()

/////////////////////////////////////////////////////////////////////////////
// CWzdDialogBar message handlers

BOOL CWzdDialogBar::InitDialog()
{
    UpdateData(FALSE);
    SetWindowText("Wzd Dialog Bar");
    return TRUE;
}

void CWzdDialogBar::OnWzdButton1()
{
    int i=0;
}

void CWzdDialogBar::OnWzdButton3()
{
    // TODO: Add your control notification handler code here

}
```

# Control Windows

Control windows are the buttons, list boxes, and scroll bars that allow a user to interact with your application. When a dialog box is created, the control windows you defined using the Dialog Editor are created for you. However, some control windows might require that you fill them in at run time, such as combo boxes. Also, some control windows can't be created in the resource template.

For a complete review of the control windows available, as well as examples of their use, see Appendix A. The examples in this chapter show you how to dynamically create and fill control windows.

**Example 46     Creating a Control Window Anywhere**   We will look at how to create a control window anywhere. And we mean *anywhere*.

**Example 47   Customizing a Common Control Window with Subclassing**   We will look at a way for our class to take control of a control window by using subclassing. Subclassing is described in detail in Chapter 3.

**Example 48    Customizing a Common Control Window with Superclassing**  We will look at a way for our class to take control of a control window by using superclassing. Superclassing is also described in Chapter 3.

**Example 49    Putting Bitmaps on a Button**  We will forsake the usual text on a button for a bitmap image, instead.

**Example 50    Dynamically Filling a Combo Box**  We will look at a method to display the freshest information by filling in a combo box at the moment the user opens it.

**Example 51    Sorting a List Control**  We will look at how to react to the user clicking on the header of a list control by sorting the column it heads.

**Example 52    Line Control**  We will finally reveal how to create an etched line in a dialog box without drawing it yourself.

# Example 46   Creating a Control Window Anywhere

## Objective

You would like to create a control window anywhere (Figure 10.1).

## Figure 10.1   Place a button control in a view or anywhere.

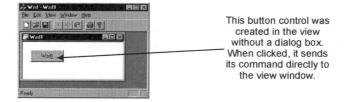

This button control was created in the view without a dialog box. When clicked, it sends its command directly to the view window.

## Strategy

Every control window, such as a button or an edit box, is simply a specialized child window that can be opened just about anywhere in your interface and report to any other window. Normally, MFC takes care of the drudgery

Example 46   Creating a Control Window Anywhere   **301**

of opening control windows when opening a dialog box. However, you can manually open control windows with two steps. First, by creating an instance of an MFC control class, like CButton, that wraps the control. Second, by calling the Create() member function of that class, which actually opens the control window.

# Steps

## Create an Instance of an MFC Control Class

1. Decide which type of control you want to create (e.g., button, edit box, etc.). Then, determine the MFC class that wraps that control (e.g., CButton wraps a button control window). See Appendix A for examples. Then, embed that class in the class you in which intend to open this control. In this example, we are embedding a CButton class in a dialog class.

```
private:
    CButton     m_ctrlButton;
```

## Create the Control Window

1. Use the ClassWizard to add a WM_CREATE message handler, or a WM_INITDIALOG message handler for a dialog box, to the class you have embedded. In this OnCreate() handler, you will call the Create() member function of the control class and give its style, extended style, caption (if any), owner, control ID, size, and position. The items you need to provide when creating a control can change from control class to control class. Styles also change. See Appendix A for more. In this example, we are creating button control window.

```
CRect rect(20,20,100,50);
m_ctrlButton.Create("Wzd&1", WS_CHILD|WS_VISIBLE, rect, this,
    IDC_WZD_CONTROL);
```

2. When a dialog box creates a control from a dialog template, it also performs tasks other than just creating the control window. For example, a

dialog box sets the font of every control window, too. Here, we set the font for our button control to that used with the 3-D Windows style.

```
CFont *pFont=CFont::FromHandle((
    HFONT)::GetStockObject(ANSI_VAR_FONT));
m_ctrlButton.SetFont(pFont);
```

## Notes

- As seen previously, creating any control window is a two-step process: first, create an instance of an MFC control class; second, call its Create() member function to create the actual window. Deleting an MFC class and control window pair can be tricky. If the MFC class instance is destroyed, it will automatically destroy its control window. However, if the control window is destroyed first, the operating system doesn't know that your MFC control class created it. It's your job to manually destroy the MFC control class. This isn't a problem in this case, since your MFC class was embedded in another class that will automatically destroy it when the dialog box is closed. See Chapter 1 for much more on this topic.

- For an example of using the ClassWizard to add a message handler, see Example 13.

## CD Notes

- When executing the project on the accompanying CD, you will notice the appearance of a button control in the middle of the view.

# Example 47 Customizing a Common Control Window with Subclassing

## Objective

You would like to add your own functionality to an existing common control by subclassing it.

## Strategy

Control windows like buttons and edit boxes are usually maintained by their very own, system-provided window process. Even when you create a control using an MFC control class, such as CButton, the system-provided function handles all messages from that control window. Subclassing is a process that allows you to cut into the message stream between a control window and its original window process by substituting the address of your own window process for the original in the window object itself. If a message comes in that you want the original window process to handle, such as drawing the control, you pass it on to the original window process address. Otherwise, you can process a message in new and exciting ways. Fortunately, MFC has taken a lot of the drudgery out of subclassing a window with CWnd::SubclassWindow(). If you create a control member variable for a control in your dialog box, you won't even need to use SubclassWindow() — it will be called automatically for you when your dialog does a data exchange.

**II**

To subclass a control window, we start by creating our very own derivation of an MFC control class. Then, we use SubclassWindow(), either directly or indirectly through DoDataExchange(), to interject the message map of our derived class into the message stream between a control window and its original window process. We can now choose to handle messages in the comfort and safety of our own class, pass them on, or handle some of them and pass the rest on. For much more on this, see Chapter 3.

**10**

# Steps

## Create a New Control Class Derived from an MFC Control Class

1. Use ClassWizard to create a new class derived from one of several control classes. In this example, we have derived CWzdEdit from CEdit and CWzdComboBox from CComboBox.

2. Use the ClassWizard to add a message handler for every message you want to intercept between your control window and its original window process.

If the control you want to subclass is in a dialog box, you can use the ClassWizard to add a member variable of the type Control to your dialog box class that will automatically subclass a control window for you as we see next.

## Subclass a Control with the ClassWizard

1. Open the ClassWizard and select the appropriate dialog class. Then, select the Member Variables tabbed page. Find and select the control ID of the control you want to subclass. Pick Control from the list of possible member variable types. In this example we have added edit and combo box control member variables to our class, which resulted in the Class-Wizard adding the following lines to that class's .h file.

```
//{{AFX_DATA(CWzdDialog)
enum {IDD = IDD_WZD_DIALOG};
CEdit m_ctrlEdit;
CComboBox     m_ctrlComboBox;
//}}AFX_DATA
```

2. Pull these definitions below the {{}} brackets so that the ClassWizard can't change them and change these standard MFC class names to your new derived class names.

```
//{{AFX_DATA(CWzdDialog)
enum {IDD = IDD_WZD_DIALOG};
//}}AFX_DATA
CWzdEdit        m_ctrlEdit;        <<<<
CWzdComboBox    m_ctrlComboBox;    <<<<
```

3. Your new classes will now have complete access to all the messages your control window generates. In this example, we have modified our derivation of CEdit to override PreTranslateMessage(). There, we translate the up and down arrow keys to Tab and Shift-Tab so that keyboard focus will change between control windows in a dialog box when using these key strokes. For details, see "Listings — Edit Class" on page 306.

## Subclass a Control with SubclassWindow()

1. Create a new derived class as before and then embed it in the dialog, view, or some other class in which this control will appear.

```
CWzdEdit m_ctrlEdit;
```

2. Use `CWnd::GetDlgItem()` to get the window handle of the control you want to subclass. You can then use `SubclassWindow()` to subclass the conrol.

```
HWND hWnd;
GetDlgItem(IDC_WZD_EDIT2, &hWnd);
m_ctrlEdit.SubclassWindow(hWnd);
```

This method of using `SubclassWindow()` will work with all control windows except one: combo boxes. A combo box is actually a control that is itself composed of two other control windows: an edit box and a list box. You can still subclass a combo box using `SubclassWindow()`, but depending on your application, you may also need to subclass its edit box and list box, too. For example, if you wanted to customize your combo box's handling of a mouse click over the edit box part, you would need to subclass its edit box control, because that's what receives the actual mouse click message.

## Subclass a Combo Box Control with `SubclassWindow()`

1. To subclass the edit box of a combo box, you first need to locate its control window from within the combo box window. The edit box and list box are child windows of the combo box, so we simply scan the child windows of the original combo box using `CWnd::GetWindow()`. We then look for the appropriate window class name, which is `EDIT` when looking for the edit box.

```
char className[8];
// examine all of the child windows of the combo box
CWnd* pWnd = m_ctrlComboBox.GetWindow(GW_CHILD);
do {
    ::GetClassName(pWnd->m_hWnd, className, 8);
    // if the class name is "Edit", we've found our control
    if(!stricmp("Edit", className))
    {
        // subclass as before
        m_ctrlComboEditBox.SubclassWindow(pWnd->m_hWnd);
    }
} while((pWnd = pWnd->GetWindow(GW_HWNDNEXT)));
```

As seen here, once the edit box control window is located, we subclass it as before.

2. If we are using the drop down style of the combo box, we can't look for its list box window component until the list box drops down; once it does, we subclass it as we did previously.

## Notes

- Controls are kingdoms unto themselves. They provide all of the functionality of drawing themselves and receiving messages, but at the price of being unable to easily modify them. That is, unless their creators had the foresight to provide you with just the right window style or just the right control notification to take control. Subclassing a control window provides you with even more control by actually being able to monitor messages to the control and selectively process some yourself. If you don't like the way a control looks, intercept its WM_PAINT message and draw it yourself. If a control doesn't have the notification message you need, who cares — you now have access to all of its messages.

## CD Notes

- When executing the project on the accompanying CD, click on the Test/Wzd commands to open a dialog box. You will notice that these dialog controls all have a hatched background. You can also now move between these controls by pressing the up or down arrow keys.

## Listings — Edit Class

```
#if !defined(AFX_WZDEDIT_H__1EC84996_C589_11D1_9B5C_00AA003D8695__INCLUDED_)
#define AFX_WZDEDIT_H__1EC84996_C589_11D1_9B5C_00AA003D8695__INCLUDED_

#if _MSC_VER >= 1000
#pragma once
#endif // _MSC_VER >= 1000
```

```
// WzdEdit.h : header file
//

////////////////////////////////////////////////////////////////////////////
// CWzdEdit window

class CWzdEdit : public CEdit
{

// Construction
public:
    CWzdEdit();

// Attributes
public:

// Operations
public:

// Overrides
    // ClassWizard generated virtual function overrides
    //{{AFX_VIRTUAL(CWzdEdit)
    public:
    virtual BOOL PreTranslateMessage(MSG* pMsg);
    //}}AFX_VIRTUAL

// Implementation
public:
    virtual ~CWzdEdit();

    // Generated message map functions
protected:
    //{{AFX_MSG(CWzdEdit)
    afx_msg HBRUSH CtlColor(CDC* pDC, UINT nCtlColor);
    //}}AFX_MSG

    DECLARE_MESSAGE_MAP()
private:
    CBrush    m_brush;
};
```

II

10

```
//////////////////////////////////////////////////////////////////////////

//{{AFX_INSERT_LOCATION}}
// Microsoft Developer Studio will insert additional declarations immediately
// before the previous line.

#endif // !defined(
    AFX_WZDEDIT_H__1EC84996_C589_11D1_9B5C_00AA003D8695__INCLUDED_)

// WzdEdit.cpp : implementation file
//

#include "stdafx.h"
#include "wzd.h"
#include "WzdEdit.h"

#ifdef _DEBUG
#define new DEBUG_NEW
#undef THIS_FILE
static char THIS_FILE[] = __FILE__;
#endif

//////////////////////////////////////////////////////////////////////////
// CWzdEdit

CWzdEdit::CWzdEdit()
{
    m_brush.CreateHatchBrush(HS_BDIAGONAL, RGB(0,255,0));
}

CWzdEdit::~CWzdEdit()
{
}
```

```
BEGIN_MESSAGE_MAP(CWzdEdit, CEdit)
    //{{AFX_MSG_MAP(CWzdEdit)
    ON_WM_CTLCOLOR_REFLECT()
    //}}AFX_MSG_MAP
END_MESSAGE_MAP()


/////////////////////////////////////////////////////////////////////////
// CWzdEdit message handlers

HBRUSH CWzdEdit::CtlColor(CDC* pDC, UINT nCtlColor)
{
        return m_brush;
}

BOOL CWzdEdit::PreTranslateMessage(MSG* pMsg)
{
    if (pMsg->message == WM_KEYDOWN &&
        (pMsg->wParam == VK_UP || pMsg->wParam == VK_DOWN ||
        pMsg->wParam == VK_RETURN))
    {
        int nextprev=GW_HWNDNEXT;
        int firstlast=GW_HWNDFIRST;
        if (pMsg->wParam == VK_UP)
        {
            nextprev=GW_HWNDPREV;
            firstlast=GW_HWNDLAST;
        }
        HWND hWnd = m_hWnd;
        while (hWnd)
        {
            HWND hWndx=hWnd;
            if ((hWnd = ::GetWindow(hWnd, nextprev))==NULL)
                hWnd = ::GetWindow(hWndx, firstlast);
            long style=::GetWindowLong(hWnd,GWL_STYLE);
            if ((style&WS_TABSTOP && !(style&WS_DISABLED) &&
                style&WS_VISIBLE) || hWnd==m_hWnd)
            break;
        }
```

II

10

```
        ::SetFocus(hWnd);

    return TRUE;
  }


  return CEdit::PreTranslateMessage(pMsg);
}
```

# Example 48   Customizing a Common Control Window with Superclassing

## Objective

You would like to intercept and process the WM_CREATE and WM_NCCREATE messages from a control window. You can't do this with subclassing because you can only subclass an existing control (i.e., a control that's already been created and has already processed the WM_CREATE and WM_NCCREATE messages).

## Strategy

To superclass a control, we simply create our own Window Class using the Window Class of a system-provided common control, as our template. We then stick our own Window Process into this new Window Class, saving the address of the original so that we can pass on all other messages. Since this change will then effect every control window created with our new class, we will make this substitution in our Application Class before our application is even initialized.

> **NOTE:** A Window Class is not an MFC class or even a C++ class. It predates and exists outside of C++ and is the template used when any window is created by your system. See Chapter 1 for much more on this topic.

Example 48　Customizing a Common Control Window with Superclassing　**311**

## Steps

### Create a New Window Class

1. In your Application Class's `InitInstance()` function, call a helper function to do the actual work. This keeps our `InitInstance()` function from looking too messy. In this example, we will be superclassing all button controls, so our helper function will be called `SuperclassButtons()`.

```
BOOL CWzdApp::InitInstance()
{
    SuperclassButtons();

    :    :    :
}
```

2. In `SuperclassButtons()`, start by using `::GetClassInfo()` to fill a WND-CLASS structure with the current information on the window class we intend to superclass (in this case, `BUTTON`).

```
void CWzdApp::SuperclassButtons()
{
    WNDCLASS wClass;

    // get existing class information
    ::GetClassInfo(AfxGetInstanceHandle(),"BUTTON",&wClass);
```

3. Next, save the current window process for the BUTTON Window Class in a static variable. Then, stuff the instance and address of our own window process into this class structure.

```
// at the top of your source
WNDPROC lpfnButtonWndProc;
    :    :    :
    // save old window process and substitute our own
    lpfnButtonWndProc=wClass.lpfnWndProc;
    wClass.hInstance=AfxGetInstanceHandle();
    wClass.lpfnWndProc=ButtonWndProc;
```

II

10

4. Finally, we reregister this class using the same name as before (i.e., BUTTON).

```
    // register this class
    ::RegisterClass(&wClass);
}
```

The system allows us to do this because we are registering a local Window Class name here. From now on, when we tell the system to create a control window using the BUTTON Window Class, it starts by searching the local Window Classes. When the system finds our new class, it creates the window using ours instead of the original.

## Create the New Window Process

1. Make the following declarations at the top of your Application Class.

```
LRESULT CALLBACK ButtonWndProc(HWND hWnd, UINT msg,
    WPARAM wParam, LPARAM lParam);
```

2. Create your new window process using the following syntax. Your process need only handle the WM_CREATE and WM_NCCREATE messages. All others can be passed on to the original windows process.

```
LRESULT CALLBACK ButtonWndProc(HWND hWnd, UINT msg,
    WPARAM wParam, LPARAM lParam)
{
    switch (msg)
    {
        case WM_CREATE:
            break;
        case WM_NCCREATE:
            break;
    }
    return (lpfnButtonWndProc)(hWnd,msg,wParam,lParam);
}
```

# Notes

- See Chapter 3 for more on Subclassing and Superclassing.

Example 49   Putting Bitmaps on a Button   **313**

- In this example, we are creating a local Window Class. We could have created a global class by including the CS_SYSTEMGLOBAL class style and registering the class. However, a global Window Class doesn't have the same power that it once did. With Windows 3.1, a system global class could affect the entire system. However starting with Win 95 and NT 4.0, you can only affect your own application. See Chapter 1 for more on Windows Classes.

- This example can be applied to any Window Class. See "Factory Installed Window Classes" on page 23 for a list of Window Classes provided by the Windows operating system.

- In this example, all of the button controls created by your application will now come to your new window process. If you would like to be more selective, you can instead rename the Window Class before you register it — perhaps calling it MYBUTTON. Then, whenever you want to create a new button control using this new Window Class, you can use CWnd::CreateEx() and the new Window Class name. To create a control using this new Window Class with the Dialog Editor, you can add a Custom Control to your dialog box template. A Custom Control allows you to specify the Window Class name that the system will use when creating the control.

- I've included Superclassing for completeness, only. You rarely need to intercept WM_CREATE or WM_NCCREATE. If you want to change the style of a window before it is visible, just create it invisible, change the style, then make it visible. Superclassing is very powerful, but also very messy.

## CD Notes

- When executing the project on the accompanying CD, set a breakpoint on ButtonWndProc() at the bottom of Wzd.cpp. From then on, you will notice that your application will break there anytime any message is sent to any button control in your application.

# Example 49   Putting Bitmaps on a Button

## Objective

You would like to replace the text on a button control with a bitmap or you would like to supply all of the bitmaps with which a button control will be drawn (Figure 10.2).

## Figure 10.2  Use bitmaps on buttons.

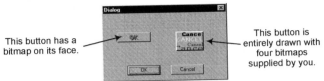

This button has a bitmap on its face.

This button is entirely drawn with four bitmaps supplied by you.

## Strategy

To put a bitmap on a button, we will simply set that button's style to BS_BITMAP and then use CButton::SetBitmap() to tell the button which bitmap to display. We can also create an owner-drawn button that will not only allow us to put a bitmap on its face, but also allow us to draw its border. We could simply set the style of the button to BS_OWNERDRAWN and override the DrawItem() member function of CButton to draw the control ourselves. However, to take the work out of handling DrawItem(), we will use the CBitmapButton class instead, which will load and draw the button for us using up to four bitmaps that we supply.

## Steps

### Add a Bitmap to a Button

1. To create a button in a dialog template with a bitmap face, use the Dialog Editor to open its Properties and give it a Bitmap property.
2. Embed a bitmap class in the dialog class that uses this template.

```
CWzdBitmap m_bitmap;
```

We are using a bitmap class we will create in Example 57 that allows our bitmap to have the same background color as our button and will blend in better. We need to embed this class because even though we will be using CButton::SetBitmap() in the next step to assign this bitmap to a button, it is still our responsibility to maintain this bitmap's object. If you didn't do this, you won't get any compile, link, or run-time errors. You also won't get a bitmap on your button.

Example 49   Putting Bitmaps on a Button   **315**

3. Use the ClassWizard to add a `WM_INITDIALOG` message handler to the dialog class that uses this template. There, you will tell the button which bitmap to display.

```
m_bitmap.LoadBitmapEx(IDB_WZD_BUTTON,TRUE);
((CButton *)GetDlgItem(IDC_WZD_BUTTON1))->SetBitmap(m_bitmap);
```

## Use `CBitmapButton` **to Create a Button**

1. Use the Dialog Editor to give the appropriate button an owner-drawn property.
2. Also use the Dialog Editor to give the button a unique caption, such as `MYBUTTON`.

   We will be using `CBitmapButton` to draw this button by supplying it with up to four bitmaps: one for the button when it's up, one for when it's down, etc. The method `CBitmapButton` uses to determine what bitmaps are in your application's resources to draw this button is somewhat kludgy. First, it gets the caption you gave the button using the Dialog Editor (in this case, `MYBUTTON`). It then looks for four bitmaps in your application's resources with text IDs that are permutations of that caption. A bitmap with the text ID `MYBUTTONU` identifies the bitmap to draw when the button is up. `MYBUTTOND` will be drawn when the button is down. `MYBUTTONF` will be drawn when the button has focus. `MYBUTTONX` will be drawn when the button is disabled. You can give a resource a text ID, rather than a numeric ID, simply by enclosing it in double quotes when defining it.

3. Use the Bitmap Editor to create up to four buttons and give each the appropriate text ID. Note that only up and down are required. Also note that when drawing a button yourself, it's up to you to give a button a 3-D look with some well-placed highlight and shadow lines along its border.
4. Embed a `CBitmapButton` class variable in your dialog class.
5. Use the ClassWizard to add a `WM_INITDIALOG` message handler to your dialog class. There, you will use `CBitmapButton::AutoLoad()` to load up your button's bitmaps and take it from there. The ID you supply Auto-Load is the ID of the button control you want it to draw.

```
CBitmapButton m_bitmapButton;
m_bitmapButton.AutoLoad(IDC_WZD_BUTTON2,this);
```

You will notice that system-drawn buttons seem more responsive than owner-drawn buttons. If you click several times quickly on a system-drawn button, it reacts to each click, while an owner-drawn button reacts only to every other click. For whatever reason, system-drawn buttons will process a double-click, while owner-drawn buttons will not. What does double-click processing have to do with it? Remember that clicking anything in rapid succession is bound to generate some double-clicks (a double-click is simply two rapid clicks in a row). A system-drawn button treats a double-click as if it were a single-click, while owner-drawn buttons ignore them altogether. To process double-clicks too, we will handle our button's WM_LBUTTONDBLCLK message.

### Make a CBitmapButton **More Responsive**

1. Use the ClassWizard to create your own derivation of the CBitmapButton class.
2. Use the ClassWizard to add a WM_LBUTTONDBLCLK message handler to this class. There you will send an additional WM_LBUTTONDOWN message.

```
void CWzdButton::OnLButtonDblClk(UINT nFlags, CPoint point)
{
    SendMessage(WM_LBUTTONDOWN, (WPARAM)nFlags,
        (LPARAM)MAKELONG(point.x, point.y));
}
```

And presto chango, your button is responsive again.

## Notes

- The CBitmapButton class takes a lot of the work out of an owner-drawn button. However, it also takes some of the control out of your hands. For example, if you wanted to create a button with the system color for a button face, you can't do it with CBitmapButton because you don't have access to the button bitmaps at run time. In this case, you need to override DrawItem() and draw the button yourself.

Example 50   Dynamically Filling a Combo Box   **317**

## CD Notes

- When executing the project on the accompanying CD, click on the Test/Wzd commands to open a dialog box that displays two bitmap buttons.

# Example 50   Dynamically Filling a Combo Box

## Objective

You would like to fill a combo box with data just before it opens (Figure 10.3).

**Figure 10.3   Fill in a combo box just before it opens to keep the data fresh.**

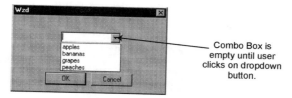

## Strategy

Since the list in a drop down combo box is only visible once the user has clicked on the drop down button, you can fill it at any time. You can add items to a combo box when it's still in a dialog template by using the Dialog Editor. You can also add items when the combo box is first created in the OnInitDialog() member function of your dialog class. Or, if the data must be very fresh, you can even add items to a combo box just before it opens its drop down list. This example covers this last case by processing two control notifications from a combo box control.

# Steps

## Dynamically Fill a Combo Box

1. Use the Dialog Editor to add a combo box to a dialog template.
2. Use the ClassWizard to create a dialog class to use this template.
3. Use the ClassWizard to add a member variable to this dialog class, which will exchange a CString value with this combo box (Example 39). In this example, we are using the name m_sSelection.
4. When using this dialog class, initialize this member variable to some valid value.
5. Use the ClassWizard to add a message handler for your combo box's CBN_DROPDOWN control notification (Example 59). There, you will fill this combo box control with text using CComboBox::AddString(). Then, you will select the line that matches what is currently in the combo box's edit box, if any.

```
void CWzdDialog::OnDropdownWzdCombo()
{
    CComboBox *pCombo=(CComboBox *)GetDlgItem(IDC_COMBO_BOX);
    pCombo->ResetContent();
    pCombo->AddString("apples");
    pCombo->AddString("peaches");
    pCombo->AddString("bananas");
    pCombo->AddString("grapes");
    pCombo->SelectString(-1,m_sSelection);
}
```

6. Use the ClassWizard to add a CBN_CLOSEUP control notification handler to your class. There, you will grab the selection your user has made (if any) and stuff it into your combo box's edit box.

```
void CWzdDialog::OnCloseupWzdCombo()
{
    int    nSel;
    if ((nSel=m_ctrlCombo.GetCurSel())!=CB_ERR)
    {
        m_ctrlCombo.GetLBText(nSel, m_sSelection);
    }
}
```

Example 50   Dynamically Filling a Combo Box   **319**

```
    else
    {

        m_ctrlCombo.ResetContent();
        m_ctrlCombo.AddString(m_sSelection);
        m_ctrlCombo.SelectString(-1,m_sSelection);
    }
}
```

Notice that if nothing was selected, this handler simply restores the original contents of the edit box.

7. When your dialog box closes, the value in m_sSelection will contain whatever selection the user has made.

**II**

## Notes

- The Dialog Editor has a couple of tricks when editing a combo box: you can increase the size of a combo box's list when it drops down.  You can also add data items to its list. To increase the size of a combo box's drop down list, first add the combo box to the dialog box template. Then, click on the drop-down button. A new outline appears that you can drag to make bigger. To add data to the list, open the Property box for the combo box and select the second tab. Make sure you press Control-Enter between items.

- There are three combo box styles. A combo box using the Simple style has a list that is always visible. A Dropdown style only displays the list when the user clicks on the drop down button. The Dropdown style also allows the user to enter their own data into a combo box's edit box. A Drop List style is similar to the Dropdown style, except that it does not allow the user to enter their own data.

- The combo box class, CComboBox, has a member function, called CComboBox::Dir(), that will automatically fill its list with the contents of a subdirectory on your disk. You can use Dir() in this example as a source of strings for CComboBox::AddString(), although you might want to reformat the output of Dir() considering it's somewhat stark.

- The CBN_DROPDOWN message is a Control Notification sent by the combo box to allow the parent to perform processing that the control cannot. As in this example, the control knows how to draw itself and accept input, but only the parent knows how to fill it with data. For more on Control Notifications, see Chapter 3.

## CD Notes

- When executing the project on the accompanying CD, click on the Test/Wzd menu commands to open a dialog box with an empty combo box. You can set a breakpoint in WzdDialog.cpp to watch this combo box being filled whenever the user clicks on the drop down button.

# Example 51    Sorting a List Control

## Objective

You would like to sort a list control's column when your user clicks on its header (Figure 10.4). Note that a list control only has columns when using its Report window style.

## Figure 10.4    Sort a list control by processing its LVN_COLUMNCLICK message.

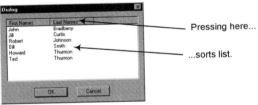

## Strategy

A list control sends out a LVN_COLUMNCLICK Control Notification whenever someone clicks on one of its headers. We will process this notification by using the CListCtrl::SortItems() function.

## Steps

The list control allows you to store one DWORD variable with each item in the list. This variable is totally user-defined and typically is a pointer to an instance of a data class or structure that more fully defines this entry in the list. In fact, when sorting a list control, this DWORD variable is all that we will have to compare one item to another. We start this example by defining our own data class that simply contains a member variable for each column in this list item.

Example 51   Sorting a List Control   **321**

> **NOTE:** If the items in this list already exist in some data collection in the document, we could forego creating a special data class here and use the other classes instead.

## Create a Data Class

1. Create a data class that will describe each item in the list. In this example, the data class contains a member variable for the first name and last name.

```
class CWzdInfo : public CObject
{
public:
    DECLARE_SERIAL(CWzdInfo)
    CWzdInfo();
    CWzdInfo(CString sFirst,CString sLast);

    // misc info
    CString m_sFirst;
    CString    m_sLast;
};
```

2. Add an instance of one of these data classes to each line of your list control as a DWORD pointer.

```
m_ctrlList.SetItemData(0,(DWORD)new CWzdInfo("Bill","Smith"));
```

## Sort a List Control

1. Use the ClassWizard to add a LVN_COLUMNCLICK message handler to the parent window of your list control. The parent window can be either a

dialog box or a view. In this handler, call the `SortItems()` member function of your list control.

```
void CWzdDialog::OnColumnclickWzdList(NMHDR* pNMHDR,
    LRESULT* pResult)
{

    NM_LISTVIEW* pNMListView = (NM_LISTVIEW*)pNMHDR;
    m_ctrlList.SortItems((PFNLVCOMPARE)CompareColumnItems,
        pNMListView->iSubItem);


    *pResult = 0;

}
```

`CompareColumnItems()` is supplied by us to tell `SortItems()` which list item comes before which line item. We create the `CompareColumnItems()` function next.

2. Create the `CompareColumnItems()` function using the following syntax.

```
int CALLBACK CompareColumnItems(CWzdInfo* pItem1,
    CWzdInfo* pItem2, LPARAM lCol)
{

    int nCmp = 0;
    switch(lCol)
    {
    case 0: //column 1
        nCmp = pItem1->m_sFirst.CompareNoCase(
            pItem2->m_sFirst);
        break;
    case 1: //column 2
        nCmp = pItem1->m_sLast.CompareNoCase(pItem2->m_sLast);
        break;
    }
    return nCmp;

}
```

The list Control class's `SortItems()` function passes the `DWORD` variable of two list control items to this function. Since we have stuck a pointer to our own data class here, we are in fact being passed the pointer to two data classes to compare. The returned value should be negative if the first item

Example 51   Sorting a List Control   **323**

should come before the second, positive if the opposite is true, and zero if they equal each other.

### Destroy the Data Classes

1. If you created a new data class instance to sort your list, rather than use an existing document data collection, you will need to delete these instances when the list control is destroyed. You can do this in the destructor of the dialog or view class in which it lives.

```
void CWzdDialog::OnDestroy()
{
    CDialog::OnDestroy();

    // destroy all data items
    int i= m_ctrlList.GetItemCount();
    while(i>-1)
    {
        delete (CWzdInfo *)m_ctrlList.GetItemData(i--);
    }
}
```

## Notes

- The list control also has a window style that will cause it to automatically sort the items in the first column alphabetically, but with this approach you can sort any column any way you see fit.
- The LVN_COLUMNCLICK message is a Control Notification sent by the list control to allow the parent to perform processing that the control cannot. The control knows how to move lines around, but only the parent window knows what comes before what. For more on Control Notifications, see Chapter 3.

## CD Notes

- When executing the project on the accompanying CD, click on the Test/Wzd menu commands to open a dialog with a list control. Then, click on either column header in the list control to sort that column alphabetically.

# Example 52  Line Control

## Objective

You would like to add an etched vertical or horizontal line to a dialog box (Figure 10.5).

## Figure 10.5  Lines in a Dialog Box.

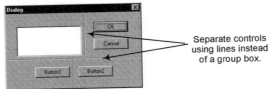

## Strategy

Sounds easy doesn't it? Well, actually it is — it's just not very obvious. You won't find an etched line control among the controls that accompany the Dialog Editor. The closest you come to a line tool is the group box control. You could use a group box control and blank out its name, but then all you would have would be a less then perfect rectangle. If you hunt around some more, you'll even find out that the Picture Control will also create an etched rectangle, but again, no line. After even more hunting around, you'll discover that you can create a line control programmatically by creating a static control with a SS_ETCHEDHORZ or SS_ETCHEDVERT window style. But there's no way to set either of these window styles using the Dialog Editor and it can be a pain to always have to create your line control using Create(), especially when precision is required to place such a control.

So how do you do it? Actually, I lied when I said you couldn't create a line with the Picture Control.

## Steps

### Create an Etched Line in a Dialog Box

1. Use the Dialog Editor to add a Picture Control to your dialog template.
2. Bring up that control's properties. In the Color combo box, select Etched. The Type should be Frame.

Example 52   Line Control   **325**

3. Now pick one of the grabber handles on the control and make the control smaller until (you guessed it) it's a single line.

## Notes

- You can use the same approach to create a black, white, or gray line. Just select a different color in the Picture Control's properties.

- The controls created using the Picture Control are actually static Control windows. Normally, you might think of a static control as displaying only text in a dialog box. However, a static control has several other window styles that cause it to create all sorts of nontextual displays, including an etched line. See Appendix A for other styles available from a static control.

- Button controls also possess a wide variety of window styles. There are push buttons, obviously, but also check boxes and radio buttons. More surprisingly, a group box control is, in fact, a button control with the BS_GROUPBOX style. See Appendix A for more available button styles.

## CD Notes

- When executing the project on the accompanying CD, click on the Test/Wzd menu commands to open a dialog box with two etched lines.

**II**

**10**

# 11

# Drawing

Bitmaps and icons allow you to add color and style to your application. Because all Windows interfaces are essentially alike, logos and splash screens are really your only way to distinguish the look of your application from someone else's. Drawing is obviously also important for creating your own controls and displaying figures in a CAD application.

You can use the examples in this chapter to give your application some distinguishing characteristics.

**Example 53**    **Drawing Figures**    We review some MFC drawing tools.

**Example 54**    **Drawing Text**    We review how to draw text.

**Example 55**    **Loading an Icon from Anywhere and Drawing**    We will look at the methods available for loading an icon from disk.

**Example 56**    **Loading a Bitmap from Anywhere**    We will look at the methods available for loading a bitmap from disk.

**Example 57**    **Creating a Bitmap From a File**    We will take over the bitmap loading process to exert our own control over it to include the ability to create a palette for the bitmap and even substitute bitmap colors for our own on demand.

**Example 58** **Creating a Bitmap from Scratch** We will create a bitmap entirely in memory using the MFC drawing tools available to us.

# Example 53 Drawing Figures

## Objective

You would like to draw one of the figures seen here in Figure 11.1.

## Figure 11.1 You can draw figures like these with MFC.

## Strategy

We will explore the different drawing tools available to us through MFC's CDC class.

## Steps

Drawing in a Windows application is done using a device context. This context defines where you will be drawing, with which tool you will be drawing, and in which mode you will be drawing. A device context helps to simplify the other Windows drawing tools by eliminating repetitive calling arguments. See Chapter 4 for much more on this topic.

### Create a Device Context

1. If you are processing a WM_PAINT message or some other similar message, chances are the device context will be provided for you. If not, you must create one yourself. If you are drawing to the screen, you can use the following to create a device context, where pWnd is a pointer to a CWnd class

Example 53   Drawing Figures   **329**

instance. This class instance should own the window in which you are drawing.

```
CDC *pDC=pWnd->GetDC();
```

2. If you create your own device context, you must destroy it after you use it. Otherwise, a different kind of memory leak occurs, called a resource leak. To destroy a device context, use

```
pWnd->ReleaseDC(pDC);
```

---

**NOTE:** The class from which you call these functions should control the window into which you are drawing. If no such window class instance is available, you can use `AfxGetMainWnd()->GetDC()` or `::GetDC(NULL)`, which will return a device context initialized using the characteristics of the desktop. The latter function returns a device context handle that you must wrap in a `CDC` class instance for use here.

---

A device context comes predefined with several drawing characteristics. One of these characteristics is the width and in line color to be drawn. This characteristic is actually contained in its own object that is referred to by the device context. This object is called a Pen and it defaults to a black line that is one pixel wide. If you need something else, you will need to create your own pen object.

## Create a Pen

1. Create an instance of the `CPen` class with the characteristics you need to draw your line(s).

```
CPen pen(
    PS_SOLID,                // solid line also
                             // PS_DASH,PS_DOT,PS_DASHDOT,
                             // PS_DASHDOTDOT, PS_INSIDEFRAME and
                             // PS_NULL
    2,                       // width in pixels
    RGB(128,128,128));       // color
```

2. Tell the device context to refer to this new pen object, while also saving a pointer to the old pen so that you can restore it later.

```
CPen *pPen=pDC->SelectObject(&pen);    // save old pen
```

Another characteristic predefined in your device context is the fill color (the color used to paint the interior of a figure) used with a function that draws a closed figure. The default is white, which you can change by telling the device context to us a new Brush object.

## Create a Brush

1. Create an instance of the CBrush class using the desired color.

```
CBrush(RGB(128,128,128));    // color
```

2. Tell the device context to refer to this new brush object, while also saving a pointer to the old brush so that you can restore it later.

```
CBrush *pBrush=pDC->SelectObject(&brush);    // save old brush
```

---

NOTE: The last two sections represent the most basic ways of creating a CPen and CBrush class instance. For much more variety, see Chapter 4 and your MFC documentation.

---

## Draw Figures with Member Functions of the CDC class

1. To draw a line with this device context, use

```
pDC->MoveTo(5,5);
pDC->LineTo(25,25);
```

The start and end coordinates of a line are split into two function calls, so that multiple contiguous lines can be drawn with a minimal of pushing and popping of arguments off the stack.

The numbers here, as well as in the rest of this example, are in logical units. When drawing to the screen, logical units equate to screen pixels.

---

NOTE: This and the other examples here use sample calling arguments. You obviously can use your own.

---

Example 53   Drawing Figures   **331**

2. To draw a rectangle, use

```
pDC->Rectangle(CRect(5,55,50,85));
```

3. To draw an arc, use

```
pDC->Arc(CRect(5,115,50,145),    // area encompassing the arc
    CPoint(5,115),               // starting point of the arc
    CPoint(50,115));             // end point
```

4. To draw a rectangle with rounded corners, use

```
pDC->RoundRect(CRect(5,185,50,215),
    CPoint(15,15));       // distance from corner to draw arc
```

5. To draw an ellipse or circle, use

```
pDC->Ellipse(CRect(250,5,305,25));
```

6. To draw a pie chart, use

```
pDC->Pie(CRect(250,55,305,85),    // area encompassing chart
CPoint(250,55),                   // starting point
CPoint(305,55));                  // end point
```

7. To draw a window frame identical in appearance to those you find around control, popup, and overlapped windows, use

```
pDC->DrawEdge(CRect(250,115,305,145),
    EDGE_BUMP,     // also EDGE_ETCHED,EDGE_RAISED,EDGE_SUNKEN
    BF_RECT);      // also BF_LEFT,BF_BOTTOM,BF_RIGHT,BF_TOP
```

8. To draw a series of contiguous lines, use

```
POINT pt[8];
pt[0].x=495;
pt[0].y=5;
pt[1].x=510;
pt[1].y=10;
pt[2].x=515;
pt[2].y=12;
pt[3].x=495;
```

```
pt[3].y=15;
pt[4].x=550;
pt[4].y=25;
pDC->Polyline(pt, 5);
```

9. To draw an enclosed figure with multiple, contiguous sides, use

```
pt[0].x=495;
pt[0].y=55;
pt[1].x=550;
pt[1].y=55;
pt[2].x=530;
pt[2].y=65;
pt[3].x=550;
pt[3].y=85;
pt[4].x=520;
pt[4].y=70;
pt[5].x=495;
pt[5].y=85;
pt[6].x=510;
pt[6].y=65;
pt[7].x=495;
pt[7].y=55;
pDC->Polygon(pt, 8);
```

## Notes

- A device context is simply a chuck of memory that's initialized with all of the characteristics of the device to which you plan to draw. Using a device context saves on performance — instead of passing 10 or 15 arguments to `CDC::LineTo()`, you only need to pass two. The start of the line, its color, its thickness, valid places it can be drawn, and the type of device to which it's being drawn are already defined in the context.

- There are four types of device contexts. In additional to the screen context we've seen, there's also a memory device context, a printer device context, and an informational device context. The information device context is much more compact then the other three and is used only to retrieve information on a device, not for drawing. We will be using a

Example 54 Drawing Text **333**

memory device context later in this chapter to create a bitmap image in memory. A printer device context is used for what you might expect, although it may be created differently than what you might expect.

- There's much more to drawing figures than this example can show. See Chapter 4 and your MFC documentation for more.

## CD Notes

- When executing the project on the accompanying CD, you will notice that the view is filled with several basic shapes you can draw using MFC and the Windows API.

# Example 54    Drawing Text

## Objective

You would like to draw text in your view (Figure 11.2).

### Figure 11.2    Two Samples of Drawn Text.

This is drawn text.

This is drawn text

## Strategy

We will be using the `CDC::TextOut()` and `CDC::DrawText()` member functions of the `CDC` class to draw our text. See the section "Notes" on page 336 for other text drawing functions.

## Steps

1. If a device context isn't available to you, create one using the techniques found in the previous example. One common method is as follows.

```
CDC *pDC=pWnd->GetDC();     // where pWnd is a pointer to an
                            // MFC window class
```

## Work with TextOut()

1. To draw a text string, use the following.

```
CString str("This is drawn text");
pDC->TextOut(
x,y,                        // device location (ex: screen pixels)
str,                        // text as either a string pointer or
                            // a CString variable
str.getLength());           // text length
```

The x and y arguments define the upper-left corner of the location at which the text will be drawn. If you would like x and y to indicate something else, such as the middle of the location at which the text will be drawn, you can use CDC::SetTextAlign() to change its meaning.

2. To make x and y refer to the middle of the the text to be drawn, call the following before you call TextOut().

```
pDC->SetTextAlign(TA_CENTER);    // (also TA_RIGHT)
```

You can also change the y-alignment with

```
pDC->SetTextAlign(TA_BOTTOM);    // (also TA_ BASELINE)
```

3. To draw your text using a different standard font, use the following before drawing. See the MFC documentation for other stock fonts.

```
pDC->SelectStockObject(ANSI_VAR_FONT);
```

4. To create your own font with which to draw text, you can use

```
CFont font;
font.CreateFont(
    -22,                // point size
    0, 0, 0,
    FW_NORMAL,          // weight, also FW_BOLD
    0,                  // if 1=italic
    0,                  // if 1=underline
    0,                  // if 1=strike through
    0, 0, 0, 0, 0,
    "Courier");         // typeface
    CFont *pFont = (CFont *)pDC->SelectObject(&font);
```

Example 54   Drawing Text   **335**

Or if you simply want to pick a font based on point size and typeface, use `CreatePointFont()`.

`CreateFont()` doesn't actually create a font. Instead, it scans your system for a currently installed font that most closely matches the criterion you specified in its calling arguments, and uses that font.

5. To change the default font for a window, you can use

```
pWnd->SetFont(pFont);                          // the font
```

The default font is automatically selected into any device context created from that window.

6. To change the color in which the text is drawn, use

```
pDC->SetTextColor(RGB(100,100,100));
```

7. To change the background color on which the text is drawn, use

```
pDC->SetBkColor(RGB(200,200,200));
```

The background color is ignored unless the background mode has been set to opaque. Opaque means that a background rectangle is drawn before the text is drawn. The transparent mode means the text is drawn over the current background.

8. To turn on the opaque background mode, use

```
pDC->SetBkMode(OPAQUE);
```

9. To turn on the transparent background mode, use

```
pDC->SetBkMode(TRANSPARENT);
```

`CDC::TextOut()` is the simplest of drawing functions. If you would like the system to do more work for you, you can use `CDC::DrawText()`. Draw-Text() allows you to draw multiple lines of text within a specified rectangle. It also has several styles that allow you to specify how text will be aligned in this rectangle.

II

11

### **Work with** DrawText()

1. To draw text using DrawText(), use

```
pDC->DrawText(
    str,              // a CString value
    rect,             // a bounding rectangle
    DT_CENTER);       // an alignment style-text will be centered
```

## Notes

- Other text drawing functions include:
  - ExtTextOut(), which clips the drawn text outside of a given rectangle;
  - TabbedTextOut(), which expands tabs embedded in your text using a table of tab positions you supply to this function; and
  - DrawState(), which you can use to draw disabled text, which appears etched.

## CD Notes

- When executing the project on the accompanying CD, you will notice that the view is filled with two drawn text lines.

# Example 55   Loading an Icon from Anywhere and Drawing

## Objective

You would like to load an icon from your resource file or directly from an icon file to draw in your application.

## Strategy

We will load an icon using three different methods. In the first method, we will use a member function of the application class called LoadIcon(), which will load a icon from our application's resources. The next method will load an icon directly from a disk file using the Windows API function

Example 55   Loading an Icon from Anywhere and Drawing   **337**

`LoadImage()`. The last method will pull an icon out of another application's executable file using the Windows API function `ExtractIcon()`.

# Steps

## Load an Icon from Your Application's Resources

1. To load an icon that's defined in your application's resources, use

```
HICON hicon;
hicon=AfxGetApp()->LoadIcon(IDR_MAINFRAME);
```

**II**

## Load an Icon Directly from an `.ico` Disk File

1. To load an icon from an `.ico` file, use the following. In this example, we will load an icon from `Wzd.ico`.

```
hicon = (HICON)LoadImage(
    NULL,              // handle of the instance that contains
                       // the image
    "Wzd.ico",         // name or identifier of image
    IMAGE_ICON,        // type of image—
                       // can also be IMAGE_CURSOR or IMAGE_ICON
    0,0,               // desired width and height
    LR_LOADFROMFILE);  // load flags
```

**11**

## Load an Icon from a DLL or `.exe` File

1. To extract an icon from another application's executable file, you can use the following. In this example, we are extracting the second icon found in `wzd.exe`.

```
HINSTANCE hinst=AfxGetInstanceHandle();
hicon=ExtractIcon(hinst,"Debug\\wzd.exe",1);
```

To determine how many icons an executable or DLL file holds, call `ExtractIcon()` with a -1 index. The number will be returned in `hIcon`.

## Draw an Icon

1. You can draw an icon to any window using the following, where 0,0 are the x,y coordinates of the upper-left corner of the drawn icon.

```
pDC->DrawIcon(0,0,hicon);
```

## Destroy an Icon

1. You must manually destroy any icons loaded or extracted with LoadImage() or ExtractIcon() to avoid resource memory leaks.

```
DestroyIcon(hicon);
```

# Notes

- Icons that you load using LoadIcon() actually hang around until your application terminates. Subsequent loads of the same icon resource merely returns the handle of the currently resident icon object.

# CD Notes

- When executing the project on the accompanying CD, you will notice that an icon has been drawn in the view.

# Example 56   Loading a Bitmap from Anywhere

## Objective

You would like to load a bitmap from your resource file or any bitmap file.

## Strategy

First, we will use the CBitmap class to load a bitmap that's defined in your application's resources. Then, we will use the Windows API function Load-Image() to load a bitmap from a .bmp file.

Example 56   Loading a Bitmap from Anywhere   **339**

# Steps

## Add Bitmaps to Your Application's Resources

1. You put the bitmap you want to load into your project in one of two ways. You can create a bitmap using the Developer Studio's Bitmap Editor or you can use the Import command under the Insert/Insert Resource Developer Studio menu commands. Note the ID you assign your bitmap.

## Load Bitmaps from Your Application's Resources

1. To load this bitmap into your application for drawing, use the following lines, where you would substitute IDB_WZD with the ID of your bitmap.

```
CBitmap bitmap;
bitmap.LoadBitmap(IDB_WZD);
```

## Load Bitmaps from a .bmp File

1. You can also load a bitmap at run time with LoadImage().

```
CBitmap bitmap;
HBITMAP hbitmap = (HBITMAP)::LoadImage(
    NULL,                  // handle of the instance that contains
                           // the image
    "Wzd2.bmp",            // name or identifier of image
    IMAGE_BITMAP,          // type of image—
                     // can also be IMAGE_CURSOR or IMAGE_ICON
    0,0,                   // desired width and height
    LR_LOADFROMFILE);      // load from file

// attach this bitmap object to our bitmap class
bitmap.Attach(hbitmap);
```

### Draw a Bitmap

1. To draw a bitmap, you can then use the following. Notice that to use BitBlt(), you need two device contexts, instead of one.

```
CDC dcComp;
dcComp.CreateCompatibleDC(pDC);
dcComp.SelectObject(&bitmap);

// get size of bitmap for BitBlt()
BITMAP bmInfo;
bitmap.GetObject(sizeof(bmInfo),&bmInfo);

// use BitBlt() to draw bitmap
pDC->BitBlt(0,0,bmInfo.bmWidth,bmInfo.bmHeight,
    &dcComp, 0,0,SRCCOPY);
```

## Notes

- One of the big differences between bitmaps and icons is that you can create an icon with a transparent color. Wherever you use a transparent color with an icon, the background on which the icon is drawn will show through. You can fake this effect by using LoadImage() and ORing the LR_LOADFROMFILE flag with LR_LOADTRANSPARENT. LoadImage() will substitute one color in your bitmap with the background color of your application (e.g., the color of a button). LoadImage() chooses which color to substitute based on the color of the pixel in the upper-left corner of your bitmap.

- Some earlier versions of Windows NT won't allow you to use LoadImage() with the LR_LOADFROMFILE flag. If you can't upgrade, you can use the next example to load your bitmap. In that example, rather than use the application's resources, you can instead load the bitmap file directly from disk using the CFile class. Make sure, however, to start your file read after the 12-byte file header.

- For more on BitBlt(), bitmaps and device contexts, see Chapter 4.

Example 57   Creating a Bitmap From a File   **341**

## CD Notes

- When executing the project on the accompanying CD, you will notice two bitmaps drawn in the view.

# Example 57   Creating a Bitmap From a File

## Objective

You would like to load a bitmap object from a `.bmp` file and potentially modify it.

## Strategy

We will load a raw bitmap file into memory from our application's resources and call the Windows API `CreateDIBitmap()` to create a bitmap from it. However, before we pass the bitmap to `CreateDIBitmap()`, we will have an opportunity to modify the palette that came with it. We will encapsulate this functionality in a member function of our very own derivation of `CBitmap`.

## Steps

### Create a New Bitmap Class

1. Use the ClassWizard to derive a new class from `CBitmap`. Use the text editor to add a `LoadBitmapEx()` member function to this class.

### Create a Bitmap Object from a Bitmap File

1. Start `LoadBitmapEx()` by getting a device context. To avoid having to require a device context in the calling arguments of `LoadBitmapEx()`, we will simply grab a device context from the desktop.

```
CDC dcScreen;
dcScreen.Attach(::GetDC(NULL));
```

2. Next, we'll load our bitmap file from our application's resources into memory.

```
HRSRC hRsrc = FindResource(AfxGetResourceHandle(),
MAKEINTRESOURCE(nID),RT_BITMAP);
HGLOBAL hglb = LoadResource(AfxGetResourceHandle(), hRsrc);
LPBITMAPINFOHEADER lpBitmap =
    (LPBITMAPINFOHEADER)LockResource(hglb);
```

The `lpBitmap` variable now points to a chunk of memory in the heap that contains your resource. Incidentally, this is how you would load any of your application's resources into memory, varying `RT_BITMAP` for the type of resource you intend to load.

We now need to create pointers into this blob of data that we can pass to `CreateDIBitmap()`.

3. Create three pointer variables into this bitmap structure that you can pass to `CreateDIBitmap()`.

```
// get pointers into bitmap structures
// (header, color table and picture bits)
LPBITMAPINFO pBitmapInfo = (LPBITMAPINFO)lpBitmap;
LPBITMAPINFOHEADER pBitmapInfoHeader =
    (LPBITMAPINFOHEADER)lpBitmap;
// determine number of colors in bitmap's palette now because
// bitmap pixel is right after it
int nNumberOfColors=0;
if (lpBitmap->biClrUsed)
    nNumberOfColors = lpBitmap->biClrUsed;
else if (pBitmapInfoHeader->biBitCount <= 8)
    nNumberOfColors = (1<<pBitmapInfoHeader->biBitCount);
// else there IS no color table
LPBYTE pBitmapPictureData = (LPBYTE)lpBitmap+lpBitmap->biSize+
    (nNumberOfColors*sizeof(RGBQUAD));
```

Notice that if the number of bits required to specify a pixel color is 24, the bitmap file has no color table because each pixel entry itself contains the full definition of an RGB color.

Example 57   Creating a Bitmap From a File   **343**

4. Since we now know the size of this bitmap from its header, we will save it for later reference. These values are not needed to create a bitmap object.

```
m_Width = lpBitmap->biWidth;
m_Height = lpBitmap->biHeight;
```

Unfortunately, we can't simply pass a pointer to our bitmap's color table to CreateDIBitmap(). We must first create an application palette with our bitmap's color table and then select the application palette into a device context, which we then pass to CreateDIBitmap().

5. To create an application palette, stuff your bitmap's color table into a logical palette and then use the CPalette class to create your application palette.

```
// create a logical palette from the color table in this bitmap
    if (nNumOfColors)
    {
        LOGPALETTE *pLogPal = (LOGPALETTE *) new BYTE[
            sizeof(LOGPALETTE) + (nNumberOfColors *
            sizeof(PALETTEENTRY))];
        pLogPal->palVersion    = 0x300;
        pLogPal->palNumEntries = nNumberOfColors;

        for (int i = 0;  i < nNumberOfColors;  i++)
        {
        // if flag set, replace grey color with window's
        // background color
            if (bTransparent &&
                pBitmapInfo->bmiColors[i].rgbRed==192 &&
                pBitmapInfo->bmiColors[i].rgbGreen==192 &&
                pBitmapInfo->bmiColors[i].rgbBlue==192)
            {
                pBitmapInfo->bmiColors[i].rgbRed=
                    GetRValue(::GetSysColor(COLOR_BTNFACE));
                pBitmapInfo->bmiColors[i].rgbGreen=
                    GetGValue(::GetSysColor(COLOR_BTNFACE));
                pBitmapInfo->bmiColors[i].rgbBlue=
                    GetBValue(::GetSysColor(COLOR_BTNFACE));
            }
```

II

11

```
            pLogPal->palPalEntry[i].peRed   =
                pBitmapInfo->bmiColors[i].rgbRed;
            pLogPal->palPalEntry[i].peGreen =
                pBitmapInfo->bmiColors[i].rgbGreen;
            pLogPal->palPalEntry[i].peBlue  =
                pBitmapInfo->bmiColors[i].rgbBlue;
            pLogPal->palPalEntry[i].peFlags = 0;

        }
        m_pPalette=new CPalette;
        m_pPalette->CreatePalette(pLogPal);
        delete []pLogPal;
        dcScreen.SelectPalette(m_pPalette,TRUE);
        dcScreen.RealizePalette();

    }
```

Notice that when creating a logical palette, we have an opportunity to play with the colors in our bitmap. Specifically, we are replacing any gray colors we find here with the color of a button face, so our bitmap is seemingly transparent when drawn on a button face background. We use this to great advantage throughout this book.

6. We can now pass our three pointers into our bitmap structure along with the realized palette to `::CreateDIBitmap()` to create a bitmap object.

```
HBITMAP bitmap = ::CreateDIBitmap(dcScreen.m_hDC,
    pBitmapInfoHeader,CBM_INIT, pBitmapPictureData,
    pBitmapInfo, DIB_RGB_COLORS);
```

7. Having created a bitmap object, we can now attach it to our bitmap class object.

```
Attach(bitmap);
```

8. For a complete listing of this bitmap class, see "Listings — Bitmap Class" on page 346.

Example 57   Creating a Bitmap From a File   **345**

## Use the New Bitmap Class

1. To draw this bitmap, first select its palette into the device context, then use BitBlt() or StretchBlt() to draw it to the screen.

```
// load bitmap with our new function
m_bitmap.LoadBitmapEx(IDB_WZD,TRUE);

// select our palette into the device context
CPalette *pOldPal =
    pDC->SelectPalette(m_bitmap.GetPalette(),FALSE);
pDC->RealizePalette();

// get device context to select bitmap into
CDC dcComp;
dcComp.CreateCompatibleDC(pDC);
dcComp.SelectObject(&m_bitmap);

// draw bitmap
pDC->BitBlt(0,0,m_bitmap.m_Width,m_bitmap.m_Height, &dcComp,
    0,0,SRCCOPY);

// reselect old palette
pDC->SelectPalette(pOldPal,FALSE);
```

## Notes

- Every dot in a bitmap is simply a pointer into a color table. A bitmap file has its very own color table. However, when a bitmap is drawn on the screen, it must contend with other bitmaps and other color tables on a system that might have a limited number of individual colors it can show at one time. These color tables are converted into palettes that are selected into the system palette as needed. ::CreateDIBitmap() converts the dot pointers in a bitmap to point to this new palette. For much more on this topic, see Chapter 4.

- To load this file directly from disk instead of through your application's resources, make sure to first advance your file pointer past the first 12-bytes of the file, which contain the file header. Then, use the CFile class to read the file into a CShareMem file, which can then be passed to ::CreateDIBitmap().

## CD Notes

- When executing the project on the accompanying CD, you will notice the view filled with a bitmap displayed in nondiffused colors using the original colors defined in the bitmap.

## Listings — Bitmap Class

```
#ifndef WZDBITMAP_H
#define WZDBITMAP_H


class CWzdBitmap : public CBitmap
{
public:
    DECLARE_DYNAMIC(CWzdBitmap)

// Constructors
    CWzdBitmap();

    void LoadBitmapEx(UINT nID, BOOL bIconBkgrd);
    CPalette *GetPalette(){return m_pPalette;};

// Implementation
public:
    virtual ~CWzdBitmap();

// Attributes
    int    m_Width;
    int    m_Height;
// Operations

private:
    CPalette *m_pPalette;
};
#endif


// WzdBitmap.cpp : implementation of the CWzdBitmap class
```

Example 57   Creating a Bitmap From a File   **347**

```
//

#include "stdafx.h"
#include "WzdBitmap.h"

///////////////////////////////////////////////////////////////////////////
// CWzdBitmap

IMPLEMENT_DYNAMIC(CWzdBitmap, CBitmap)


CWzdBitmap::CWzdBitmap()
{
    m_pPalette=NULL;
}


CWzdBitmap::~CWzdBitmap()
{
    if (m_pPalette)
    {
        delete m_pPalette;
    }
}

void CWzdBitmap::LoadBitmapEx(UINT nID, BOOL bTransparent)
{
// can only load once
    ASSERT(!m_pPalette);

    CDC dcScreen;
    dcScreen.Attach(::GetDC(NULL));

// find and lock bitmap resource
    HRSRC hRsrc = FindResource(AfxGetResourceHandle(),
        MAKEINTRESOURCE(nID),RT_BITMAP);
    HGLOBAL hglb = LoadResource(AfxGetResourceHandle(), hRsrc);
    LPBITMAPINFOHEADER lpBitmap = (LPBITMAPINFOHEADER)LockResource(hglb);

// get pointers into bitmap structures (header, color table and picture bits)
    LPBITMAPINFO pBitmapInfo = (LPBITMAPINFO)lpBitmap;
```

II

11

```
    LPBITMAPINFOHEADER pBitmapInfoHeader = (LPBITMAPINFOHEADER)lpBitmap;
    int nNumberOfColors=0;
    if (lpBitmap->biClrUsed)
        nNumberOfColors = lpBitmap->biClrUsed;
    else if (pBitmapInfoHeader->biBitCount <= 8)
        nNumberOfColors = (1<<pBitmapInfoHeader->biBitCount);
    LPBYTE pBitmapPictureData = (LPBYTE)lpBitmap+lpBitmap->biSize+
        (nNumberOfColors*sizeof(RGBQUAD));

    // get width and height
    m_Width = lpBitmap->biWidth;
    m_Height = lpBitmap->biHeight;

// create a logical palette from the color table in this bitmap
    if (nNumberOfColors)
    {
        LOGPALETTE *pLogPal   = (LOGPALETTE *) new BYTE[sizeof(LOGPALETTE) +
            (nNumberOfColors * sizeof(PALETTEENTRY))];
        pLogPal->palVersion    = 0x300;
        pLogPal->palNumEntries = nNumberOfColors;

        for (int i = 0;  i < nNumberOfColors;  i++)
        {
// if flag set, replace grey color with window's background color
            if (bTransparent &&
                pBitmapInfo->bmiColors[i].rgbRed==192 &&
                pBitmapInfo->bmiColors[i].rgbGreen==192 &&
                pBitmapInfo->bmiColors[i].rgbBlue==192)
            {
                pBitmapInfo->bmiColors[i].rgbRed=
                    GetRValue(::GetSysColor(COLOR_BTNFACE));
                pBitmapInfo->bmiColors[i].rgbGreen=
                    GetGValue(::GetSysColor(COLOR_BTNFACE));
                pBitmapInfo->bmiColors[i].rgbBlue=
                    GetBValue(::GetSysColor(COLOR_BTNFACE));
            }
```

Example 58  Creating a Bitmap from Scratch  **349**

```
            pLogPal->palPalEntry[i].peRed   = pBitmapInfo->bmiColors[i].rgbRed;
            pLogPal->palPalEntry[i].peGreen =
                pBitmapInfo->bmiColors[i].rgbGreen;
            pLogPal->palPalEntry[i].peBlue  = pBitmapInfo->bmiColors[i].rgbBlue;
            pLogPal->palPalEntry[i].peFlags = 0;
        }
        m_pPalette=new CPalette;
        m_pPalette->CreatePalette(pLogPal);
        delete []pLogPal;
        dcScreen.SelectPalette(m_pPalette,TRUE);
        dcScreen.RealizePalette();
    }

// create device dependant bitmap
    HBITMAP bitmap = ::CreateDIBitmap(dcScreen.m_hDC, pBitmapInfoHeader,
        CBM_INIT, pBitmapPictureData, pBitmapInfo, DIB_RGB_COLORS);

// attach this new bitmap object to our CBitmap class
    Attach(bitmap);

// release dc
    ::ReleaseDC(NULL, dcScreen.Detach());
}
```

**II**

**11**

# Example 58   Creating a Bitmap from Scratch

## Objective

You would like to create an empty bitmap in memory and then draw into it using MFC's drawing tools.

## Strategy

We will create a memory device context and draw into it. All drawing functions in Windows involve a device context, which is an object that keeps track of the characteristics of the device to which you're drawing. A memory device context allows you to draw straight into memory. We will encapsulate the functionality of this example in a class we derive from CBitmap.

## Steps

### Create a New Bitmap Class

1. Use the ClassWizard to create a new class derived from CBitmap.
2. Use the Text Editor to create a CreateBitmapEx() member function to this class.

### Create a Bitmap Object from Scratch

Although we will be drawing into memory, we need to initialize our device context with the characteristics of the device on which it will finally appear. First, we need to get a device context to a real device. The desktop is as good a device as any.

1. Start CreateBitmapEx() by wrapping a device context from the desktop in a CDC class instance.

```
CDC dcScreen;
dcScreen.Attach(::GetDC(NULL));
```

2. Create your memory device context and initialize it with the desktop's characteristics.

```
CDC dcMem;
dcMem.CreateCompatibleDC(&dcScreen);
```

3. Create an empty bitmap object and tell your memory device context that that's where it will be drawing.

```
CreateCompatibleBitmap(&dcScreen, size.cx, size.cy);
dcMem.SelectObject(this);
```

Example 58   Creating a Bitmap from Scratch   **351**

4. Draw to the memory device context as if you were drawing to the screen. In this example, we will draw our application's icon inside a circle on top of a square.

```
CBrush bluebrush,greenbrush;
bluebrush.CreateSolidBrush(RGB(0,0,255));
greenbrush.CreateSolidBrush(RGB(0,255,0));
dcMem.FillRect(CRect(0,0,size.cx,size.cy), &bluebrush);
dcMem.SelectObject(&greenbrush);
dcMem.Ellipse(0,0, size.cx, size.cy);
HICON hicon=AfxGetApp()->LoadIcon(IDR_MAINFRAME);
dcMem.DrawIcon((size.cx-32)/2,(size.cy-32)/2,hicon);
```

5. Release your screen and memory contexts when you're done.

```
dcMem.DeleteDC();
::ReleaseDC(NULL, dcScreen.Detach());
```

For a complete listing of this bitmap class, see "Listings — Derived Bitmap Class" on page 352.

## Notes

- Creating a bitmap from scratch is one way of drawing a figure once that you can use repeatedly. You can also use a metafile or a path to accomplish the same task in different ways. See Chapter 4 for a description of metafiles and paths.

- If your application is using its own palette, you must select that palette into the memory device context before you start drawing into it.

## CD Notes

- When executing the project on the accompanying CD, you will notice a bitmap drawn in the view. This bitmap was created using the standard MFC icon and a drawn circle and square.

**II**

**11**

# Listings — Derived Bitmap Class

```cpp
#ifndef WZDBITMAP_H
#define WZDBITMAP_H


class CWzdBitmap : public CBitmap
{
public:
    DECLARE_DYNAMIC(CWzdBitmap)

// Constructors
    CWzdBitmap();

    void CreateBitmapEx(CSize size);

// Implementation
public:
    virtual ~CWzdBitmap();

// Attributes
    int    m_Width;
    int m_Height;

// Operations

};
#endif

// WzdBitmap.cpp : implementation of the CWzdBitmap class
//

#include "stdafx.h"
#include "WzdBitmap.h"
#include "resource.h"


////////////////////////////////////////////////////////////////////
// CWzdBitmap

IMPLEMENT_DYNAMIC(CWzdBitmap, CBitmap)
```

Example 58   Creating a Bitmap from Scratch   **353**

```
CWzdBitmap::CWzdBitmap()
{
    m_Width=0;
    m_Height=0;
}


CWzdBitmap::~CWzdBitmap()
{
}


void CWzdBitmap::CreateBitmapEx(CSize size)
{
    CDC dcMem;
    CDC dcScreen;
    dcScreen.Attach(::GetDC(NULL));

    // create our bitmap in memory
    dcMem.CreateCompatibleDC(&dcScreen);
    CreateCompatibleBitmap(&dcScreen, size.cx, size.cy);
    dcMem.SelectObject(this);

    // do our drawing
    CBrush bluebrush,greenbrush;
    bluebrush.CreateSolidBrush(RGB(0,0,255));
    greenbrush.CreateSolidBrush(RGB(0,255,0));
    dcMem.FillRect(CRect(0,0,size.cx,size.cy), &bluebrush);
    dcMem.SelectObject(&greenbrush);
    dcMem.Ellipse(0,0, size.cx, size.cy);
    HICON hicon=AfxGetApp()->LoadIcon(IDR_MAINFRAME);
    dcMem.DrawIcon((size.cx-32)/2,(size.cy-32)/2,hicon);

    // delete and release device contexts
    dcMem.DeleteDC();
    ::ReleaseDC(NULL, dcScreen.Detach());

    m_Width=size.cx;
    m_Height=size.cy;

}
```

II

11

# Section III

# Internal Processing Examples

Your application is not all user interface. A great deal goes on under the hood, from reading and writing files to timing events. Even though MFC is mostly known as an interface development system, there are several classes in the MFC library that provide support for the noninterface part of your application, as well.

The examples in this section relate to a broad range of processing inside your application. This includes sending messages within and outside of your application. Also included is reading and writing to files on your disk and files in memory. There are also examples of maintaining data within your application in lists and arrays, cutting and pasting screen data, and playing with the time.

## Messaging

The examples in this chapter all relate to sending data and messages within and outside of your application. There are examples of subclassing, superclassing, and creating your own message types, among others.

# Files, Serialization, and Databases

All things persistent are covered in this chapter, from flat files to accessing the databases of all major database providers from your application. Also included is serializing your data, such that it can be easily organized and updated.

# Potpourri

This last chapter is the repository of "everything else internal". In this case, that turns out to be cut and paste, lists and arrays, and time, among others.

# 12

# Messaging

Given the fact that a windows application is itself made up of dozens of potentially autonomous windows and also given the requirements of C++ encapsulation and messaging, in an MFC application becomes a complicated issue. Technically, messages in a Windows application can only be sent from one window to another. How then to allow these messages into the inner sanctum of a C++ class without bogging down the system? For the long story, see Chapter 3.

**Example 59     Adding a Message Handler or MFC Class Override**
We will use the Class Wizard to tie into the MFC messaging system and direct a particular message to one of our class member functions. The Class Wizard also allows you to automatically override any virtual functions that can be found in the MFC class from which you derive.

**Example 60     Adding a Command Range Message Handler**   We will manually add a message handler to our class that processes a whole range of Command messages and, therefore, saves us the time of adding each one using the ClassWizard.

**Example 61** **Rerouting Command Messages** We will look at how to direct command messages to classes for which they aren't normally routed.

**Example 62** **Creating Your Own Windows Message** We will create a message that we can use within our application for control.

# Example 59 Adding a Message Handler or MFC Class Override

## Objective

You would like to add a message handler or override an MFC member function in one of your classes.

## Strategy

First, we will use the ClassWizard to add a message handler or override automatically. Then, we will look at how to do this manually for those times when the handler or override we want is outside the repertoire of the ClassWizard.

## Steps

### Add a Message Handler or Class Override Using the ClassWizard

1. Click on the Developer Studio's View/ClassWizard menu commands to open the ClassWizard. Determine the class to which you would like to add an override or message handler and select it from the Class name combo box. Then, in the Object IDs list box, select the name of the class. (The other entries are menu and toolbar command IDs, and are discussed in Example 13.) The Messages list box will now display most, if not all, of the overrides and windows messages available to that class. Class overrides top the list and are shown in upper and lower case as the actual virtual member function names. Windows messages complete the list, come in all capitals, and depict the actual windows message ID. To add the override or handler you want, click the Add Function button. The

Example 59   Adding a Message Handler or MFC Class Override   **359**

ClassWizard will then add this function to the tables it maintains in your source and append a slightly commented function to your class's source (Figure 12.1).

## Figure 12.1   Use the ClassWizard to add a message handler.

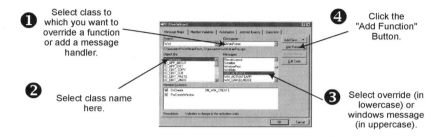

① Select class to which you want to override a function or add a message handler.

② Select class name here.

④ Click the "Add Function" Button.

③ Select override (in lowercase) or windows message (in uppercase).

2. If you can't find the message or override you're looking for in the Messages list, you can sometimes expand the list by selecting the Class info tab of the ClassWizard. In this page, you will find the Message filter combo box, which can change the selections in the Messages list box on the first page. For example, selecting Not a Window here will eliminate all windows messages from the Messages list. (Technically, only windows can receive messages, so there is no need to add a message handler.) Selecting Window will give you the full complement of window messages (Figure 12.2).

## Figure 12.2   Pick a new message filter to change the selections in the Messages list box.

III   12

If an override or windows message isn't listed in the "Message Maps" tab, try selecting a different filter in this combo box .

---

NOTE: If you can't find the appropriate handler or override in the Messages list box, it might not be available to that class. However, sometimes the ClassWizard simply ignores some messages because they're rarely used by a class. Experiment for yourself.

---

3. To delete a handler or override, select it from the Messages list and click the Delete Function button. Actually, all the ClassWizard deletes is any entries in the tables it maintains in your source between the {{}} brackets. You are responsible for deleting everything else, which turns out to be the function, itself. Even if you accidently add a handler and then immediately delete it using the ClassWizard, you must locate the empty function it created and manually delete it.

## Manually Add a Message Handler

If the ClassWizard doesn't list the message you want to process, either by omission or because you created your very own window message, you must manually add the handler yourself. A message handler requires three edits to your class: two to declare and implement the handler and one to enter the handler function in your class's message map.

1. Add the declaration for your handler to your class's .h file. Add your declaration immediately following the //}}AFX_MSG line. This is the message map area of your class's .h file, which, with your addition, should now look something like this. (In this example, our message handler is called, "OnDestroy().")

```
// Generated message map functions
protected:
    //{{AFX_MSG(CWzdClass)
    //}}AFX_MSG
    afx_msg void OnDestroy(); <<<< add declarations after "}}"
    DECLARE_MESSAGE_MAP()
```

By convention, the best place to add your handler declaration is underneath the tables that the ClassWizard maintains in your source, but outside of the {{}} brackets that it uses to demark its territory. Anything within these brackets can and will be destroyed by the ClassWizard.

Example 59   Adding a Message Handler or MFC Class Override   **361**

2. Next, add a message map macro to the message map in your class's `.cpp` file. Locate the `//}}AFX_MSG_MAP` line and add your macro immediately following it.

```
BEGIN_MESSAGE_MAP(CWzdClass, CClass)
    //{{AFX_MSG_MAP(CWzdClass)
    //}}AFX_MSG_MAP
    ON_WM_DESTROY()  //<<<< add macros after "}}"
END_MESSAGE_MAP()
```

For a complete list of message map macros, see Appendix B.

3. You can then add the body of your handler to this file.

```
void CWzdClass::OnDestroy()
{

}
```

See Appendix B for the precise calling syntax to use for a particular message map macro.

## Manually Add a Class Override

1. To manually add an override, you only need to make two edits. Nothing is added to the message map because you are getting control directly from the base class, which itself might be processing a message. First, add your override declaration to the `.h` table. (In this example, we are overriding `PreCreateWindow()`.)

```
// Overrides
    // ClassWizard generated virtual function overrides
    //{{AFX_VIRTUAL(CWzdClass)
    public:
        :    :    :
    //}}AFX_VIRTUAL
    public:                          << make sure to match virtual
                                     << function's protection
    virtual BOOL PreCreateWindow(CREATESTRUCT& cs);
```

2. Then add the body of your override to the `.cpp` file, as seen here. You can optionally call the base class's implementation, too.

```
BOOL CWzdClass::PreCreateWindow(CREATESTRUCT& cs)
{
    return CClass::PreCreateWindow(cs);
}
```

## Notes

- There are actually several ways to invoke the ClassWizard. You can usually right-click on the source file of the class and the ClassWizard will open with that class. You can also create a toolbar button by clicking on the Tools/Customize menu commands to open the Studio's Customize dialog box. There, under the Commands tab, is a Categories combo box, from which you can select View. The ClassWizard is one of the buttons listed here that you can drag into an existing toolbar or into empty space to create a new toolbar.

- When manually adding your own message handler or when using the filter to expand the list of message handlers you can add, the possibility remains that this class might not be in a position to receive that message anyway. Heck, even when the ClassWizard says you can add the handler, your class might not be eligible. See Chapter 3 for more. Also, see Example 61 for a way of adding this class to the path of Command Messages.

- When manually adding an override to a class, the base class must have this function in order to override it. Fortunately, if it doesn't, the compiler will tell you immediately. The same is not true with a message handler. If you manually add a message handler to a class that never receives that message, you won't find out until you test it.

- The ClassWizard won't list an available class override when it's undocumented. You can find these undocumented overrides by searching the source code provided with MFC. However, since they aren't documented, Microsoft has no motivation to maintain these overrides in future releases. If you can't proceed without overriding an undocumented override, you might have to revisit whatever code you write when porting to the next version of MFC.

Example 59   Adding a Message Handler or MFC Class Override   **363**

- When you use the ClassWizard to add a message handler to your class, it will also typically add the code necessary to call your base class's function, passing it the same arguments your message handler receives.

```
void CAppView::OnLButtonDown(UINT nFlags, CPoint point)
{
    // TODO: Add your message handler code here and/or call default

    CView::OnLButtonDown(nFlags, point);
}
```

Don't bother trying to change these arguments to affect the base class however. These arguments are put here merely to torment you. As soon as the base class gets these arguments, it throws them away and chuckles. Instead, it uses the arguments it initially passed you.

If you would really like to modify a message before it gets to the base class and the default window process, you will need to reconstitute the original message into a wParam and lParam argument and, instead of calling the base class's function, call DefWindowProc() directly. To determine how to reconstitute your message into arguments for DefWindowProc(), see your MFC Documentation for the message that you're processing. In this example, you would use

```
DefWindowProc(WM_LBUTTONDOWN,nFlags,point);
```

- To retrieve the message that the base class is working with, you can use

```
MSG msg=CWnd::GetCurrentMessage();
```

where MSG is a message structure that looks like the following.

```
typedef struct tagMSG {
    HWND    hwnd;
    UINT    message;
    WPARAM  wParam;
    LPARAM  lParam;
    DWORD   time;
    POINT   pt;
} MSG;
```

**III** 12

- You could then call `DefWindowProc()` with these values.

```
MSG msg=CWnd::GetCurrentMessage();
msg.wParam = ...changed...
DefWindowProc(msg.message,msg.wParam,msg.lParam);
```

## CD Notes

- There is no accompanying project on the CD for this example.

# Example 60    Adding a Command Range Message Handler

## Objective

You would like to add a handler to your class that can process a range of command messages or update a range of command interfaces.

## Strategy

We will be manually adding two message map macros, `ON_COMMAND_RANGE` and `ON_UPDATE_COMMAND_UI_RANGE`, to our class's message map. The Class-Wizard doesn't currently handle either of these macros.

## Steps

### Use the Command Range Macro

The `ON_COMMAND_RANGE` macro intercepts all command messages (`WM_COMMAND` messages) that have an ID in a specified range. Using this macro allows you to handle several commands with one function.

1. Define the command handler in your class's `.h` file map below the `{{}}` brackets so that the ClassWizard won't disturb it.

```
protected:
    //{{AFX_MSG(CWzdView)
```

Example 60   Adding a Command Range Message Handler   **365**

```
    //}}AFX_MSG
    afx_msg void OnWzdCommandRange(UINT nID);<<<<<<
    DECLARE_MESSAGE_MAP()
};
```

2. Add the ON_COMMAND_RANGE macro to your class's message map, again below the {{}} brackets. The first two arguments in this macro define a range of command IDs to process. These IDs must be in a sequential range with the last ID being greater then the first. The last argument is the name of the command handler you defined in step 1.

```
BEGIN_MESSAGE_MAP(CWzdView, CView)
    //{{AFX_MSG_MAP(CWzdView)
    //}}AFX_MSG_MAP
    ON_COMMAND_RANGE(ID_TEST_WZD1,ID_TEST_WZD4,OnWzdComman-
dRange)
END_MESSAGE_MAP()
```

3. Add the command handler with the following syntax. The nID argument is the ID of the command.

```
void CWzdView::OnWzdCommandRange(UINT nID)
{
    switch (nID)
    {
    case ID_TEST_WZD1:
        break;
    case ID_TEST_WZD2:
        break;
    case ID_TEST_WZD3:
        break;
    case ID_TEST_WZD4:
        break;
    }
}
```

**III** **12**

## Use the User Interface Command Range Macro

The ON_UPDATE_COMMAND_UI_RANGE macro intercepts all queries to update the user interface for a range of commands. Using this macro as an example, you can enable an entire range of menu commands or process a group of toolbar buttons.

1. Define the user interface command handler in your class's .h file map below the {{}} brackets so that the ClassWizard won't disturb it.

```
protected:
    //{{AFX_MSG(CWzdView)
    //}}AFX_MSG
    afx_msg void OnUpdateWzdCommandRange(CCmdUI* pCCmdUI);
    DECLARE_MESSAGE_MAP()
};
```

2. Add the ON_UPDATE_COMMAND_UI_RANGE macro to your class's message map, again below the {{}} brackets. The first two arguments in this macro define a range of command IDs to process. These IDs must be in a sequential range with the last ID being greater then the first. The last argument is the name of the user interface command handler you defined in step 1.

```
BEGIN_MESSAGE_MAP(CWzdView, CView)
    //{{AFX_MSG_MAP(CWzdView)
    //}}AFX_MSG_MAP
    ON_UPDATE_COMMAND_UI_RANGE(ID_TEST_WZD1,ID_TEST_WZD4,
        OnUpdateWzdCommandRange)
END_MESSAGE_MAP()
```

3. The syntax for processing an interface command message is as follows. As seen here, you can determine what menu item, toolbar button, etc., to update based on the m_nID member variable of the CCmdUI class.

```
void CWzdView::OnUpdateWzdCommandRange(CCmdUI *pCmdUI)
{

    switch (pCmdUI->m_nID)
    {
        case ID_TEST_WZD1:
            break;
```

Example 61   Rerouting Command Messages   **367**

```
        case ID_TEST_WZD2:
            pCmdUI->SetRadio();
            break;
        case ID_TEST_WZD3:
            break;
        case ID_TEST_WZD4:
            break;
    }
}
```

## Notes

- Command range macros are typically used to handle a collection of command messages whose processing differs only by the command that called it. These macros can greatly simplify a class that would otherwise be filled with line after line of dummy message handlers calling the same function.

- Two other less frequently used message macros are ON_CONTROL_RANGE and ON_NOTIFY_RANGE for handling the older and newer type of Control Notifications, respectively. For details, see Chapter 3 and Appendix B.

- Another way to process a range of any type of message would be to override one of the message functions of your CWnd class (WindowProc(), OnCommand(), OnNotify(), etc.). See Chapter 3 for more on these message functions.

## CD Notes

- When executing the project on the accompanying CD, you will notice that the four menu commands in the Test menu are all processed by the same OnWzdCommandRange() function in WzdView.cpp.

# Example 61   Rerouting Command Messages

## Objective

You would like to direct main menu commands to a dialog bar, a modeless dialog box, or another class that normally doesn't receive command messages.

## Strategy

We will use the ClassWizard to override the `OnCmdMsg()` function in the `CMainFrame` Class, where we will call the `OnCmdMsg()` of the class to which we want to direct command messages.

## Steps

### Reroute Command Messages

1. Use the ClassWizard to override the `OnCmdMsg()` function in `CMainFrame`.
2. Fill this override with the following, where `m_WzdDialogBar` is the class object of the target class.

```
BOOL CMainFrame::OnCmdMsg(UINT nID, int nCode, void* pExtra,
    AFX_CMDHANDLERINFO* pHandlerInfo)
{
    if (m_WzdDialogBar.OnCmdMsg(nID,nCode,pExtra,pHandlerInfo))
        return TRUE;

    return CMDIFrameWnd::OnCmdMsg(nID, nCode, pExtra,
        pHandlerInfo);
}
```

## Notes

- Main menu and toolbar commands are automatically routed to only the Frame, Document, View, and Application classes. You must use this method to reroute commands to other classes.
- Only classes derived from `CCmdTarget` are eligible to receive command messages. This class does the actual message map processing required to handle a command message. The `OnCmdMsg()` function you call in the new target class is actually a member function of `CCmdTarget`.
- For more on the automatic routing of command messages, as well as messaging in general, see Chapter 3. For an example of using the Class-Wizard to add a class override, see Example 59.

Example 62   Creating Your Own Windows Message   **369**

## CD Notes

- When executing the project on the accompanying CD, you will notice that the Wzd menu command in the Test menu is now being processed by `OnWzdOptions()` in the `CWzdDialogBar` class.

# Example 62   Creating Your Own Windows Message

## Objective

You would like to create your own window message to use within your application.

## Strategy

We will define a new windows message, actually just a numeric ID. We will then look at how to send and receive this message between our application's classes.

## Steps

### Create a New Windows Message

1. To define the new message, use the following. For more messages, just use higher increments of `WM_USER`.

```
#define WM_WZD_MESSAGE WM_USER+1
```

### Send a New Windows Message

1. To send our message, use `SendMessage()`.

```
AfxGetMainWnd()->SendMessage(WM_WZD_MESSAGE, wParam, lParam);
```

Notice that you can also define what to store in `wParam` and `lParam`, which are both `DWORD` variables.

There are two ways to receive this new windows message. The first approach is to add the generic `ON_MESSAGE` macro to your receiving class's message map, along with the handler itself. The second approach would be to create your very own message map macro for this message. The advantage to the first approach is that it's quick and simple. The advantage of the second is that wParam and 1Param can be customized before they reach your handler. This can be an important factor if you will be using this message thoughout your application. Neither of these approaches allows you the luxury of using the ClassWizard, since your new message will obviously not be in its vocabulary.

### Receive a New Windows Message with ON_MESSAGE()

1. Use the Text Editor to add the definition of your message handler to your receiving class. Use the following syntax and make sure to include it below the ClassWizard's {{}} brackets to prevent it from deleting your work.

```
//{{AFX_MSG(CMainFrame)
//}}AFX_MSG
afx_msg LRESULT OnMyMessage(WPARAM wParam, LPARAM 1Param);<<<<
DECLARE_MESSAGE_MAP()
```

2. Then, we add our message map macro to the message map in the implementation file.

```
BEGIN_MESSAGE_MAP(CMainFrame, CMDIFrameWnd)
    //{{AFX_MSG_MAP(CMainFrame)
    //}}AFX_MSG_MAP
    ON_MESSAGE(WM_WZD_MESSAGE,OnMyMessage)<<<<<
END_MESSAGE_MAP()
```

3. And finally, we add our message handler.

```
LRESULT CMainFrame::OnWzdMessage(WPARAM wParam, LPARAM 1Param)
{
    //process
    return 0;
}
```

Just as you can determine what to pass in wParam and 1Param, what you return from this handler is also your call.

Example 62   *Creating Your Own Windows Message*   **371**

## Create a New Message Map Macro

1.  Define a new message map macro for our message as follows, ideally in an include file that will be shared by all of your application's files.

```
#define ON_WZD_MESSAGE() \
{WM_WZD_MESSAGE, 0, 0, 0, AfxSig_vW, (AFX_PMSG)(AFX_PMSGW)(void \
(AFX_MSG_CALL CWnd::*)(CWnd *))&OnWzdMessage},
```

The value `AfxSig_vW` tells MFC that our message comes with a pointer to a `CWnd` class and returns a `void`. It then converts the normal `wParam` and `lParam` that comes with a window message into these calling arguments. For other potential values for `AfxSig_vW`, see the `AFXMSG_.H` file in the MFC subdirectories. Changing the `AfxSig_vW` value would also means changing the calling arguments of the function that this macro defines.

2.  Use the Text Editor to add the definition of your new message handler to your receiving class. Use the following syntax and make sure to include it below the ClassWizard's {{}} brackets to prevent it from deleting your work.

```
//{{AFX_MSG(CMainFrame)
//}}AFX_MSG
afx_msg void OnMyMessage(CWnd *pWnd);<<<<
DECLARE_MESSAGE_MAP()
```

3.  Then, we add our message map macro to the message map in the implementation file.

```
BEGIN_MESSAGE_MAP(CMainFrame, CMDIFrameWnd)
    //{{AFX_MSG_MAP(CMainFrame)
    //}}AFX_MSG_MAP
    ON_WZD_MESSAGE(WM_WZD_MESSAGE,OnMyMessage)<<<<<
END_MESSAGE_MAP()
```

4.  And finally, we add our message handler.

```
void CMainFrame::OnWzdMessage(CWnd *pWnd)
{
//process
}
```

III   **12**

Note that rather than being passed generic `wParam` and `lParam` arguments, our message handler is now passed a more user-friendly `CWnd` pointer.

## Notes

- Windows messages are intended more for internal messaging between your windows, than for processing command messages, such as those coming from the menu. To add a new command message, you should instead use `WM_COMMAND` and a new command ID.

- Notice that when manually adding anything to the message map, you should avoid putting anything between the `{{}}` brackets. Otherwise, the ClassWizard can, and will, blow your stuff away!

```
//{{AFX_MSG_MAP(CMainFrame)
                              <---- NOT HERE!!!

//}}AFX_MSG_MAP
ON_WZD_MESSAGE()             <---- HERE!!!!
```

- For more information on messages in general, see Chapter 3. Also, see Example 8 for a practical example.

## CD Notes

- When executing the project on the accompanying CD, notice that the view class sends a custom windows message (`WM_WZD_MESSAGE`) to the main frame class when you click on the Options/Wzd menu commands.

# Files, Serialization, and Databases

The examples in this chapter relate to all things persistent, from simple binary files to accessing third-party database management systems. Unique to MFC is the concept of file serialization. Serialization allows you to organize how you save your class objects to disk so that they can be retrieved and even upgraded with limited effort on your part. Serialization saves and loads from an archive file, which is actually a binary file.

**Example 63**     **Accessing Binary Files**    We will review accessing and manipulating a simple binary file using MFC classes.

**Example 64**     **Accessing Standard I/O Files**    We will review accessing text files using MFC classes.

**Example 65**     **Accessing Memory File Files**    We will review the creation and manipulation of binary files in memory. Essentially, memory files give you the ability to access chucks of heap memory using a file system approach.

### Example 66    Implementing Serialization in Your Data Classes

We will add the ability for our data classes to be serialized. We don't actually serialize anything in this example — that's left for the next five examples.

### Example 67    Serializing Your SDI or MDI Document

We will activate MFC's built-in functionality, which will automatically serialize the data classes we prepared in the last example as one cohesive document.

### Example 68    Serialization on Command

We will serialize one or more data classes ourselves for reasons other than saving and loading a document.

### Example 69    Transparently Upgrading Serialized Documents

We will see how we can still use old documents created with serialization with new data classes by slightly modifying those data classes to convert the old data.

### Example 70    Serializing Polymorphous Classes

We will serialize a list of data classes whose only similarity is that they're all derived from the same base class.

### Example 71    Serializing Data Collections

We will review the serialization functionality built right into the MFC Collection Classes.

### Example 72    Accessing an ODBC Database

We will review how to access an ODBC-compliant database using MFC classes.

### Example 73    Accessing a DAO Database

We will review how to access a DAO-compliant database using MFC classes.

# Example 63    Accessing Binary Files

## Objective

You would like to maintain binary data in a flat disk file.

Example 63   Accessing Binary Files   **375**

# Strategy

We will use the MFC `CFile` class, which wraps the Windows API for handling binary files.

# Steps

## Check for the Existence of a File

1. To determine if a file already exists on disk, you can use the static `CFile` function, `GetStatus()`.

```
CFileStatus status;
if (!CFile::GetStatus(sFile,status))
{
    msg.Format("%s does not exist",sFile);
    AfxMessageBox(msg);
}
```

## Create a Binary File

1. To create a binary file for writing, you first create a `CFile` class object and then use its `Open()` member function to open and create a file object.

```
CFile file;
CString msg;
CString sFile("Wzd.tmp");
if (!file.Open(sFile, CFile::modeCreate|CFile::modeWrite))
{
    msg.Format("Failed to create %s.",sFile);
    AfxMessageBox(msg);
}
```

III

13

## Work with a Binary File

1. To write binary data to this file, you can use

```
file.Write(buffer,sizeof(buffer));
```

2. To close this file and destroy the file object, you can use

```
file.Close();
```

3. To open a binary file for reading, you can use

```
if (!file.Open(sFile, CFile::modeRead))
{
    msg.Format("Failed to open %s.",sFile);
    AfxMessageBox(msg);
}
```

4. To get the length of a binary or any type of file, you can use

```
UINT nBytes = file.GetLength();
```

5. To change the place where file reading or writing occurs, you can use one of these functions.

```
file.SeekToEnd();
file.Seek(20,           // file offset in bytes
    CFile::begin);      // also CFile::end and CFile::current
file.SeekToBegin();
```

6. To read from a binary file, you would use the following, where `nBytes` is the number of bytes actually read. If `nBytes` is less than the number of bytes you wanted to read or is zero, you have reached the end of the file.

```
nBytes=file.Read(buffer,sizeof(buffer));
```

7. To make a file read-only, you would use the static `CFile` function, `SetStatus()`. In this example, we grab whatever status a file currently has and add the read-only status bit to it.

```
if (CFile::GetStatus(sFile, status))
{
    status.m_attribute|=0x01;
    CFile::SetStatus(sFile, status);
}
```

8. To delete a file, you would use another static `CFile` function called `Remove()`.

```
CFile::Remove(sFile);
```

Example 63   Accessing Binary Files   **377**

## Notes

- CFile follows in the tradition of most MFC classes, in that it wraps a bunch of Windows API calls in a C++ class. The CWnd class does the same for windows and CMenu does the same for menus.
- For manipulating text files, see the next example.

## CD Notes

- When executing the project on the accompanying CD, set a breakpoint on the OnTestWzd() function in WzdView.cpp. Then click on Test/Wzd in the menu and step through as a binary file is accessed.

# Example 64   Accessing Standard I/O Files

## Objective

You would like to manipulate a text file.

## Strategy

We will use the MFC CStdioFile class to manipulate text files. CStdioFile is derived from the binary CFile class and wraps the Windows API for handling text files.

## Steps

### Check for the Existence of a File

1. To see if a text file exists, you would use the GetStatus() static function of CFile.

```
CFileStatus status;
CString sFile("Wzd.txt");
if (!CFile::GetStatus(sFile,status))
{
    msg.Format("%s does not exist",sFile);
    AfxMessageBox(msg);
}
```

III

13

## Create a Text File

1. To open a text file for writing text strings, you would use the following. In this example, if the named file can't be opened, a new file is created.

```
CStdioFile file;
if(!file.Open(sFile, CFile::modeWrite | CFile::typeText))
{
    if(!file.Open(sFile, CFile::modeCreate |
        CFile::modeWrite | CFile::typeText))
    {
        CString msg;
        msg.Format("Failed to create %s.",sFile);
    }
}
```

## Work with Text Files

1. To write a text string to this file, you can use

```
file.WriteString(sRecord);
```

2. To close this file and destroy the file object created, you can use

```
file.Close();
```

3. To open a text file for reading, you would use

```
if (!file.Open(sFile, CFile::modeRead | CFile::typeText))
{
    msg.Format("Failed to open %s.",sFile);
    AfxMessageBox(msg);
}
```

4. To move to the front or end of a text file, you can use

```
file.SeekToEnd();
file.SeekToBegin();
```

5. To read a string of text from a text file, you can use

```
file.ReadString(sRecord);
```

Example 65   Accessing Memory File Files   **379**

6. To delete a text file, you can use

```
CFile::Remove(sFile);
```

## Notes

- Unlike a binary file that treats every byte the same, a standard I/O file uses the `fgets()` and `fputs()` functions to get and put text strings to the file.

- Since `CStdioFile` is derived from `CFile`, every function found in `CFile` can also be used here.

## CD Notes

- When executing the project on the accompanying CD, set a breakpoint on the `OnTestWzd()` function in `WzdView.cpp`. Then click on Test/Wzd in the menu and step through as a standard I/O file is accessed.

# Example 65   Accessing Memory File Files

## Objective

Rather than create a temporary file on disk, you would like to instead create it in memory.

## Strategy

We will use the MFC `CMemFile` class, which is derived from the binary `CFile` class and wraps the Windows API for allocating memory.

**III**

**13**

## Steps

### Create a Memory File

1. To create a memory file, just create a `CMemFile` class object.

```
CMemFile file;
```

Creating a `CMemFile` class object automatically opens a memory file.

## Work with a Memory File

1. To write to a memory file, you can use

```
file.Write(buffer,sizeof(buffer));
```

2. To determine the current length of a memory file, you can use

```
UINT nBytes = file.GetLength();
```

3. To position your memory file for the next read or write operation, you can use one of the following.

```
file.SeekToEnd();
file.Seek(20,            // file offset in bytes
    CFile::begin);       // also CFile::end and CFile::current
file.SeekToBegin();
```

4. To read data from your memory file, you can use

```
nBytes=file.Read(buffer,sizeof(buffer));
```

5. To delete a memory file, you can either just destroy the CMemFile class object or, if you can't destroy the class object, you can just set the file length to zero.

```
file.SetLength(0);
```

# Notes

- Memory files provide a very convenient way to manipulate data in memory without having to worry about allocating and deallocating blocks of memory in heap space. They are also recommended over temporary (.tmp) files since they make your application faster and also solve the eternal problem of naming and deleting temporary files.

- Memory files created with CMemFile are created in your application's heap space. If you would like your file to be created in the global heap, you should use CSharedFile instead. The clipboard in particular requires that the data it gets be in the global heap. For an example of this, see Example 76 on cutting and pasting.

- Since CMemFile is derived from CFile and not CStdioFile, you will need to write your own ReadString() and WriteString() functions to manipulate text data easily in a memory file.

Example 66  Implementing Serialization in Your Data Classes  **381**

## CD Notes

- When executing the project on the accompanying CD, set a breakpoint on the `OnTestWzd()` function in `WzdView.cpp`. Then, click on Test/Wzd in the menu and step through as a memory file is accessed.

# Example 66   Implementing Serialization in Your Data Classes

## Objective

You would like to be able to serialize your data classes.

## Strategy

We will implement serialization in our data classes by deriving them from `CObject`. This MFC class adds a virtual override to your class called `Serialize()`. We will then override this function with our own `Serialize()` function that will individually load and save our member variables to disk.

---

NOTE: In this example, we will be giving our data classes the ability to serialize themselves. How to actually serialize your classes to disk is covered in the following examples.

---

## Steps

III

13

### Add `Serialize()` to Your Data Classes

1. Derive your data classes from MFC's `CObject` class. Classes not derived from `CObject` cannot be serialized. Since the ClassWizard does not offer `CObject` as a base class, you will need to add this base class yourself.

```
class CMyData : public CObject
{
};
```

Add a `Serialize()` member function to each of your data classes with the following format.

```
void CWzdInfo2::Serialize(CArchive& ar)
{
    CObject::Serialize(ar);
    int version=1;
    if(ar.IsStoring())
    {
        // store data
    }
    else
    {
        // load data
    }
}
```

We will now use this `Serialize()` function to load and save every member variable of our class to the archive device. How we serialize a member variable depends on the variable type. There are three variable types: simple variables that aren't derived from `CObject` (e.g., integers, floats), variables that are derived from `CObject`, and data collections, which include lists and arrays.

## Serialize Simple Variables

1. To store each of the member variables of your data class that is not derived from `CObject`, use the << overloaded operator in the "store data" section of `Serialize()`.

```
CObject::Serialize(ar);
    if(ar.IsStoring())
    {
    // store member variables not derived from CObject
        ar << m_sName;
        ar << m_sComment;
```

Example 66 Implementing Serialization in Your Data Classes **383**

2. To load simple member variables back into your data classes, use the >> overloaded operator in the "load data" section of `Serialize()`.

```
CObject::Serialize(ar);
if(ar.IsStoring())
{
// store data
else
{
// load member variables not derived from CObject
    ar >> m_sName;
    ar >> m_sComment;
}
```

## Serialize Variables Derived from `CObject`

1. To store member variables that *are* derived from `CObject`, call that member variable's `Serialize()` function.

```
// load and store member variables derived from CObject
m_wzdInfo.Serialize(ar);
```

You can call this `Serialize()` function within the load and save sections of your class's `Serialize()` function, or you can call it once outside of the `if{}else{}` statement. If `Serialize()` isn't yet implemented in this variable's class, it's time to add one.

## Serialize Data Collections

III

13

1. To store a data collection, use the following syntax.

```
nCount = m_WzdInfoList.GetCount();      // get number of
                                        // items in
ar << nCount;                           // and store
for (POSITION pos = m_WzdInfoList.GetHeadPosition(); pos;)
{
    CWzdInfo2 *pInfo = m_WzdInfoList.GetNext(pos);
    pInfo->Serialize(ar);
}
}
```

Essentially, all we are doing here is first saving the count of the number of items in this collection and then saving each item one-by-one in the order in which they were stored.

2. To load a data collection, use the following syntax.

```
// load a list of data
ar >> nCount;
while (nCount-- > 0)
{
    CWzdInfo2* pInfo = new CWzdInfo2;
    pInfo->Serialize(ar);
    m_WzdInfoList.AddTail(pInfo);
}
}
```

You can find the complete listing for an example data class with serialization in "Listings — First Data Class" on page 385 or "Listings — Second Data Class" on page 387.

## Notes

- Serialize takes each member variable in your data class and stores it sequentially on disk. Loading and storing a document is simply a matter of serializing potentially thousands of data classes. As long as a document is loaded in the same sequence it was stored, there is no other file control necessary. As we will see in a future example, this method lends itself to quickly converting older document formats to the latest format. For more on serialization, and the CObject class, see Chapter 2.

- The archive device you serialize can be a disk file, a memory file, or even a file on another system. As we will see in Example 76, serialization to a memory file provides a quick and easy way to cut and paste user proprietary data classes.

- Some MFC classes can also use the << and >> overloaded operators to serialize their contents, but only because they have individually implemented this operator. As an example, CString elements can use this operator. You might want to experiment to see what other MFC classes support it.

- This ability to store a class object such that it can be restored later is called object *persistence*.

Example 66   Implementing Serialization in Your Data Classes   **385**

## CD Notes

- The project for this example is called "Serialization". Start by setting a breakpoint on the Serialize() function in WzdDoc.cpp. Then click on the Test/Wzd menu commands to fill the document. Then, click on File/Save to start the serialization function.

## Listings — First Data Class

```
#ifndef WZDINFO1_H
#define WZDINFO1_H


#include "afxtempl.h"
#include "WzdInfo2.h"


class CWzdInfo1 : public CObject
{
public:
    CWzdInfo1();
    ~CWzdInfo1();


    //misc info
    CString                     m_sGroupName;
    CString                     m_sComment;
    CList<CWzdInfo2*,CWzdInfo2*> m_WzdInfo2List;
    void Serialize(CArchive& archive);


};
#endif


// WzdInfo1.cpp : implementation of the CWzdInfo class
//


#include "stdafx.h"
#include "WzdInfo1.h"


/////////////////////////////////////////////////////////////////////////////
// CWzdInfo
```

III

13

```
CWzdInfo1::CWzdInfo1()
{
    m_sGroupName=_T("");
    m_sComment=_T("");
}


CWzdInfo1::~CWzdInfo1()
{
    while (!m_WzdInfo2List.IsEmpty())
    {
        delete m_WzdInfo2List.RemoveHead();
    }
}



void CWzdInfo1::Serialize(CArchive& ar)
{
    CObject::Serialize(ar);
    int nCount;
    if (ar.IsStoring())
    {
        // name and comment
        ar << m_sGroupName;
        ar << m_sComment;

        // other list
        nCount = m_WzdInfo2List.GetCount();
        ar << nCount;
        for (POSITION pos = m_WzdInfo2List.GetHeadPosition(); pos;)
        {
            CWzdInfo2 *pInfo = m_WzdInfo2List.GetNext(pos);
            ar << pInfo;
        }
    }
```

Example 66   Implementing Serialization in Your Data Classes   **387**

```
    else
    {
        // name and comment
        ar >> m_sGroupName;
        ar >> m_sComment;

        // other list
        ar >> nCount;
        CObject* pInfo;
        while (nCount-- > 0)
        {
            ar >> pInfo;
            m_WzdInfo2List.AddTail((CWzdInfo2*)pInfo);
        }
    }
}
```

## Listings — Second Data Class

```
#ifndef WZDINFO2_H
#define WZDINFO2_H


class CWzdInfo2 : public CObject
{
public:

enum STATES {
    OLD,
    NEW,
    MODIFIED,
    DELETED
    };

    DECLARE_SERIAL(CWzdInfo2)

    CWzdInfo2();
    CWzdInfo2(CString sName,int nVersion);
```

III

13

```
    void Set(CString sName,CString sComment,int nVersion, int nState);


    // misc info
    CString    m_sName;
    CString    m_sComment;
    int        m_nState;


    void Serialize(CArchive& archive);


    CWzdInfo2& operator=(CWzdInfo2& src);

};
#endif

// WzdInfo2.cpp : implementation of the CWzdInfo class
//

#include "stdafx.h"
#include "WzdInfo2.h"

//////////////////////////////////////////////////////////////////////////////
// CWzdInfo

IMPLEMENT_SERIAL(CWzdInfo2, CObject, 0)


CWzdInfo2::CWzdInfo2()
{
    m_sName=_T("");
    m_sComment=_T("");
    m_nState=CWzdInfo2::NEW;
}


CWzdInfo2::CWzdInfo2(CString sName,int nVersion) :
    m_sName(sName),m_nVersion(nVersion)
{
    m_sComment=_T("");
    m_nState=CWzdInfo2::OLD;
}
```

Example 66  Implementing Serialization in Your Data Classes  **389**

```
void CWzdInfo2::Set(CString sName,CString sComment,int nVersion, int nState)
{
    m_sName=sName;
    m_sComment=sComment;
    m_nState=nState;
}


void CWzdInfo2::Serialize(CArchive& ar)
{
    CObject::Serialize(ar);
    if(ar.IsStoring())
    {
        // data
        ar << m_sName;
        ar << m_sComment;
        ar << m_nState;

    }
    else
    {
        // data
        ar >> m_sName;
        ar >> m_sComment;
        ar >> m_nState;

    }
}


CWzdInfo2& CWzdInfo2::operator=(CWzdInfo2& src)
{
    if(this != &src)
    {
        m_sName = src.m_sName;
        m_sComment = src.m_sComment;
        m_nState = src.m_nState;
    }
    return *this;
}
```

III

13

# Example 67 Serializing Your SDI or MDI Document

## Objective

You would like to serialize your SDI or MDI document to disk using the built-in features of the MFC framework.

## Strategy

We will implement two functions in our Document Class to serialize documents to and from a disk file. The first is our Document Class's `Serialize()` function, which we will program to systematically call the `Serialize()` member function of every data class instance in our document. (We added this member function in the previous example.) The second function is `DeleteContents()`, which we will use to initialize our document. We will also edit a string resource to customize the File Open dialog box for our document's extension.

## Steps

### Use the Document Class's `Serialize()`

1. Locate the `Serialize()` member function of your Document Class and fill it in the same way you filled in the `Serialize()` functions in your data classes in the previous example. As before, you will simply serialize each member variable of your Document Class to the supplied archive class reference.

### Use the Document Class's `DeleteContents()`

1. Use the ClassWizard to add a `DeleteContents()` override to your Document Class. This function is automatically called by your application

Example 67   Serializing Your SDI or MDI Document   **391**

whenever the user selects the File/New or File/Open menu command and gives you an opportunity to reinitialize the document.

```
void CWzdDoc::DeleteContents()
{
    // called by new and open document
    // opportunity to initialize the data collections
    // that make up your document
    while (!m_WzdInfoList.IsEmpty())
    {
        delete m_WzdInfoList.RemoveHead();
    }
    CDocument::DeleteContents();
}
```

No other overrides or message handlers need to be added to your Document Class. When the user clicks on File/New or File/Open, the framework will automatically open an Open File dialog box that will prompt your user to enter a filename. Once selected, your application will open a file and an archive and then call your Document Class's DeleteContents(). Once the document has been initialized, your application will call the Serialize() function. The same is true when your user uses File/Save or File/Save As.

If your user clicks on the File/New menu command, only the DeleteContents() function is called.

The File Dialog that automatically opens uses a file extension you provide to filter out all nondocument files from its file list. You can provide this file extension when creating your application using the AppWizard (Example 2). However, if you weren't so foresightful, you can directly edit a string in your application's string table to provide this extension.

## Specify a Document's File Extension

1. To customize the File Dialog box that your application opens to prompt the user for a document name, you can edit a string in your resource's string table. Locate the string that defines your document type. In this example, that ID is IDC_WZDTYPE. The changes you need to make to this string are shown in boldface and underlined in the following example.

```
\nWzd\nWzd\nWzd Files (*.wzd)\n.wzd\nWzd.Document\nWzd Document
```

## Notes

- Strictly speaking, even the `DeleteContents()` function is only required in an SDI application. Whenever an MDI application opens a new or existing document, a new Document Class is created. Therefore, any document initialization can take place in the Document Class's constructor. But for convention's sake, you should use `DeleteContents()`, instead.

## CD Notes

- The project for this example is called "Serialization". Start by setting a breakpoint on the `Serialize()` function in `WzdDoc.cpp`. Then, click on the Test/Wzd menu commands to fill the document. Then, click on File/Save to start the serialization function.

## Listings — Document Class

```cpp
// WzdDoc.h : interface of the CWzdDoc class
//
/////////////////////////////////////////////////////////////////////////////

#if !defined(AFX_WZDDOC_H__CA9038EE_B0DF_11D1_A18C_DCB3C85EBD34__INCLUDED_)
#define AFX_WZDDOC_H__CA9038EE_B0DF_11D1_A18C_DCB3C85EBD34__INCLUDED_

#if _MSC_VER >= 1000
#pragma once
#endif // _MSC_VER >= 1000

#include "afxtempl.h"
#include "WzdInfo1.h"
#include "WzdInfo2.h"

class CWzdDoc : public CDocument
{
protected:                              // create from serialization only
    CWzdDoc();
    DECLARE_DYNCREATE(CWzdDoc)

// Attributes
public:
    CList<CWzdInfo1*,CWzdInfo1*> *GetInfo1List(){return &m_WzdInfo1List;};
```

Example 67    Serializing Your SDI or MDI Document    **393**

```
// Operations
public:

// Overrides
    // ClassWizard generated virtual function overrides
    //{{AFX_VIRTUAL(CWzdDoc)
    public:
    virtual void Serialize(CArchive& ar);
    virtual void DeleteContents();
    virtual BOOL OnNewDocument();
    //}}AFX_VIRTUAL

// Implementation
public:
    virtual ~CWzdDoc();
#ifdef _DEBUG
    virtual void AssertValid() const;
    virtual void Dump(CDumpContext& dc) const;
#endif

protected:

// Generated message map functions
protected:
    //{{AFX_MSG(CWzdDoc)
        // NOTE - the ClassWizard will add and remove member functions here.
        // DO NOT EDIT what you see in these blocks of generated code!
    //}}AFX_MSG
    DECLARE_MESSAGE_MAP()
private:
    CList<CWzdInfo1*,CWzdInfo1*> m_WzdInfoList;
};


/////////////////////////////////////////////////////////////////////////////
//{{AFX_INSERT_LOCATION}}
// Microsoft Developer Studio will insert additional declarations immediately
// before the previous line.
```

III

13

```
#endif // !defined(
    AFX_WZDDOC_H__CA9038EE_B0DF_11D1_A18C_DCB3C85EBD34__INCLUDED_)


// WzdDoc.cpp : implementation of the CWzdDoc class
//

#include "stdafx.h"
#include "Wzd.h"

#include "WzdDoc.h"

#ifdef _DEBUG
#define new DEBUG_NEW
#undef THIS_FILE
static char THIS_FILE[] = __FILE__;
#endif

/////////////////////////////////////////////////////////////////////
// CWzdDoc

IMPLEMENT_DYNCREATE(CWzdDoc, CDocument)

BEGIN_MESSAGE_MAP(CWzdDoc, CDocument)
    //{{AFX_MSG_MAP(CWzdDoc)
        // NOTE - the ClassWizard will add and remove mapping macros here.
        // DO NOT EDIT what you see in these blocks of generated code!
    //}}AFX_MSG_MAP
END_MESSAGE_MAP()

/////////////////////////////////////////////////////////////////////
// CWzdDoc construction/destruction

CWzdDoc::CWzdDoc()
{
    // TODO: add one-time construction code here

}

CWzdDoc::~CWzdDoc()
{
}
```

Example 67   Serializing Your SDI or MDI Document   **395**

```cpp
BOOL CWzdDoc::OnNewDocument()
{
    if (!CDocument::OnNewDocument())
        return FALSE;

    return TRUE;
}


//////////////////////////////////////////////////////////////////////////////
// CWzdDoc serialization

void CWzdDoc::Serialize(CArchive& ar)
{
    int nCount;
    if (ar.IsStoring())
    {
        nCount = m_WzdInfolList.GetCount();
        ar << nCount;
        for (POSITION pos = m_WzdInfolList.GetHeadPosition(); pos;)
        {
            CWzdInfol *pInfo = m_WzdInfolList.GetNext(pos);
            pInfo->Serialize(ar);
        }
    }
    else
    {
        ar >> nCount;
        while (nCount-- > 0)
        {
            CWzdInfol* pInfo = new CWzdInfol;
            pInfo->Serialize(ar);
            m_WzdInfolList.AddTail(pInfo);
        }
    }
}
```

III

13

```
/////////////////////////////////////////////////////////////////////////////
// CWzdDoc diagnostics

#ifdef _DEBUG
void CWzdDoc::AssertValid() const
{
    CDocument::AssertValid();
}

void CWzdDoc::Dump(CDumpContext& dc) const
{
    CDocument::Dump(dc);
}
#endif //_DEBUG


/////////////////////////////////////////////////////////////////////////////
// CWzdDoc commands

void CWzdDoc::DeleteContents()
{
    // called with new and open document
    // opportunity to initialize the data collections that make up our document
    while (!m_WzdInfolList.IsEmpty())
    {
        delete m_WzdInfolList.RemoveHead();
    }


    CDocument::DeleteContents();
}
```

# Example 68   Serialization on Command

## Objective

You would like to manually serialize your data classes at any time.

Example 68   Serialization on Command   **397**

## Strategy

Assuming you have added a `Serialize()` function to each of your data classes, as seen in Example 66, we will simply open a file and an archive and serialize these data classes to that archive.

## Steps

### Save Serialized Data Classes

1. To save any data classes that have a `Serialize()` function to disk, start by opening a file for creation. Then, open an instance of the `CArchive` class.

```
CFile file;
if (file.Open("filename.ext", CFile::modeCreate|
    CFile::modeWrite))
{
    CArchive ar(&file, CArchive::store);
```

   Notice that we use the `CArchive::store` flag when creating our archive instance.

2. You can now loop through the appropriate data classes, calling their `Serialize()` function using this archive. In this example, we serialize a `CList` variable.

```
CList<CWzdInfo1*,CWzdInfo1*> *pList =
    GetDocument()->GetInfo1List();
nCount = pList->GetCount();
ar << nCount;
for (POSITION pos = pList->GetHeadPosition(); pos;)
{
    CWzdInfo1 *pInfo = pList->GetNext(pos);
    pInfo->Serialize(ar);
}
```

III

13

3. When finished, first close the archive, then the file.

```
    ar.Close();
    file.Close();
}
```

## Load Serialized Data Classes

1. To reload these data classes, you again open a file, open an archive and load the data.

```
CFile file;
if (file.Open("filename.ext", CFile::modeRead))
{
    CArchive ar(&file, CArchive::load);
    CList<CWzdInfo1*,CWzdInfo1*> *pList =
        GetDocument()->GetInfo1List();
    ar >> nCount;
    while (nCount-- > 0)
    {
        CWzdInfo1* pInfo = new CWzdInfo1;
        pInfo->Serialize(ar);
        pList->AddTail(pInfo);
    }
    ar.Close();
    file.Close();
}
```

Notice that we use the CArchive::load flag when creating our archive instance.

## Notes

- This approach is particularly useful if you want to deviate from the framework's automatic document loading and saving. For example, if your application is saving both a proprietary database and a log. We will use this approach later on to implement cut and paste. Instead of a CFile class object however, we will be serializing to a shared memory file.

Example 69   Transparently Upgrading Serialized Documents   **399**

## CD Notes

- The project for this example is called "Serialization". Start by setting a breakpoint on the Serialize() function in WzdDoc.cpp. Then, click on the Test/Wzd menu commands to fill the document. Then, click on File/Save to start the serialization function.

# Example 69   Transparently Upgrading Serialized Documents

## Objective

You would like to be able to allow your user to access and convert older versions of your application's documents without creating a conversion utility.

## Strategy

We will add version control to the Serialize() function in each of our data classes that will invisibly convert older documents to the current format. This example appears separately from Example 66 only for the sake of clarity. In reality, you should never add a Serialize() function to your data class without also adding version control.

## Steps

## Add Data Class Version Numbers

III

13

1. Prepare your data classes as seen in Example 66. However, also serialize one additional variable that will keep track of the version of this data class and initialize this variable to one (1).

```
void CWzdInfo1::Serialize(CArchive& ar)
{
    int version=1;
    CObject::Serialize(ar);
    int nCount;
    if (ar.IsStoring())
```

```
    {
        // version
        ar << version;

        :    :    :

    }
    else
    {
        // version
        ar >> version;

        :    :    :

    }
}
```

## Upgrade Older Data Classes

1. Weeks have passed and you have decided to add a new variable to one of your data classes. After you have added this variable, you must now modify the `Serialize()` function of that data class. Start by incrementing the version number.

```
void CWzdInfo::Serialize(CArchive& ar)
{
    int version=2; //<<<<<<<< incremented from 1
    CObject::Serialize(ar);
    if(ar.IsStoring())
    {
        :    :    :
```

2. In the storing part of your `Serialize()` function, simply add your new variable at the end. Don't mix it up with the other variables.

```
// version
ar << version;

// data
```

Example 69   Transparently Upgrading Serialized Documents   **401**

```
   :     :     :

// new with version 2
ar << m_nModNum;
```

3. In the loading part of your `Serialize()` function, you will only conditionally load this variable if the version of the document matches the current version of this class. Again, you will only do this test at the end without mixing it up with the other variables. This example depends on each data item being loaded and stored in the exact same order each time.

```
    // version
    ar >> version;

    // data
    :     :     :

    // new with version 2
    if (version>=2)
    {
        ar >> m_nModNum;
    }
}
```

4. And that's it. If the document is an older version, this variable won't get its value from the document and will instead wind up with whatever you initialized it with in the constructor. If the document *is* the current version, the variable gets its value from the archive. To see a complete listing of a data class using version control, see "Listings — Data Class" on page 403.

III

13

## Notes

- Notice that data class is responsible for its own version number. Your serialized document can now be made of several versions rather than just one.

- The method presented in this example will only allow you to add new member variables to the end of your data class's serialization. You can neither delete a variable nor mix it with existing variables. To be able to delete old variables, use the following case statement approach, instead.

```
switch (version)
case 1:
    // old way
case 2:
    // new way, with the sky being the limit
case 3:
```

With this approach, there can be little in common between formats. However, this approach may be harder for another programmer to follow and understand.

## CD Notes

- The project for this example is called "Serialization". Start by setting a breakpoint on the `Serialize()` function in `WzdDoc.cpp`. Then, click on the Test/Wzd menu commands to fill the document. Then click on File/Save to start the serialization function.

Example 69   Transparently Upgrading Serialized Documents   **403**

## Listings — Data Class

```cpp
#ifndef WZDINFO2_H
#define WZDINFO2_H


class CWzdInfo2 : public CObject
{
public:

enum STATES {
    OLD,
    NEW,
    MODIFIED,
    DELETED
    };

    CWzdInfo2();
    CWzdInfo2(CString sName,int nVersion);

    void Set(CString sName,CString sComment,int nVersion, int nState);

    // misc info
    CString m_sName;
    CString m_sComment;
    int m_nVersion;
    int m_nState;

    // new with version 2
    int m_nModNum;

    void Serialize(CArchive& archive);

    CWzdInfo2& operator=(CWzdInfo2& src);

};
#endif
```

```cpp
// WzdInfo2.cpp : implementation of the CWzdInfo class
//

#include "stdafx.h"
#include "WzdInfo2.h"


/////////////////////////////////////////////////////////////////////////////
// CWzdInfo

CWzdInfo2::CWzdInfo2()
{
    m_sName=_T("");
    m_sComment=_T("");
    m_nVersion=1;
    m_nState=CWzdInfo2::NEW;

    // new with version 2
    m_nModNum = 0;
}

CWzdInfo2::CWzdInfo2(CString sName,int nVersion) :
    m_sName(sName),m_nVersion(nVersion)
{
    m_sComment=_T("");
    m_nState=CWzdInfo2::OLD;

    // new with version 2
    m_nModNum = 0;
}

void CWzdInfo2::Set(CString sName,CString sComment,int nVersion, int nState)
{
    m_sName=sName;
    m_sComment=sComment;
    m_nVersion=nVersion;
    m_nState=nState;
    m_nModNum = 0;
}
```

Example 69    Transparently Upgrading Serialized Documents    **405**

```
void CWzdInfo2::Serialize(CArchive& ar)
{
    int version=2;
    CObject::Serialize(ar);
    if(ar.IsStoring())
    {
        // version
        ar << version;

        // data
        ar << m_sName;
        ar << m_sComment;
        ar << m_nVersion;
        ar << m_nState;

        // new with version 2
        ar << m_nModNum;

    }
    else
    {
        // version
        ar >> version;

        // data
        ar >> m_sName;
        ar >> m_sComment;
        ar >> m_nVersion;
        ar >> m_nState;

        // new with version 2
        if (version>=2)
        {
            ar >> m_nModNum;
        }
    }
}
```

III

13

```
CWzdInfo2& CWzdInfo2::operator=(CWzdInfo2& src)
{
    if(this != &src)
    {
        m_sName = src.m_sName;
        m_sComment = src.m_sComment;
        m_nVersion = src.m_nVersion;
        m_nState = src.m_nState;
    }
    return *this;
}
```

# Example 70   Serializing Polymorphous Classes

## Objective

You would like to serialize a list of data class instances, but the class types are unknown.

## Strategy

In a previous example, we serialized a list of classes, but they were all of the same type. Because they were all of the same type, we could recreate each data instance with a simple new operator.

```
while (nCount-- > 0)
{
    CWzdInfo2 pInfo=new CWzdInfo2; <<all instances of CWzdInfo2
    pInfo->Serialize(ar);
    m_WzdInfo2List.AddTail((CWzdInfo2*)pInfo);
}
```

But if our data includes polymorphous instances of some base class, we wouldn't know what class to construct.

**Example 70 Serializing Polymorphous Classes** **407**

To serialize polymorphous classes, we will again derive each class from CObject as before. Then, we will add two macros to our class called DECLARE_SERIAL() and IMPLEMENT_SERIAL(). These macros add a new overloaded operator to our data classes that will allow the class to be serialized without knowing their derived class.

---

**NOTE:** This approach only works if your data class does not have a member variable that is itself derived from CObject. For example, if your data class uses CList to maintain a list, you cannot use this method.

---

# Steps

## Serialize a Polymorphous Class

1. Derive the data class from CObject and add a Serialize() function (Example 66).

2. Include the DECLARE_SERIAL() macro in each data class's .h file.

```
class CWzdInfo : public CObject
{
public:


    DECLARE_SERIAL(CWzdInfo)
        :    :    :
```

3. Include the IMPLEMENT_SERIAL() macro in this data class's .cpp file.

```
///////////////////////////////////////////////////////////
// CWzdInfo
IMPLEMENT_SERIAL(CWzdInfo, CObject, 0)
CWzdInfo::CWzdInfo()
{
}
```

III

13

4. Make sure this data class also has a constructor that takes no arguments. An instance of this class will be created automatically by MFC with `CreateObject()`, which looks for a constructor with no arguments.

5. When storing this data class to disk, you should now use the `<<` overloaded operator instead of using the class's `Serialize()` function.

```
for (POSITION pos = m_WzdInfoList.GetHeadPosition(); pos;)
{
    CWzdInfo2 *pInfo = m_WzdInfoList.GetNext(pos);
    ar << pInfo;
}
```

The `<<` operator not only calls your data class's `Serialize()` function, but it also serializes the class's name so that it can be used to recreate the correct class in the next step. This operator was added with the `DECLARE_SERIAL` macro.

6. When loading this data class from disk, you should now use the `>>` overloaded operator.

```
CObject* pInfo;
while (nCount-- > 0)
{
    ar >> pInfo;
    m_WzdInfoList.AddTail(pInfo);
}
```

The `>>` operator creates the object for you using the information stored with the `<<` operator. Therefore, you don't need to know what class derivation to create yourself.

## Notes

- `IMPLEMENT_SERIAL` has an argument called a schema number that you can fill with the current version of a data class. You can then query this schema number using

```
UINT version = ar.GetObjectSchema();
```

However, this schema number is only available if you serialize the class using the `>>` or `<<` overloaded operators. A much more universal way of saving a data class's version number was presented in Example

Example 71   Serializing Data Collections   **409**

69. If you use that method, just set this schema number to zero (0) and never touch it again. However, if you do intend to use it, make sure you OR the version number with VERSIONABLE_SCHEMA in the macro, as seen in the following example. Otherwise, changing the version number will cause an archive exception error.

```
IMPLEMENT_SERIAL(CWzdInfo, CObject, VERSIONABLE_SCHEMA|0)
```

- The CObject class adds the Serialize() function that every data class must override. The DECLARE_SERIAL and IMPLEMENT_SERIAL macros on the other hand just add the >> and << overloaded member functions, which you only need in this example.

## CD Notes

- The project for this example is called "Serialization". Start by setting a breakpoint on the Serialize() function in WzdDoc.cpp. Then, click on the Test/Wzd menu commands to fill the document. Then, click on File/Save to start the serialization function.

# Example 71   Serializing Data Collections

## Objective

You would like to take advantage of the built-in serialization capabilities of MFC's collection classes (i.e., CList, CArray, etc.) to make serializing data collections easier.

## Strategy

In previous examples, when serializing a data collection, we looped through the collection and individually serialized each element. However, each MFC collection class has its very own Serialize() function that will automatically serialize its elements. In this example, we will use that Serialize() function.

III

13

# Steps

## Serialize a Data Collection

1. To serialize a collection class, just call its `Serialize()` function.

```
void CWzdInfo1::Serialize(CArchive& ar)
{
    CObject::Serialize(ar);
    if (ar.IsStoring())
    {
        m_WzdInfo2List.Serialize(ar);
    }
    else
    {
        m_WzdInfo2List.Serialize(ar);
    }
}
```

Without any further modifications, your data class will serialize itself to disk. However, each data class object in this collection will simply be copied as a block of memory — the `Serialize()` function you carefully crafted for the classes in this list will not be called. If you would like these `Serialize()` functions to be called, you will need to override a member function of your data collection class, `SerializeElement()`.

## Override `SerializeElement()`

1. To override the `SerializeElement()` of your data collection, start by adding the following declaration to your data class's `.h` file.

```
void AFXAPI SerializeElements(CArchive& ar,
    CWzdInfo2** ppElements, int nCount);
```

Example 71   Serializing Data Collections   **411**

2. Your `SerializeElement()` function should then be implemented with the following syntax.

```
void AFXAPI SerializeElements(CArchive& ar,
    CWzdInfo2** ppElements, int nCount)
{

    CWzdInfo2 *pWzdInfo2;
    for (int i = 0; i < nCount; i++)
    {

        if (ar.IsStoring())
        {

            pWzdInfo2 = *(ppElements + i);
        }
        else
        {

            pWzdInfo2 = new CWzdInfo2;
            *(ppElements + i) = pWzdInfo2;
        }
        pWzdInfo2->Serialize(ar);
    }
}
```

For your application, simply substitute `CWzdInfo2` for the class name of the data class stored in your data collection.

3. If you are also working with a list of polymorphised data classes, as discussed in the previous example, your version of `SerializeElement()` should look like the following example.

```
void AFXAPI SerializeElements(CArchive& ar,
    CWzdInfo2** ppElements, int nCount)
{

    CWzdInfo2 *pWzdInfo2;
    for (int i = 0; i < nCount; i++)
    {

        if (ar.IsStoring())
        {

            pWzdInfo2 = *(ppElements + i);
            ar << pWzdInfo2;
        }
```

III

13

```
        else
        {
            ar >> pWzdInfo2;
            *(ppElements + i) = pWzdInfo2;
        }
    }
}
```

The only difference between this function and the function in the last step is the use of the << and >> overloaded operators.

For a complete listing of a data class that implements SerializeElement(), see "Listings — Data Class" on page 413.

## Notes

- SerializeElement() is a template override. If the calling arguments don't exactly match what your collection class is looking for, you won't get a syntax error from the compiler — however, the collection class call will not call your function, either.
- The example presented here is for a template collection class. The only difference there should be for a regular collection class (e.g., CObList) is the syntax for SerializeElement(), which I will leave to you to determine.

## CD Notes

- The project for this example is called "Serialization". Start by setting a breakpoint on the Serialize() function in WzdDoc.cpp. Then, click on the Test/Wzd menu commands to fill the document. Then, click on File/Save to start the serialization function.

Example 71   Serializing Data Collections   **413**

# Listings — Data Class

```
#ifndef WZDINFO1_H
#define WZDINFO1_H

#include "afxtempl.h"
#include "WzdInfo2.h"

void AFXAPI SerializeElements(CArchive& ar, CWzdInfo2** pWzdInfo2, int nCount);

class CWzdInfo1 : public CObject
{
public:
    CWzdInfo1();
    ~CWzdInfo1();

    // misc info
    CString                         m_sGroupName;
    CString                         m_sComment;
    CList<CWzdInfo2*,CWzdInfo2*> m_WzdInfo2List;
    void Serialize(CArchive& archive);

};
#endif

// WzdInfo1.cpp : implementation of the CWzdInfo class
//

#include "stdafx.h"
#include "WzdInfo1.h"
```

III

13

```
///////////////////////////////////////////////////////////////////////
// CWzdInfo

void AFXAPI SerializeElements(CArchive& ar, CWzdInfo2** ppElements, int nCount)
{
    CWzdInfo2 *pWzdInfo2;
    for (int i = 0; i < nCount; i++)
    {
        if (ar.IsStoring())
        {
            pWzdInfo2 = *(ppElements + i);
        }
        else
        {
            pWzdInfo2 = new CWzdInfo2;
            *(ppElements + i) = pWzdInfo2;
        }
        pWzdInfo2->Serialize(ar);
    }
}

CWzdInfo1::CWzdInfo1()
{
    m_sGroupName=_T("");
    m_sComment=_T("");
}

CWzdInfo1::~CWzdInfo1()
{
    while (!m_WzdInfo2List.IsEmpty())
    {
        delete m_WzdInfo2List.RemoveHead();
    }
}
```

Example 72    Accessing an ODBC Database    **415**

```
void CWzdInfo1::Serialize(CArchive& ar)
{
    int version=1;
    CObject::Serialize(ar);
    if (ar.IsStoring())
    {
        // version
        ar << version;

        // name and comment
        ar << m_sGroupName;
        ar << m_sComment;

        // other list
        m_WzdInfo2List.Serialize(ar);
    }
    else
    {
        // version
        ar >> version;

        // name and comment
        ar >> m_sGroupName;
        ar >> m_sComment;

        // other list
        m_WzdInfo2List.Serialize(ar);
    }
}
```

III

13

# Example 72    Accessing an ODBC Database

## Objective

You would like to access an ODBC-compliant database from your application.

## Strategy

We will use MFC's CDatabase and CRecordSet classes to access an ODBC database. The CDatabase class will allow us to open the database and the CRecordSet class will allow us to open table(s) in that database and read its records.

## Steps

### Set Up Your Application

1. Start by making sure the following includes are in your Stdafx.h file. These are the include files that the AppWizard adds to your project if you select Header files only for your database support when creating your application.

```
#ifndef _AFX_NO_DB_SUPPORT
#include <afxdb.h>              // MFC ODBC database classes
#endif                          // _AFX_NO_DB_SUPPORT


#ifndef _AFX_NO_DAO_SUPPORT
#include <afxdao.h>             // MFC DAO database classes
#endif                          // _AFX_NO_DAO_SUPPORT
```

### Work with an ODBC Database

1. You can create a database class instance anywhere, but by convention, we will embed ours in the Document Class.

```
CDocument m_WzdDocument;
```

Example 72   Accessing an ODBC Database   **417**

2. You can then open this database using its `Open()` member function in response to a New or Open file command from your user, which you can handle in your Document Class.

```
if(!m_WzdDatabase.Open(
    NULL,                                   // Data source name,
                                            // NULL if defined in
                                            // DSN below
    FALSE,                                  // exclusive access,
                                            // NOT SUPPORTED,
                                      // should always be FALSE
    FALSE,                           // TRUE = read only access
    "ODBC;DSN=MS Access 97 Database")
// connect string where DSN= is the data source name
// found in ODBC32 utility, UID= is user id, PSW= is password
    )
    {
        AfxMessageBox("Failed to open database.");
    }
```

The connect string we use here is based on this database entry's name in the ODBC32 Setup utility. You can find the ODBC32 Setup utility in your system's Control Panel. Using ODBC32, you can add and delete other databases. You can even use ODBC to access a DAO database.

3. It's also customary to create a wrapper function in your Document Class to retrieve a pointer to the database.

```
// Attributes
public:
    CDatabase *GetDatabase(){return &m_WzdDatabase;};
```

4. You can close this database in your Document Class's `DeleteContents()`.

```
// close database
m_WzdDatabase.Close();
```

## Open an ODBC Record Set

1. Use the ClassWizard to create a new record set class derived from CRecordSet. Selecting CRecordSet will cause the ClassWizard to enter a mode you might not have seen before, effectively turning the ClassWizard into a Record Set Wizard. You are first prompted for the type of record set you want to create: ODBC or DAO. Take the default response of ODBC. Then you select from the current list of ODBC databases for the one this record set class will access. The ClassWizard will then attempt to open that database and, if successful, prompt you for which table(s) to include in this record set class. Selecting multiple classes will allow you to perform operations on multiple tables at one time, such as when doing joins. The ClassWizard will then create your derived record set class. You will notice that there is a member variable in this class for each column in each table you selected. When the record set is open and you are scrolling through it, you will be retrieving your column values through these member variable.

2. To open a database record set class, you can use

```
CWzdRecordSet wzdSet(GetDocument()->GetDatabase());
wzdSet.Open();
```

## Work with an ODBC Record Set

1. To scroll through the records in this record set, you can use

```
while (! wzdSet.IsEOF())
{
    // column values can be accessed from record
    // set member variables like these:
    //      wzdSet.m_CustomerID
    //      wzdSet.m_CompanyName
    //      wzdSet.m_ContactName
    //      etc.

    wzdSet.MoveNext();
}
```

Example 72   Accessing an ODBC Database   **419**

2. To move back to the first record, you can use

```
wzdSet.MoveFirst();
```

3. To determine if this record set can be updated or appended, you can use

```
// if we can't update or append records, leave now
    if (!wzdSet.CanUpdate() || !wzdSet.CanAppend())
        return;
```

4. To add a record to this record set, you can use

```
try
{
    // add new record
    wzdSet.AddNew();

    // initialize each field
    wzdSet.m_CustomerID="ABCDE";
    wzdSet.m_CompanyName="ABC Inc.";
    //etc.

    // update database
    wzdSet.Update();
}
catch (CDBException *e)
{
    // AddNew failed
    AfxMessageBox(e->m_strError);
    e->Delete();
}
```

Notice that a record isn't actually added until you call CRecord-Set::Update().

5. To edit a record in the database, you can use

```
try
{
    wzdSet.Edit();

    // update affected fields
    wzdSet.m_CompanyName="ABCEF Inc.";
    wzdSet.m_ContactName="Frank";
    //etc.

    // update database
    wzdSet.Update();
}
catch (CDBException *e)
{
    // Edit failed
    AfxMessageBox(e->m_strError);
    e->Delete();
}
```

6. To delete a record from the database, you can use

```
try
{
    // delete record to which we're currently opened
    wzdSet.Delete();
}
catch (CDBException *e)
{
    // Delete failed
    AfxMessageBox(e->m_strError);
    e->Delete();
}
```

7. To close a record set, use

```
wzdSet.Close();
```

Example 72   Accessing an ODBC Database   **421**

## Open an ODBC Record Set with WHERE

1. To open a record set class with a simple WHERE SQL filter that filters out all records that don't match your criteria, you can use the following. In this example, only records which have a [Country] column value of UK are accessed when scrolling through the record set.

```
wzdSet.m_strFilter = "[Country]='UK'";
if(!wzdSet.Open() || wzdSet.IsEOF())
{

    AfxMessageBox("Cannot find records.");

}


// scroll through records as above...
while (! wzdSet.IsEOF())
{

    wzdSet.MoveNext();

}
wzdSet.Close();
```

## Open an ODBC Record Set with SELECT

1. To open a record set class using a complete SQL SELECT statement that can allow you to specify joins, sorts, etc., you can use the following code. Be aware, however, that anything we leave in m_strFilter is appended to this SQL command.

```
wzdSet.m_strFilter = ""; // appended to the following!
if (!wzdSet.Open(AFX_DB_USE_DEFAULT_TYPE,
    //>>>>>>>>>> SELECT STARTS
    "SELECT [CustomerID], [CompanyName], [ContactName], \
    [ContactTitle], [Address], [City], [Region], [PostalCode], \
    [Country], [Phone], [Fax] \FROM [Customers] \
    WHERE [Country] = 'Mexico'") ||
    //<<<<<<<<<<<< SELECT ENDS
    wzdSet.IsEOF())
{

    AfxMessageBox("Cannot find records.");

}
```

**III**

**13**

```
// scroll through records as above...
while (! wzdSet.IsEOF())
{
    wzdSet.MoveNext();
}
wzdSet.Close();
```

## Create ODBC Transactions

1. Some database operations may require accessing several database tables at once, where a change to one requires changes in others. In this situation, if an error occurs when making a change to one table and you have to undo the change, you must also undo any changes you made in the other tables. In database terminology, this is called rolling back a transaction and is supported by the `CDatabase` class's `BeginTrans()`, `CommitTrans()`, and `RollbackTrans()` member functions.

```
// Create a transaction we can rollback
GetDocument()->GetDatabase()->BeginTrans();
try
{
    // perform transactions on several database tables/records
    wzdSet.Open();
    wzdSet.Update();
    :    :    :
    // No exception occurred so we can commit
    // the previous transactions
    GetDocument()->GetDatabase()->CommitTrans();
}
catch (CDBException *e)
{
    // An exception occurred, rollback the transaction
    GetDocument()->GetDatabase()->Rollback();
    AfxMessageBox(e->m_strError);
    e->Delete();
}
```

Example 72   Accessing an ODBC Database   **423**

## Notes

- You can also use `CDatabase::OpenEx()` to open a database. However, when doing so, make sure you omit the `ODBC;` part of your connection string or it will fail. With databases that require a user ID and password, you can either specify those values in the connection string or, if omitted, the system will prompt your user for those values. The system will also prompt your user if it can't find the specified database. However if you use the `OpenEx()` function, you can set an option that will either always display this prompt or never display this prompt. The latter can be useful if you simply want to check for the existence of a database without alarming your user.

- When using `CDatabase::OpenEx()`, errors are reported solely through database exceptions, so you must use the try/catch syntax to catch any errors.

- ODBC drivers allow your application to access any third-party vendor's database management system (DBMS). However, this flexibility comes at a performance hit since your database requests are encoded in an SQL text command that the driver must first translate into a native command before it can process it. If portability is important to your application, you may have no choice. However, if you don't mind being tied to one vendor for the near future, you might consider using their own database API directly, if available. You give up portability for speed. The DAO database is much more efficient then an ODBC database, simply because it doesn't use SQL to talk to its management system and instead directly wraps the API for Microsoft's Jet Engine Database Management System.

## CD Notes

- When executing the project on the accompanying CD, set a breakpoint on the `OnTestWzd()` function in `WzdView.cpp`. Then click on Test/Wzd in the menu and step through as an ODBC database is accessed.

III

13

# Example 73   Accessing a DAO Database

## Objective

You would like to access a database created for Microsoft's Jet Database Management System. This is the database that Microsoft's Access application creates.

## Strategy

We will be using MFC's `CDaoDatabase` and `CDaoRecordSet` classes to open a database and access its records. A great deal of the functionality in these two classes is identical to the MFC classes that access an ODBC database, but with better performance and flexibility.

## Steps

### Set Up Your Application

1. Start by making sure the following include files are in your `Stdafx.h` file. These are the include files that the AppWizard adds to your project if you select Header files only for your database support.

```
#ifndef _AFX_NO_DB_SUPPORT
#include <afxdb.h>                    // MFC ODBC database classes
#endif                                // _AFX_NO_DB_SUPPORT


#ifndef _AFX_NO_DAO_SUPPORT
#include <afxdao.h>                   // MFC DAO database classes
#endif                                // _AFX_NO_DAO_SUPPORT
```

### Work with a DAO Database

1. You can create your `CDaoDocument` class instance anywhere, but by convention, we will embed it in our Document Class.

```
CDaoDocument m_WzdDocument;
```

Example 73   Accessing a DAO Database   **425**

2. You can then open this database using its `Open()` member function in response to either a New or Open file command from your user, which you can handle in your Document Class.

```
try
{
    m_WzdDatabase.Open(
    "C:\\jeswanke\\examples\\Sampdata.mdb",    // Data source name,
                                               // NULL if defined
                                               // in DSN = below
    FALSE,                                     // TRUE = exclusive
                                               // access to database
    FALSE);                                    // TRUE = read only access
}
catch (CDaoException *e)
{
    AfxMessageBox(e->m_pErrorInfo->m_strDescription);
    e->Delete();
    return FALSE;
}
```

Notice that you can directly specify the name of the Access *.mdb file to open and, unlike an ODBC open, exclusive access is supported.

3. It's also customary to create a wrapper function in your Document Class to retrieve a pointer to the database.

```
// Attributes
public:
    CDaoDatabase *GetDatabase(){return &m_WzdDatabase;};
```

4. To close this database, use the following code.

```
if (m_WzdDatabase.IsOpen())
    m_WzdDatabase.Close();
```

If you attempt to close a DAO database that isn't currently open, it will throw an exception.

## Work with a DAO Record Set

1. Use the ClassWizard to create a new record set class derived from `CDaoRecordSet`. Selecting `CDaoRecordSet` will cause the ClassWizard to

enter a mode you might not have seen before, effectively turning the ClassWizard into a Record Set Wizard. You are first prompted for the type of record set you want to create: ODBC or DAO. Take the default response of DAO. Then, you are prompted to explore your disk for a *.mdb database file. The ClassWizard will then attempt to open that database and, if successful, prompt you for which table(s) to include in this record set class. Selecting multiple classes will allow you to perform operations on multiple tables at one time, such as when doing joins. The ClassWizard will then create the record set class. You will notice that there is a member variable in this class for each column in each table you selected. When the record set is open and you are scrolling through it, you will be retrieving your column values through these member variable.

2. To open this record set, use

```
CWzdRecordSet wzdSet(GetDocument()->GetDatabase());
wzdSet.Open();
```

3. To scroll through all of the records in this record set, you can use

```
while (! wzdSet.IsEOF())
{
    // values can be accessed from record set member variables:
    //      wzdSet.m_CustomerID
    //      wzdSet.m_CompanyName
    //      wzdSet.m_ContactName
    //      etc.

    wzdSet.MoveNext();
}
```

4. To scroll back to the first record, you can use

```
wzdSet.MoveFirst();
```

5. To find records with a column value that matches a particular criteria, you can use CDaoRecordSet's FindFirst(), and FindNext() functions. In

Example 73   Accessing a DAO Database   **427**

this example, we are looking for all records which have a [Country] column equal to UK.

```
if (wzdSet.FindFirst("[Country]='UK'"))
{
    while (wzdSet.FindNext("[Country]='UK'"))
    {
    }
}
```

6. To determine if the table(s) in this record set can be updated or appended, you can use the following.

```
// if we can't update or append records, leave now
if (!wzdSet.CanUpdate() || !wzdSet.CanAppend())
    return;
```

7. To add a record to this table, you can use the following.

```
try
{
    // add new record
    wzdSet.AddNew();

    // initialize each field
    wzdSet.m_CustomerID="ABCDX";
    wzdSet.m_CompanyName="ABC Inc.";
    //etc.

    // update database
    wzdSet.Update();
}
catch (CDaoException *e)
{
    // AddNew failed
    AfxMessageBox(e->m_pErrorInfo->m_strDescription);
    e->Delete();
}
```

Notice that you must update the record for it to be added.

8. To edit a record in this table, you can use

```
try
{
    wzdSet.Edit();

    // update affected fields
    wzdSet.m_CompanyName="ABCEF Inc.";
    wzdSet.m_ContactName="Frank";
    //etc.

    // update database
    wzdSet.Update();
}
catch (CDaoException *e)
{
    // Edit failed
    AfxMessageBox(e->m_pErrorInfo->m_strDescription);
    e->Delete();
}
```

9. To delete a record from this record set, you can use

```
try
{
    // delete record to which we're currently opened
    wzdSet.Delete();
}
catch (CDaoException *e)
{
    // Delete failed
    AfxMessageBox(e->m_pErrorInfo->m_strDescription);
    e->Delete();
}
```

Example 73   Accessing a DAO Database   **429**

10.To close a record set, you can use

```
wzdSet.Close();
```

If your record set class was created on the stack, simply returning will destruct the class instance and close the record set.

## Open a DAO Record Set with WHERE

1. To open a database table with a simple SQL WHERE statement, you can use the following. In this example, only records whose [Country] column value equals UK are accessed when scrolling through the record set.

```
wzdSet.m_strFilter = "[Country]='UK'";
wzdSet.Open();
if(wzdSet.IsEOF())
{
    AfxMessageBox("Cannot find records.");
}

// scroll through records as above...
while (! wzdSet.IsEOF())
{
    wzdSet.MoveNext();
}
wzdSet.Close();
```

## Open a DAO Record Set with SORT

1. To open a table with a simple SQL SORT statement, you can use the following. In this example, when scrolling through the table, records are returned alphabetically based on the values in the [ContactName] column.

```
wzdSet.m_strSort = "[ContactName]";
wzdSet.Open();
    :    :    :
```

III

13

## Open a DAO Record Set with SELECT

1. To open a database record set using a complete SQL SELECT statement, you can use the following. Notice that anything left in either m_strFilter or m_strSort is appended to the end of your SQL statement.

```
wzdSet.m_strFilter = ""; // appended to the following!
wzdSet.m_strSort = "";
wzdSet.Open(AFX_DB_USE_DEFAULT_TYPE,
    "SELECT [CustomerID], [CompanyName], [ContactName], \
    [ContactTitle], [Address], [City], [Region], [PostalCode], \
    [Country], [Phone], [Fax] \
    FROM [Customers] WHERE [Country] = 'Mexico'");
if (wzdSet.IsEOF())
{
    AfxMessageBox("Cannot find records.");
}


// scroll through records as above...
while (! wzdSet.IsEOF())
{
    wzdSet.MoveNext();
}
wzdSet.Close();
```

Example 73   Accessing a DAO Database   **431**

## Create DAO Transactions

1. Some database operations may require accessing several tables at once, where a change to one requires changes in others. In this situation, if an error occurs when making a change to one table that you must then undo, you must also undo any changes you made in the other tables. In database terminology, this is called rolling back a transaction and is supported by the CDaoDatabase class's BeginTrans(), CommitTrans(), and RollbackTrans() member functions.

```
// Create a transaction we can rollback
GetDocument()->GetDatabase()->m_pWorkspace->BeginTrans();
try
{
    // perform transactions on several database tables/records
    wzdSet.Open();
    wzdSet.Update();
    // Success
    GetDocument()->GetDatabase()->m_pWorkspace->CommitTrans();
}
catch (CDaoException *e)
{
    // An exception occurred, rollback the transaction
    GetDocument()->GetDatabase()->m_pWorkspace->Rollback();
    AfxMessageBox(e->m_pErrorInfo->m_strDescription);
    e->Delete();
}
```

## Notes

- For frequent use of a fully defined SQL open, you might consider encapsulating a tailored Open() member function in your record set class. Such a function might be called SelectByCountry(arg) and would always open the record set class with WHERE '[COLUMN]' = arg.
- The MFC class library also includes CDaoQueryDef, which allows you to predefine an SQL query in your database for future use.

- The MFC class library also includes `CDaoTableDef`, which allows you to create a new DAO database table.

## CD Notes

- When executing the project on the accompanying CD, set a breakpoint on the `OnTestWzd()` function in `WzdView.cpp`. Then click on Test/Wzd in the menu and step through as a DAO database is accessed.

# 14

# Potpourri

The only criteria an example had to pass to make it into this chapter was that it didn't have enough accompanying examples to warrant a chapter in this book. Cut and paste, as well as MFC's data collection classes, almost had enough examples for their own chapter, but time and the rest were simply leftovers. And yet, this chapter contains several useful examples.

**Example 74**     **Cutting, Copying, and Pasting Text Data**    We will review the built-in cut and paste functionality of an Edit control window.

**Example 75**     **Cutting, Copying, and Pasting Rich Text Data**    We will review the built-in cut and paste functionality of a Rich Edit Control window.

**Example 76**     **Cutting, Copying, and Pasting Binary Data**    We will use serialization to cut and paste our own proprietary data classes.

**Example 77**     **Array Functions**    We will review the MFC classes that can maintain arrays of identical classes or pointers.

**Example 78**     **List Functions**    We will review the MFC classes that can maintain linked lists of identical classes or pointers.

**Example 79    Map Functions**  We will review the MFC classes that allow us to access identical classes or pointers using a binary, or even text key, rather than an index.

**Example 80    System Keys**  We will direct some system keys, such as the Delete key, to our view class's `Delete()` function.

**Example 81    Time**  We will review the different time formats we can access using MFC classes.

# Example 74   Cutting, Copying, and Pasting Text Data

## Objective

You would like to cut or copy text from your edit control and then paste it back at another point in your control or in another control. You would also like to cut and paste using the popup menu (Figure 14.1).

## Figure 14.1   Add your own menu commands to the cut and paste text popup menu.

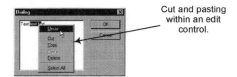

Cut and pasting within an edit control.

## Strategy

Actually, an edit control already provides this functionality, right down to the popup menu that appears when you right-click on the control. The purpose, therefore, of this example is to show how you can implement this popup menu yourself so that you might add your own menu commands to it in the future. We can also review the member functions of an edit control class that support cut and paste. We will be encapsulating this functionality in our own derivation of the `CEdit` control class.

Example 74   Cutting, Copying, and Pasting Text Data   **435**

# Steps

## Create a New Edit Control Class

1. Use the ClassWizard to create a new edit control class and derive it from CEdit. Use the ClassWizard again to add a WM_RBUTTONDOWN message handler to this new class.

## Load and Enable the Popup Menu

1. Use the Menu Editor to create a menu resource identical to the one in Figure 14.1.
2. Edit the WM_RBUTTONDOWN message handler to create a popup menu from this menu resource.

```
void CWzdEdit::OnRButtonDown(UINT nFlags, CPoint point)
{
CMenu menu;
    // load a menu from the resources
    menu.LoadMenu(IDR_SELECTION_MENU);

    // get a pointer to actual popup menu
    CMenu* pPopup = menu.GetSubMenu(0);
```

Once loaded, all menu items in our popup menu will be enabled. We need to possibly disable some items to reflect their unavailability. For example, if there is nothing to undo, we must disable the Undo command.

3. To enable or disable the Undo command, use CEdit's CanUndo().

```
UINT nUndo=(CanUndo() ? 0 : MF_GRAYED);
pPopup->EnableMenuItem(ID_EDIT_UNDO,  MF_BYCOMMAND|nUndo);
```

4. For commands that depend on text in the edit control being selected, we can use CEdit::GetSel() to get the beginning and end of the current

**III**

**14**

selection. If the beginning and end are equal, nothing is selected and the Cut, Copy, and Delete commands are unavailable.

```
int beg,end;
GetSel(beg,end);
UINT nSel=((beg!=end) ? 0 : MF_GRAYED);
pPopup->EnableMenuItem(ID_EDIT_CUT, MF_BYCOMMAND|nSel);
pPopup->EnableMenuItem(ID_EDIT_COPY, MF_BYCOMMAND|nSel);
pPopup->EnableMenuItem(ID_EDIT_CLEAR, MF_BYCOMMAND|nSel);
```

5. For the Paste command, we need to check the clipboard to see if there's any text available.

```
UINT nPaste=(::IsClipboardFormatAvailable(CF_TEXT) ? 0 :
    MF_GRAYED);
pPopup->EnableMenuItem(ID_EDIT_PASTE, MF_BYCOMMAND|nPaste);
```

6. We can now display this popup menu and wait for the user to click something.

```
    // pop up the menu
    CPoint pt;
    GetCursorPos(&pt);
    pPopup->TrackPopupMenu(TPM_RIGHTBUTTON, pt.x, pt.y, this);
    pPopup->DestroyMenu();

    CEdit::OnRButtonDown(nFlags, point);
}
```

## Handle the Popup Menu Commands

1. If the user clicks a command, it will return to this window as a WM_COMMAND message with the command ID we assigned with the Menu Editor. To handle these commands we must manually add the following message macros to our message map.

```
ON_COMMAND(ID_EDIT_UNDO, OnUndo)
ON_COMMAND(ID_EDIT_CUT, OnCut)
ON_COMMAND(ID_EDIT_COPY, OnCopy)
ON_COMMAND(ID_EDIT_PASTE, OnPaste)
ON_COMMAND(ID_EDIT_CLEAR, OnDelete)
ON_COMMAND(ID_EDIT_SELECT_ALL, OnSelectAll)
```

Example 74  Cutting, Copying, and Pasting Text Data  **437**

2. Because the CEdit member functions that correspond to these commands usually have no arguments, we can simply implement them as inline functions in this edit class's .h file.

```
//{{AFX_MSG(CWzdEdit)
afx_msg void OnRButtonDown(UINT nFlags, CPoint point);
//}}AFX_MSG
afx_msg void OnUndo(void){Undo();};
afx_msg void OnCut(void){Cut();};
afx_msg void OnCopy(void){Copy();};
afx_msg void OnPaste(void){Paste();};
afx_msg void OnDelete(void){Clear();};
afx_msg void OnSelectAll(void){SetSel(0,-1);};
DECLARE_MESSAGE_MAP()
```

As you can see, only the Select All command required any sophistication.

3. To see a complete listing of this edit class, see "Listings — Edit Control Class" on page 438.

## Notes

- You can now add your own commands to this popup menu. To make this class completely encapsulated, you might also consider creating the popup menu dynamically, as shown in Example 21, rather than depend on an external menu resource.

## CD Notes

- When executing the project on the accompanying CD, click on the Test/Wzd menu commands to open a dialog box. This dialog box contains an edit box that, if you right-click on it, will display a popup menu with cut, copy, paste, etc., editing commands.

III

14

## Listings — Edit Control Class

```cpp
#if !defined(AFX_WZDEDIT_H__19B437E6_E7F5_11D1_A18D_DCB3C85EBD34__INCLUDED_)
#define AFX_WZDEDIT_H__19B437E6_E7F5_11D1_A18D_DCB3C85EBD34__INCLUDED_

#if _MSC_VER >= 1000
#pragma once
#endif // _MSC_VER >= 1000

// WzdEdit.h : header file
//

/////////////////////////////////////////////////////////////////////////////
// CWzdEdit window

class CWzdEdit : public CEdit
{
// Construction
public:
    CWzdEdit();

// Attributes
public:

// Operations
public:

// Overrides
    // ClassWizard generated virtual function overrides
    //{{AFX_VIRTUAL(CWzdEdit)
    //}}AFX_VIRTUAL

// Implementation
public:
    virtual ~CWzdEdit();
```

**Example 74   Cutting, Copying, and Pasting Text Data   439**

```
    // Generated message map functions
protected:
    //{{AFX_MSG(CWzdEdit)
    afx_msg void OnRButtonDown(UINT nFlags, CPoint point);
        //}}AFX_MSG
        afx_msg void OnUndo(void){Undo();};
        afx_msg void OnCut(void){Cut();};
        afx_msg void OnCopy(void){Copy();};
        afx_msg void OnPaste(void){Paste();};
        afx_msg void OnDelete(void){Clear();};
        afx_msg void OnSelectAll(void){SetSel(0,-1);};
        DECLARE_MESSAGE_MAP()
};


/////////////////////////////////////////////////////////////////////////////
//{{AFX_INSERT_LOCATION}}
// Microsoft Developer Studio will insert additional declarations immediately
// before the previous line.

#endif // !defined(
    AFX_WZDEDIT_H__19B437E6_E7F5_11D1_A18D_DCB3C85EBD34__INCLUDED_)


// WzdEdit.cpp : implementation file
//

#include "stdafx.h"
#include "wzd.h"
#include "WzdEdit.h"

#ifdef _DEBUG
#define new DEBUG_NEW
#undef THIS_FILE
static char THIS_FILE[] = __FILE__;
#endif
```

III

14

```
//////////////////////////////////////////////////////////////////////////
// CWzdEdit

CWzdEdit::CWzdEdit()
{
}

CWzdEdit::~CWzdEdit()
{
}



BEGIN_MESSAGE_MAP(CWzdEdit, CEdit)
    //{{AFX_MSG_MAP(CWzdEdit)
    ON_WM_RBUTTONDOWN()
    //}}AFX_MSG_MAP
    ON_COMMAND(ID_EDIT_UNDO, OnUndo)
    ON_COMMAND(ID_EDIT_CUT, OnCut)
    ON_COMMAND(ID_EDIT_COPY, OnCopy)
    ON_COMMAND(ID_EDIT_PASTE, OnPaste)
    ON_COMMAND(ID_EDIT_CLEAR, OnDelete)
    ON_COMMAND(ID_EDIT_SELECT_ALL, OnSelectAll)
END_MESSAGE_MAP()


//////////////////////////////////////////////////////////////////////////
// CWzdEdit message handlers


void CWzdEdit::OnRButtonDown(UINT nFlags, CPoint point)
{
    CMenu menu;
    // load a menu from the resources
    menu.LoadMenu(IDR_SELECTION_MENU);

    // get a pointer to actual popup menu
    CMenu* pPopup = menu.GetSubMenu(0);

    // enable/disable Undo command
    UINT nUndo=(CanUndo() ? 0 : MF_GRAYED);
    pPopup->EnableMenuItem(ID_EDIT_UNDO, MF_BYCOMMAND|nUndo);
```

Example 74   Cutting, Copying, and Pasting Text Data   **441**

```
// enable/disable selection commands
int beg,end;
GetSel(beg,end);
UINT nSel=((beg!=end) ? 0 : MF_GRAYED);
pPopup->EnableMenuItem(ID_EDIT_CUT, MF_BYCOMMAND|nSel);
pPopup->EnableMenuItem(ID_EDIT_COPY, MF_BYCOMMAND|nSel);
pPopup->EnableMenuItem(ID_EDIT_CLEAR, MF_BYCOMMAND|nSel);

// enable/disable Paste command
UINT nPaste=(::IsClipboardFormatAvailable(CF_TEXT) ? 0 : MF_GRAYED);
pPopup->EnableMenuItem(ID_EDIT_PASTE, MF_BYCOMMAND|nPaste);

// pop up the menu
CPoint pt;
GetCursorPos(&pt);
pPopup->TrackPopupMenu(TPM_RIGHTBUTTON, pt.x, pt.y, this);
pPopup->DestroyMenu();

CEdit::OnRButtonDown(nFlags, point);
}
```

# Example 75   Cutting, Copying, and Pasting Rich Text Data

## Objective

III

You would like to cut or copy text from your rich edit control and then paste it back at another point in your control or in another control. You would also like to cut and paste using the popup menu (Figure 14.2).

14

## Figure 14.2   Create a cut and paste rich text popup menu.

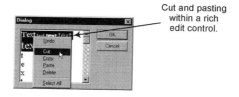

Cut and pasting within a rich edit control.

## Strategy

Unlike an edit control, a rich edit control doesn't already provide this functionality. To provide this functionality, we will be using the member functions of the `CRichEditCtrl` class along with the `CMenu` class to create the popup menu. We will be encapsulating all of this in our own derivative of the `CRichEditCtrl` class.

## Steps

### Set Up Your Application

1. Support for the rich edit control class must be initialized separately from your other MFC control classes. Add the following line to your Application Class's `InitInstance()` function.

```
BOOL CWzdApp::InitInstance()
{

    AfxInitRichEdit();


        :    :    :
```

2. Use the Menu Editor to create a menu resource identical to the one in Figure 14.2.

3. Use the ClassWizard to create a new rich edit control class derived from `CRichEditCtrl`. Then, use the ClassWizard to add a `WM_RBUTTONDOWN` message handler.

### Load and Enable the Popup Menu

1. Start the `WM_RBUTTONDOWN` handler out by creating a popup menu from the menu resource you created in the previous step.

```
void CWzdRichEditCtrl::OnRButtonDown(UINT nFlags, CPoint point)
{
    CMenu menu;
    // load a menu from the resources
    menu.LoadMenu(IDR_SELECTION_MENU);

    // get a pointer to actual popup menu
    CMenu* pPopup = menu.GetSubMenu(0);
```

Example 74   Cutting, Copying, and Pasting Text Data   **443**

Once loaded, all menu items in our popup menu will be enabled. We need to possibly disable some items to reflect their unavailability. For example, if there is nothing to undo, we must disable the Undo command.

2. To enable or disable the Undo command, use `CRichEditCtrl`'s `CanUndo()`.

```
// enable/disable Undo command
UINT nUndo=(CanUndo() ? 0 : MF_GRAYED);
pPopup->EnableMenuItem(ID_EDIT_UNDO,  MF_BYCOMMAND|nUndo);
```

3. For commands that depend on text in the control being selected, use `CRichEditCtrl::GetSelectionType()` to determine whether anything is selected. If the selection type is SEL_EMPTY the Cut, Copy and Delete commands are invalid.

```
UINT nSel=((GetSelectionType()!=SEL_EMPTY) ? 0 : MF_GRAYED);
pPopup->EnableMenuItem(ID_EDIT_CUT, MF_BYCOMMAND|nSel);
pPopup->EnableMenuItem(ID_EDIT_COPY, MF_BYCOMMAND|nSel);
pPopup->EnableMenuItem(ID_EDIT_CLEAR, MF_BYCOMMAND|nSel);
```

4. For the Paste command, we use `CRichEditCtrl::CanPaste()`.

```
UINT nPaste=(CanPaste() ? 0 : MF_GRAYED);
pPopup->EnableMenuItem(ID_EDIT_PASTE, MF_BYCOMMAND|nPaste);
```

5. We can now display this popup menu and wait for the user to click something.

```
    CPoint pt;
    GetCursorPos(&pt);
    pPopup->TrackPopupMenu(TPM_RIGHTBUTTON, pt.x, pt.y, this);
    pPopup->DestroyMenu();

    CRichEditCtrl::OnRButtonDown(nFlags, point);
}
```

## Handle the Popup Menu Commands

1. If the user clicks a command, it will return to this control window as a WM_COMMAND message with the command ID we assigned with the Menu

III

14

Editor. To handle these commands, we must manually add the following message macros to our message map.

```
ON_COMMAND(ID_EDIT_UNDO, OnUndo)
ON_COMMAND(ID_EDIT_CUT, OnCut)
ON_COMMAND(ID_EDIT_COPY, OnCopy)
ON_COMMAND(ID_EDIT_PASTE, OnPaste)
ON_COMMAND(ID_EDIT_CLEAR, OnDelete)
ON_COMMAND(ID_EDIT_SELECT_ALL, OnSelectAll)
```

2. Because the `CRichEditCtrl` member functions that correspond to these commands usually have no arguments, we can simply implement them as inline functions in this class's `.h` file.

```
//{{AFX_MSG(CWzdEdit)
afx_msg void OnRButtonDown(UINT nFlags, CPoint point);
//}}AFX_MSG
afx_msg void OnUndo(void){Undo();};
afx_msg void OnCut(void){Cut();};
afx_msg void OnCopy(void){Copy();};
afx_msg void OnPaste(void){Paste();};
afx_msg void OnDelete(void){Clear();};
afx_msg void OnSelectAll(void){SetSel(0,-1);};
DECLARE_MESSAGE_MAP()
```

As you can see, only the Select All command required any sophistication.

3. To see a complete listing of this edit class, see "Listings — Rich Edit Control Class" on page 445.

## Notes

- In some versions of MFC, you might find that a rich edit control's Cut, Copy and Paste functions cause the debug version of your application to blow up. The release version, however, has no such problem.
- Excellent menu command candidates for inclusion in this popup menu include Font and Paragraph, which are both easily implemented using the member functions of a rich edit control class.

Example 74   Cutting, Copying, and Pasting Text Data   **445**

## CD Notes

- When executing the project on the accompanying CD, click on the Test/Wzd menu commands to open a dialog box. This dialog box contains a rich text edit box that, if you right-click on it, will display a popup menu with cut, copy, paste, etc., editing commands.

## Listings — Rich Edit Control Class

```
#if !defined(
    AFX_WZDRICHEDITCTRL_H__19B437E7_E7F5_11D1_A18D_DCB3C85EBD34__INCLUDED_)
#define AFX_WZDRICHEDITCTRL_H__19B437E7_E7F5_11D1_A18D_DCB3C85EBD34__INCLUDED_

#if _MSC_VER >= 1000
#pragma once
#endif // _MSC_VER >= 1000


// WzdRichEditCtrl.h : header file
//


/////////////////////////////////////////////////////////////////////////////
// CWzdRichEditCtrl window

class CWzdRichEditCtrl : public CRichEditCtrl
{
// Construction
public:
    CWzdRichEditCtrl();

// Attributes
public:

// Operations
public:

// Overrides
    // ClassWizard generated virtual function overrides
    //{{AFX_VIRTUAL(CWzdRichEditCtrl)
    //}}AFX_VIRTUAL
```

III

14

```cpp
// Implementation
public:
    virtual ~CWzdRichEditCtrl();

    // Generated message map functions
protected:
    //{{AFX_MSG(CWzdRichEditCtrl)
    afx_msg void OnRButtonDown(UINT nFlags, CPoint point);
    //}}AFX_MSG
    afx_msg void OnUndo(void){Undo();};
    afx_msg void OnCut(void){Cut();};
    afx_msg void OnCopy(void){Copy();};
    afx_msg void OnPaste(void){Paste();};
    afx_msg void OnDelete(void){Clear();};
    afx_msg void OnSelectAll(void){SetSel(0,-1);};
    DECLARE_MESSAGE_MAP()
};


/////////////////////////////////////////////////////////////////////////////
//{{AFX_INSERT_LOCATION}}
// Microsoft Developer Studio will insert additional declarations immediately
// before the previous line.

#endif // !defined(
    AFX_WZDRICHEDITCTRL_H__19B437E7_E7F5_11D1_A18D_DCB3C85EBD34__INCLUDED_)
// WzdRichEditCtrl.cpp : implementation file
//

#include "stdafx.h"
#include "wzd.h"
#include "WzdRichEditCtrl.h"

#ifdef _DEBUG
#define new DEBUG_NEW
#undef THIS_FILE
static char THIS_FILE[] = __FILE__;
#endif
```

Example 74   Cutting, Copying, and Pasting Text Data   **447**

```
///////////////////////////////////////////////////////////////////////////
// CWzdRichEditCtrl


CWzdRichEditCtrl::CWzdRichEditCtrl()
{
}


CWzdRichEditCtrl::~CWzdRichEditCtrl()
{
}



BEGIN_MESSAGE_MAP(CWzdRichEditCtrl, CRichEditCtrl)
    //{{AFX_MSG_MAP(CWzdRichEditCtrl)
    ON_WM_RBUTTONDOWN()
    //}}AFX_MSG_MAP
    ON_COMMAND(ID_EDIT_UNDO, OnUndo)
    ON_COMMAND(ID_EDIT_CUT, OnCut)
    ON_COMMAND(ID_EDIT_COPY, OnCopy)
    ON_COMMAND(ID_EDIT_PASTE, OnPaste)
    ON_COMMAND(ID_EDIT_CLEAR, OnDelete)
    ON_COMMAND(ID_EDIT_SELECT_ALL, OnSelectAll)
END_MESSAGE_MAP()


///////////////////////////////////////////////////////////////////////////
// CWzdRichEditCtrl message handlers

void CWzdRichEditCtrl::OnRButtonDown(UINT nFlags, CPoint point)
{
    CMenu menu;
    // load a menu from the resources
    menu.LoadMenu(IDR_SELECTION_MENU);

    // get a pointer to actual popup menu
    CMenu* pPopup = menu.GetSubMenu(0);

    // enable/disable Undo command
    UINT nUndo=(CanUndo() ? 0 : MF_GRAYED);
    pPopup->EnableMenuItem(ID_EDIT_UNDO,  MF_BYCOMMAND|nUndo);
```

**III**

**14**

```
// enable/disable selection commands
UINT nSel=((GetSelectionType()!=SEL_EMPTY) ? 0 : MF_GRAYED);
pPopup->EnableMenuItem(ID_EDIT_CUT, MF_BYCOMMAND|nSel);
pPopup->EnableMenuItem(ID_EDIT_COPY, MF_BYCOMMAND|nSel);
pPopup->EnableMenuItem(ID_EDIT_CLEAR, MF_BYCOMMAND|nSel);


// enable/disable Paste command
UINT nPaste=(CanPaste() ? 0 : MF_GRAYED);
pPopup->EnableMenuItem(ID_EDIT_PASTE, MF_BYCOMMAND|nPaste);


// pop up the menu
CPoint pt;
GetCursorPos(&pt);
pPopup->TrackPopupMenu(TPM_RIGHTBUTTON, pt.x, pt.y, this);
pPopup->DestroyMenu();


CRichEditCtrl::OnRButtonDown(nFlags, point);
}
```

# Example 76   Cutting, Copying, and Pasting Binary Data

## Objective

You would like to cut or copy binary data to paste back at another point in your document or in another document.

## Strategy

This example assumes the following scenario: your document is composed of several proprietary data objects. These data objects represent graphic figures, icons, control windows, etc., which are drawn to your application's view. This example also assumes you have implemented some way for your user to select one or more of these drawn objects using the mouse cursor. The functionality that this example provides maintains a list of these selected data objects for cutting or copying to the clipboard so that they can be pasted into another part of the document or into another document.

We will be using serialization to cut or copy the selected data to the clipboard. Serialization is discussed in Chapter 13 and is primarily intended to

Example 76   Cutting, Copying, and Pasting Binary Data   **449**

load and store documents to disk. However, it also lends itself perfectly to loading and storing sections of your document to the clipboard. We will encapsulate this cutting and pasting functionality in a new selection class.

# Steps

### Create a CSelection **Class**

1. Use the Developer Studio to create a new class derived from CObject called CSelection. Click on the Insert/New Class... menu commands to open the New Class dialog box. Then, select a Class Type of Generic Class.

2. In the constructor of this new class, create a new clipboard type for your document. Documents with this type will be able to paste this clipboard type, but no others.

```
CWzdSelect::CWzdSelect()
{
    m_clipboardFormat = ::RegisterClipboardFormat("CWzdInfo1");
}
```

3. Add a Select() member function to this new class, as follows. Select() maintains a list of selected items. It's up to your View Class to call this function with new selections. This function will also keep track of the last data item selected.

```
void CWzdSelect::Select(CWzdInfo1 *pInfo)
{
    m_pActiveSelection=pInfo;
    if (!m_WzdSelectionList.Find(pInfo))
    {
        m_WzdSelectionList.AddTail(pInfo);
    }
}
```

III

14

4. Add a Serialize() function to this class. Its job will be to serialize this class's selection list to the clipboard.

```
void CWzdSelect::Serialize(CArchive& ar)
{
    int nCount;
    CObject::Serialize(ar);
    if(ar.IsStoring())
    {
        nCount = m_WzdSelectionList.GetCount();
        ar << nCount;
        for (POSITION pos =
            m_WzdSelectionList.GetHeadPosition(); pos;)
        {
            m_WzdSelectionList.GetNext(pos)->Serialize(ar);
        }
    }
    else
    {
        ar >> nCount;
        while (nCount-- > 0)
        {
            CWzdInfo1* pInfo = new CWzdInfo1;
            pInfo->Serialize(ar);
            m_WzdSelectionList.AddTail(pInfo);
        }
    }
}
```

Example 76   Cutting, Copying, and Pasting Binary Data   **451**

5. Add a `CutSelections()` function. Since cutting and copying are very similar, this function will simply call the copy version. We will use the calling argument `pList` as a flag — if `NULL`, then `CopySelections()` will only copy to the clipboard.

```
void CWzdSelect::CutSelections(CList<CWzdInfo1*,
    CWzdInfo1*> *pList)
{

    CopySelections(pList);
}
```

The `CopySelections()` function will handle both cutting and copying data objects to the clipboard.

## Implement the `CopySelections()` Helper Function

1. Add a `CopySelections()` function, whose first job will be to return if there's nothing selected.

```
void CWzdSelect::CopySelections(CList<CWzdInfo1*,
    CWzdInfo1*> *pList /*=NULL*/)
{

    if (m_WzdSelectionList.GetCount() <= 0) return;
```

2. Next, we will serialize our selection list to a shared memory file. A `CSharedFile` file is required because it's globally allocated, which is required since we're going to be sticking our selections in the clipboard.

```
CSharedFile file;
CArchive ar(&file, CArchive::store);
Serialize(ar);
ar.Close();
```

3. Now, stick this file in the clipboard. We use our registered clipboard type and we give it the handle from our memory file which is, in fact, a globally allocated memory handle.

```
COleDataSource *pDS = new COleDataSource();
pDS->CacheGlobalData(m_clipboardFormat, file.Detach());
pDS->SetClipboard();
```

**III**

**14**

4. Remember that we made the `pList` our argument to determine whether this is cut or copy? To cut the selection(s) from the document, the View Class should fill this argument with a pointer to the document's data object list. We will now remove those items from the document.

```
// if cutting, delete items in selection list from document
if (pList)
{
    for (POSITION pos=m_WzdSelectionList.GetHeadPosition();pos;)
    {
        CWzdInfo1 *pInfo=m_WzdSelectionList.GetNext(pos);

        POSITION posx;
        if (posx=pList->Find(pInfo))
        {
            pList->RemoveAt(posx);
        }
        delete pInfo;
    }
}
```

5. And finally we will empty our selection list.

```
m_WzdSelectionList.RemoveAll();
m_pActiveSelection=NULL;
```

## Paste Binary Data

1. Create a `PasteClipboard()` function that will take as its argument a pointer into the document's data object list. This function's first job is to get rid of whatever is currently selected.

```
void CWzdSelect::PasteClipboard(CList<CWzdInfo1*,
    CWzdInfo1*> *pList)
{
    // reset selections
    m_WzdSelectionList.RemoveAll();
    m_pActiveSelection=NULL;
```

**Example 76   Cutting, Copying, and Pasting Binary Data   453**

2. Next, we attach a data object to the clipboard and serialize it back into our selection list.

```
COleDataObject object;
object.AttachClipboard();
CFile* pFile = object.GetFileData(m_clipboardFormat);
if (!pFile) return;
CArchive ar(pFile, CArchive::load);
Serialize(ar);
ar.Close();
delete pFile; //deletes file
```

3. And finally, we add this selection back into our application's document.

```
for (POSITION pos=m_WzdSelectionList.GetHeadPosition();pos;)
{
    m_pActiveSelection=m_WzdSelectionList.GetNext(pos);
    pList->AddTail(m_pActiveSelection);
}
```

For a complete listing of this selection class, see "Listings — Selection Class" on page 455.

## Implement the New Selection Class

1. Embed the CSelection class in your View Class.

```
/////////////////////////////
// WzdView.h
xprivate:
    CWzdSelect m_select;
```

2. Call the different member functions of the selection class to handle your select, cut, copy, and paste menu commands:

```
void CWzdView::OnSelect()
{
    m_select.Select(GetDocument()->GetInfolList()->GetHead());
}
void CWzdView::OnEditCut()
```

**III**

**14**

```
{
    m_select.CutSelections(GetDocument()->GetInfo1List());
}
void CWzdView::OnUpdateEditCut(CCmdUI* pCmdUI)
{
    pCmdUI->Enable(m_select.SelectionCount());
}
void CWzdView::OnEditCopy()
{
    m_select.CopySelections();
}
void CWzdView::OnUpdateEditCopy(CCmdUI* pCmdUI)
{
    pCmdUI->Enable(m_select.SelectionCount());
}
void CWzdView::OnEditPaste()
{
    m_select.PasteClipboard(GetDocument()->GetInfo1List());
}
void CWzdView::OnUpdateEditPaste(CCmdUI* pCmdUI)
{
    pCmdUI->Enable(m_select.CanPasteClipboard());
}
```

## Notes

- You will probably need to further refine `PasteClipboard()` to allow your View Class to pick a particular part of the document to which you want to paste. For graphic figures, this might not be necessary, since a figure's x,y position is more important than its location in the document's data list. However, a list of database items might require special consideration when pasting data into it.

- For selecting graphic items, you can use MFC's `CTracker` class to highlight the items in the view.

Example 76   Cutting, Copying, and Pasting Binary Data   **455**

## CD Notes

- When executing the project on the accompanying CD, set breakpoints on each of the functions in `WzdSelect.cpp`. Then, fill the document by clicking on Test/Wzd in the menu. Then, click on one of the various EditZ menu commands and watch as document data is cut, copied, and pasted.

## Listings — Selection Class

```
#ifndef WZDSELECT_H
#define WZDSELECT_H


#include "afxtempl.h"
#include "WzdInfo1.h"


class CWzdSelect : public CObject
{
public:
    CWzdSelect();
    ~CWzdSelect();

    void Select(CWzdInfo1 *pInfo);
    int SelectionCount();
    void CutSelections(CList<CWzdInfo1*,CWzdInfo1*> *pList);
    void CopySelections(CList<CWzdInfo1*,CWzdInfo1*> *pList=NULL);
    BOOL CanPasteClipboard();
    void PasteClipboard(CList<CWzdInfo1*,CWzdInfo1*> *pList);
    void Serialize(CArchive& archive);

private:
    int m_clipboardFormat;
    CWzdInfo1 *m_pActiveSelection;
    CList<CWzdInfo1*,CWzdInfo1*> m_WzdSelectionList;


};
#endif
```

III

14

```
// WzdSelect.cpp : implementation of the CWzdSelect class
//

#include "stdafx.h"
#include <afxadv.h>
#include <afxole.h>
#include "WzdSelect.h"


/////////////////////////////////////////////////////////////////////////////
// CWzdSelect

CWzdSelect::CWzdSelect()
{
    m_pActiveSelection=NULL;
    m_clipboardFormat = ::RegisterClipboardFormat("CWzdInfo1");
}


CWzdSelect::~CWzdSelect()
{
}
/////////////////////////////////////////////////////////////////////////////
//CWzdSelect Methods

void CWzdSelect::Serialize(CArchive& ar)
{
    int nCount;
    CObject::Serialize(ar);
    if(ar.IsStoring())
    {
        nCount = m_WzdSelectionList.GetCount();
        ar << nCount;
        for (POSITION pos = m_WzdSelectionList.GetHeadPosition(); pos;)
        {
            m_WzdSelectionList.GetNext(pos)->Serialize(ar);
        }
    }
```

Example 76   Cutting, Copying, and Pasting Binary Data   **457**

```
        else
        {
            ar >> nCount;
            while (nCount-- > 0)
            {
                CWzdInfo1* pInfo = new CWzdInfo1;
                pInfo->Serialize(ar);
                m_WzdSelectionList.AddTail(pInfo);
            }
        }
}


void CWzdSelect::Select(CWzdInfo1 *pInfo)
{
    m_pActiveSelection=pInfo;
    if (!m_WzdSelectionList.Find(pInfo))
    {
        m_WzdSelectionList.AddTail(pInfo);
    }
}


int CWzdSelect::SelectionCount()
{
    return m_WzdSelectionList.GetCount();
}


void CWzdSelect::CutSelections(CList<CWzdInfo1*,CWzdInfo1*> *pList)
{
    CopySelections(pList);
}


void CWzdSelect::CopySelections(CList<CWzdInfo1*,CWzdInfo1*> *pList /*=NULL*/)
{
    if (m_WzdSelectionList.GetCount() <= 0) return;

    // create an archive to a memory file and selection list to it
    CSharedFile file;
    CArchive ar(&file, CArchive::store);
    Serialize(ar);
```

III

14

```cpp
    // close archive and put it in clipboard
    ar.Close();
    COleDataSource *pDS = new COleDataSource();
    pDS->CacheGlobalData(m_clipboardFormat, file.Detach());
    pDS->SetClipboard();

    // if cutting, delete items in selection list from document
    if (pList)
    {
        for (POSITION pos=m_WzdSelectionList.GetHeadPosition();pos;)
        {
            CWzdInfo1 *pInfo=m_WzdSelectionList.GetNext(pos);

            POSITION posx;
            if (posx=pList->Find(pInfo))
            {
                pList->RemoveAt(posx);
            }
            delete pInfo;
        }
    }


    // kill selections
    m_WzdSelectionList.RemoveAll();
    m_pActiveSelection=NULL;
}


BOOL CWzdSelect::CanPasteClipboard()
{
    COleDataObject object;
    return (object.AttachClipboard() &&
        object.IsDataAvailable(m_clipboardFormat));
}
```

Example 77   Array Functions   **459**

```
void CWzdSelect::PasteClipboard(CList<CWzdInfo1*,CWzdInfo1*> *pList)
{
    // reset selections
    m_WzdSelectionList.RemoveAll();
    m_pActiveSelection=NULL;

    // open archive to clipboard and serialize into selection list
    COleDataObject object;
    object.AttachClipboard();
    CFile* pFile = object.GetFileData(m_clipboardFormat);
    if (!pFile) return;
    CArchive ar(pFile, CArchive::load);
    Serialize(ar);
    ar.Close();
    delete pFile; //deletes file

    // add selection back into document
    for (POSITION pos=m_WzdSelectionList.GetHeadPosition();pos;)
    {
        m_pActiveSelection=m_WzdSelectionList.GetNext(pos);
        pList->AddTail(m_pActiveSelection);
    }
}
```

# Example 77   Array Functions

## Objective

You would like to organize some of the data in your application with MFC's array class.

## Strategy

We will use the MFC CArray class to create and maintain our array. We will use the template version of CArray because it's type-safe and easier to remember than the several type-specific versions of CArray. Our example will collect an array of CWzdInfo class objects, but CArray can collect any data type (e.g., ints, floats, etc.)

III

14

## Steps

### Declare Your Array Class

1. To use a template class, you need to include the following in your code.

```
#include <afxtempl.h>
```

2. To declare the `CArray` class, you need to supply two arguments. The first argument is the object type you will be collecting in the array. The second argument is the type statement that the template class should use when referring to objects in this collection. If the first argument specifies a class, the second argument should specify a reference to that class. If the first argument is a pointer to a class, the second argument should be the same pointer.

```
CArray<CWzdInfo,CWzdInfo&> m_WzdClassArray;
```

3. Then, tell `CArray` how big to make your array.

```
m_WzdClassArray.SetSize(10); //sets size of array
```

### Work with an Array Class

1. If you will be storing an entire data object in your array, as opposed to a pointer to an instance of your data object, you will need to add a copy constructor to your data class, if you haven't already done so, in the next two steps.

2. To the declare file of your data class, add

```
CWzdInfo& operator=(CWzdInfo& src);
```

3. To the implementation file of your data class, add

```
CWzdInfo& CWzdInfo::operator=(CWzdInfo& src)
{
    if(this != &src)
    {
        m_sName = src.m_sName;
    // repeat for every member variable in the class
        :    :    :
    }
    return *this;
}
```

Example 77   Array Functions   **461**

4. Then, to store entire objects in your array is a simple matter, use

```
m_WzdClassArray[2]=info1;
m_WzdClassArray[3]=info2;
```

where info1 and info2 are CWzdInfo class objects.

5. You can look through your array with

```
for (int i=0;i<m_WzdClassArray.GetUpperBound();i++)
{
    info=m_WzdClassArray[i];
    /////


}
```

6. You can remove items from your array with

```
info=m_WzdClassArray[3]; //remove class object at index 3
m_WzdClassArray.RemoveAt(3);
```

7. You can destroy your array with

```
m_WzdClassArray.RemoveAll();
```

Or, if you declared your array on the stack, simply returning from a routine will cause CArray's destructor to deallocate all data objects in your array.

8. If you declared an array of object pointers instead of objects, you can use the following to destroy your array and its objects.

```
while (m_WzdPtrArray.GetUpperBound()>-1)
{
    delete m_WzdPtrArray[0];
    m_WzdPtrArray.RemoveAt(0);
}
```

# Notes

- Why use CArray rather than a simple array[] declare? The CArray class includes member functions that you can call to increase or decrease the size of your array, remove selected items, and insert a new item at any point. In other words, functions that you normally would have to write yourself.

- You should use `CArray` (rather than `CList` or `CMap`) when you can maintain your objects by a sequential index. If your indexes will not be sequential or your index won't even be a number, use a `CMap` object, instead. If it doesn't make sense to refer to your objects by an index, use `CList` to collect your objects, instead.

- See the next two examples for `CList` and `CMap` collections.

## CD Notes

- When executing the project on the accompanying CD, set a breakpoint on the `OnTestWzd()` function in `WzdView.cpp`. Then, click the Test/Wzd menu commands and watch as an array class is used.

# Example 78   List Functions

## Objective

You would like to organize some of the data in your application in MFC's linked list class.

## Strategy

We will use the MFC `CList` class to create and maintain our linked list. We will use the template version of `CList` because it's type-safe and easier to remember than the several type-specific versions of `CList`. Our example will collect a list of `CWzdInfo` class objects. However, `CList` can collect any data type (e.g., `ints`, `floats`, etc.).

## Steps

### Declare Your List Class

1. To use a template class, you need to include the following in your code.

```
#include <afxtempl.h>
```

2. To declare the `CList` class, you need to supply two arguments. The first argument is the object type you will be collecting in the list. The second argument is the type statement that the template class should use when referring to objects in this collection. If the first argument specifies a

Example 78   List Functions   **463**

class, the second argument should specify a reference to that class. If the first argument is a pointer to a class, the second argument should be the same pointer.

```
CList<CWzdInfo,CWzdInfo&> m_WzdClassList;
```

## Work with a List Class

1. If you will be storing an entire data object in your list, as opposed to a pointer to an instance of your data object, you will need to add a copy constructor to your data class, if you haven't already done so, in the next two steps.
2. To the declare file of your data class, add

```
CWzdInfo& operator=(CWzdInfo& src);
```

3. To the implementation file of your data class, add

```
CWzdInfo& CWzdInfo::operator=(CWzdInfo& src)
{
    if(this != &src)
    {
        m_sName = src.m_sName;
        // repeat for every member variable in the class
        :   :   :
    }
    return *this;
}
```

4. To add class objects to your list, use

```
m_WzdList.AddTail(info1); // adds to the tail of the list
m_WzdList.AddHead(info2); // adds to the head of the list
```

where `info1` and `info2` are `CWzdInfo` class objects.

5. You can look through your list with

```
for (POSITION pos=m_WzdList.GetHeadPosition();pos;)
{
    info=m_WzdList.GetNext(pos);

    /////

}
```

III

14

6. You can remove items from your list with

```
pos=m_WzdList.FindIndex(1); // find 2nd element in list (zero based)
m_WzdList.RemoveAt(pos);    // remove from class list
```

7. You can destroy your list with

```
m_WzdClassList.RemoveAll();
```

Or, if you declared your list on the stack, you can simply return from the function. This will destroy the class, which will cause it to deallocate all data objects in the list.

8. If you declared a list of object pointers instead of objects, you can use the following to destroy your list and its objects.

```
while (!m_WzdPtrList.IsEmpty())
{
    delete m_WzdPtrList.RemoveHead();
}
```

## Notes

- Use `CList` to maintain a collection of objects that don't need to be referenced directly. Use `CArray` or `CMap`, otherwise.

## CD Notes

- When executing the project on the accompanying CD, set a breakpoint on the `OnTestWzd()` function in `WzdView.cpp`. Then click the Test/Wzd menu commands and watch as a list class is used.

# Example 79   Map Functions

## Objective

You would like to organize some of the data in your application with a data map. A data map is a collection of identical objects that can be referenced by any number or string key.

Example 79   Map Functions   **465**

## Strategy

We will use the MFC `CMap` class to create and maintain a data map. We will use the template version of `CMap` because it's type-safe and easier to remember than the several type-specific versions of `CMap`. Our example will collect two maps of `CWzdInfo` class objects — the first with an integer key, the second with a string key. However, `CMap` can collect any data type (e.g., `ints`, `floats`, etc.) with any type of key.

## Steps

### Declare Your Map Class

1. To use a template class, you need to include the following in your code.

```
#include <afxtempl.h>
```

2. To declare the `CMap` class, you need to supply four arguments. The first two arguments define the variable type of the key used to access objects in this map. The second two arguments define the type of the data object stored in the map. In each of these argument pairs, the first argument is the object type of the key or object. The second argument is the type statement that the template class should use when referring to this key or object. If the first argument specifies a class, the second argument should specify a reference to that class. If the first argument is a pointer to a class, the second argument should be the same pointer. The following declares a map that uses an integer key to reference a collection of `CWzdInfo` objects.

```
CMap<int,int,CWzdInfo,CWzdInfo&> m_WzdIntToClassMap;
```

3. Next, we declare a map that uses a `CString` for its key to again reference a collection of `CWzdInfo` objects.

```
CMap<CString,LPCSTR,CWzdInfo,CWzdInfo &> m_WzdStringToClassMap;
```

See notes below to learn why we use `LPCSTR` here.

### Work with a Map Class

1. If you will be storing entire data objects in your map, as opposed to pointers to instances of your data class, you will need to add a copy

III

14

constructor to your data class, if you haven't already done so, in the next two steps.

2. To the declare file of your data class, add

```
CWzdInfo& operator=(CWzdInfo& src);
```

3. To the implementation file of your data class, add

```
CWzdInfo& CWzdInfo::operator=(CWzdInfo& src)
{
    if(this != &src)
    {
        m_sName = src.m_sName;
        // repeat for every member variable in the class
        :    :    :
    }
    return *this;
}
```

4. To add class objects to either map we declared previously, use

```
m_WzdIntToClassMap[3]=info2;
m_WzdStringToPtrMap["these"]=info1;
```

where info2 and info1 are CWzdInfo class objects.

5. You can look through the map declared with a CString using

```
CString str;
for (POSITION pos = m_WzdStringToClassMap.GetStartPosition();
    pos;)
{

    m_WzdStringToPtrMap.GetNextAssoc(pos,str,pInfo);


    // str contains key
    // pInfo contains pointer to data


}
```

You can do the same with the integer-keyed map by using an integer key.

Example 79   Map Functions   **467**

6. You can remove items from your maps with

```
m_WzdIntToClassMap.RemoveKey(3);
m_WzdStringToClassMap.RemoveKey("them");
```

7. You can destroy your maps with

```
m_WzdIntToClassMap.RemoveAll();
m_WzdStringToClassMap.RemoveAll();
```

However, if you declared your maps on the stack, you don't need to add another line of code — the CMap objects, as well as all the objects they contain, will be automatically destroyed for you.

8. If you declared a map of object pointers instead of objects, you can use the following to destroy your map and its objects.

```
CWzdInfo *pInfo;

for (pos = m_WzdStringToPtrMap.GetStartPosition(); pos;)
{
    m_WzdStringToPtrMap.GetNextAssoc(pos,str,pInfo);
    m_WzdStringToPtrMap.RemoveKey(str);
    delete pInfo;
}
```

# Notes

- CMap doesn't actually use a string key to perform a search. Instead, it converts the string into an integer token that's not too unlike a checksum of the string. The routine that does this conversion requires a LPCSTR in the CMap declaration, even though the key is a CString.

```
CMap<CString,LPCSTR,CWzdInfo,CWzdInfo &> m_WzdStringToClassMap;
```

- Maps are automatically divided into 17 object lists. When looking for an object, a binary search is performed using the key you provide to determine which of these 17 lists contains your object. Then, that list is simply scanned for the object. You can speed up a search on larger map collections by dividing the map into more lists using the InitHashTable() member function of CMap.

```
m_WzdStringToPtrMap.InitHashTable(50);    //create 50 lists
```

III

14

However, you can only set this once before anything has been added to the map.

- MFC uses a `CMap` collection to remember which `CWnd` class object owns which window object. When you use the `FromHandle()` member function of `CWnd`, MFC looks in this map for the `CWnd` object that uses the supplied `hWnd` key. (If there is none, MFC creates a temporary `CWnd` object that wraps the `hWnd` handle you gave it. This `CWnd` object is then destroyed when your application is idle.)

## CD Notes

- When executing the project on the accompanying CD, set a breakpoint on the `OnTestWzd()` function in `WzdView.cpp`. Then click the Test/Wzd menu commands and watch as a map class is used.

# Example 80   System Keys

## Objective

You would like a combination of keys to trigger a menu command. For example, you would like the Alt+D key combination to delete selected items in the view.

## Strategy

First, we will use the Accelerator Editor provided by the Developer Studio to create a resource that will be loaded when our view is created and used when our view has the input focus. We will also look at handling the `WM_KEYDOWN` window message manually for situations when the view doesn't have input focus or when we want to customize the handling of a key.

## Steps

### Automatically Process Accelerator Keys

1. Select the Resource View tab in your Workspace View. Open the Accelerator folder and double-click on the current accelerator resource to open the Accelerator Editor.

2. To modify an existing accelerator key, click on its entry in the editor's list. To add a new key, click on the empty focus box at the bottom of the

Example 80    System Keys    **469**

list. Specify the command ID you want to trigger and the key combination that should trigger it. Clicking the close button makes the change.

3. The framework loads your accelerator table once when creating the application.

You can only have one accelerator table loaded at any given time. However, you can also manually override the accelerator table for a given window by overriding the PreTranslateMessage() function of that window.

In PreTranslateMessage(), you are given the opportunity to translate a key combination into a menu command before the accelerator table gets a crack at it (thus the name PreTranslateMessage()). In this function, you can look for the WM_KEYDOWN or WM_SYSKEYDOWN windows message. If the keyboard combination you want to translate contains the Alt key or the Shift-Alt key combination, you should look for the WM_SYSKEYDOWN message. All other key combinations, including Ctrl-Alt will appear here in a WM_KEYDOWN message.

## Manually Process Alt and Shift-Alt Key Combinations

1. Use the ClassWizard to override the PreTranslateMessage() function in the class that controls the input window which is usually the view.

2. Add the following to your PreTranslateMessage() function to process Alt and Shift-Alt keyboard combinations:

```
if (pMsg->message==WM_SYSKEYDOWN)
{
    switch(pMsg->wParam)
    {
        case VK_DELETE:
            SendMessage(WM_COMMAND,ID_EDIT_CLEAR);
            return TRUE; // translated

        case VK_INSERT:
            SendMessage(WM_COMMAND,ID_EDIT_PASTE);
            return TRUE; // translated
    }
}

return CView::PreTranslateMessage(pMsg);
```

III

14

3. If you want to check whether the Shift key is also pressed, use

```
BOOL bShift=::GetKeyState(VK_SHIFT)&0x8000;
```

## Manually Process All Other Key Combinations

1. Add the following to your PreTranslateMessage() function to process all other keyboard combinations, including Ctrl, Ctrl-Alt, and Shift-Ctrl-Alt.

```
if (pMsg->message==WM_KEYDOWN)
{
    switch(pMsg->wParam)
    {
        case VK_DELETE:
            SendMessage(WM_COMMAND,ID_EDIT_CLEAR);
            return TRUE; // translated

        case VK_INSERT:
            SendMessage(WM_COMMAND,ID_EDIT_PASTE);
            return TRUE; // translated

    }
}
return CView::PreTranslateMessage(pMsg);
}
```

2. If you want to check whether the Shift, Control, or Alt keys are also pressed, use one of the following.

```
BOOL bCtrl=::GetKeyState(VK_CONTROL)&0x8000;
BOOL bShift=::GetKeyState(VK_SHIFT)&0x8000;
BOOL bAlt=::GetKeyState(VK_MENU)&0x8000;
```

To see a complete listing of an overridden PreTranslateMessage() function, see "Listings — PreTranslateMessage() Overriding Function" on page 472.

Example 80   System Keys   **471**

## Notes

- Another approach to this example might have been to add a WM_KEYDOWN or WM_SYSKEYDOWN message handler to your window. However, the approach presented here allows you to also override any definitions in the accelerator table. As the name implies, the PreTranslateMessage() function is called before keystrokes are translated into command messages using the accelerator table. Any successfully translated keystrokes are discarded and, therefore, would never make it to your WM_KEYDOWN or WM_SYSKEYDOWN message handler.

## CD Notes

- When executing the project on the accompanying CD, set a breakpoint on the PreTranslateMessage() function in WzdView.cpp. Then, press various key combinations using the Control, Shift, and Alt keys and watch them being handled.

III

14

## Listings — PreTranslateMessage() **Overriding Function**

```
BOOL CWzdView::PreTranslateMessage(MSG* pMsg)
{

    // ALT key is not pressed or is pressed with CTRL key....
    if (pMsg->message==WM_KEYDOWN)
    {
        BOOL bCtrl=::GetKeyState(VK_CONTROL)&0x8000;
        BOOL bShift=::GetKeyState(VK_SHIFT)&0x8000;

        // only gets here if CTRL key is pressed
        BOOL bAlt=::GetKeyState(VK_MENU)&0x8000;
        switch(pMsg->wParam)
        {
            case 'N':
                if (bAlt&&bShift) //&&bCtrl assumed
                    SendMessage(WM_COMMAND,ID_EDIT_CLEAR);
                else if (bShift&&bCtrl)
                    SendMessage(WM_COMMAND,ID_EDIT_CLEAR);
                else if (bAlt) //&&bCtrl assumed
                    SendMessage(WM_COMMAND,ID_EDIT_CLEAR);
                else if (bCtrl)
                    SendMessage(WM_COMMAND,ID_EDIT_CLEAR);
                else if (bShift)
                    SendMessage(WM_COMMAND,ID_EDIT_CLEAR);
                else
                    SendMessage(WM_COMMAND,ID_EDIT_CLEAR);
                return TRUE; // translated
```

Example 80    System Keys    **473**

```
        case VK_ESCAPE:
            if (bShift)
                SendMessage(WM_COMMAND,ID_EDIT_CLEAR);
            return TRUE; // translated


        case VK_DELETE:
            SendMessage(WM_COMMAND,ID_EDIT_CLEAR);
            return TRUE; // translated

        case VK_INSERT:
            SendMessage(WM_COMMAND,ID_EDIT_PASTE);
            return TRUE; // translated
    }
}
// ALT key is pressed but not with CTRL key....
else if (pMsg->message==WM_SYSKEYDOWN)
{
    BOOL bShift=::GetKeyState(VK_SHIFT)&0x8000;
    switch(pMsg->wParam)
    {
        case 'N':
            if (bShift)
                SendMessage(WM_COMMAND,ID_EDIT_CLEAR);
            else
                SendMessage(WM_COMMAND,ID_EDIT_CLEAR);
            return TRUE; // translated


        case VK_DELETE:
            SendMessage(WM_COMMAND,ID_EDIT_CLEAR);
            return TRUE; // translated
```

III

14

```
        case VK_INSERT:
            SendMessage(WM_COMMAND,ID_EDIT_PASTE);
            return TRUE; // translated
    }
  }

  return CView::PreTranslateMessage(pMsg);
}
```

# Example 81   Time

## Objective

You would like to get, manipulate, and display the time in several exciting formats.

## Strategy

We will simply review the different time classes and functions provided by MFC.

## Steps

### Work with the CTime Class

1. To get the current time, use

```
CTime time;
time=CTime::GetCurrentTime();
```

Example 81   Time   **475**

2. To get an element of that or any time, use

```
int year = time.GetYear();
int month = time.GetMonth();
int day = time.GetDay();
int hour = time.GetHour();
int minute = time.GetMinute();
int second = time.GetSecond();
int DayOfWeek = time.GetDayOfWeek();
```

3. To get a time span, use

```
CTimeSpan timespan(0,0,1,0); // days,hours,minutes,seconds
timespan = CTime::GetCurrentTime() - time;
```

4. To print time to a string, use

```
CString sDate,sTime,sElapsedTime;
sDate = time.Format("%m/%d/%y");               //ex: 12/10/98
sTime = time.Format("%H:%M:%S");               //ex: 9:12:02
sElapsedTime = timespan.Format("%D:%H:%M:%S");
    // %D is total elapsed days
```

Please see `strftime` in your MFC documentation for more time formats.

## Work with the `COleDateTime` **Class**

1. To get the day of the year in order to create a Julian date, use

```
COleDateTime datetime;
datetime = COleDateTime::GetCurrentTime();
int DayOfYear = datetime.GetDayOfYear();
```

2. To read the time from a text string, use

```
COleDateTime datetime;
datetime.ParseDateTime("12:12:23 27 January 93");
```

III

14

## Notes

- CTime and COleDateTime have almost identical member functions. The COleDateTime class, however, allows you to get the day of the year (a great way to create a Julian date) and parse a time text string.

- The advantage to CTime is that it can be contained in a DWORD variable. COleDateTime uses twice the space with a double float. However since CTime is simply a count of how many seconds have elapsed since January 1, 1970, it will have a meltdown in 2037, when it reaches 4294967295 seconds. COleDateTime, on the other hand, is a floating-point number that represents the number of days since December 30, 1900 (where hours are represented by fractional days) and won't run out for several thousand years to come.

## CD Notes

- When executing the project on the accompanying CD, set a breakpoint on the OnWzdType() function in WzdView.cpp. Then, click the Options/Wzd menu commands and watch as the time is manipulated.

# Section IV

# Packaging Examples

The functionality you are creating with your software can be distributed as a single executable program or portions of it can be sectioned off into libraries of functions that any other application can use. Libraries can be either statically or dynamically linked to other applications. A dynamically linked library can share its functions with several applications at once, which can limit the amount of churning your hard drive has to do, even if your system has lots of memory.

The examples in this section all relate to packaging your software in several types of libraries. Examples of statically and dynamically linked libraries and resource libraries are included in this chapter.

# 15

# Libraries

Code libraries are a way of sharing the functionality you've created for your application with other applications. All of the examples in this chapter relate to libraries, including statically and dynamically linked libraries and resource libraries. Dynamically linked libraries can be shared among applications at run time.

**Example 82    Statically Linked C/C++ Libraries**  We use the Developer Studio and some simple edits to create a static C and C++ library you can access from your MFC application.

**Example 83    Dynamically Linked C/C++ Libraries**  We create a DLL that doesn't require MFC to run. This makes for a much smaller, albeit less featured, DLL.

**Example 84    Dynamically Linked MFC Extension Class Libraries**
We create a full-blown MFC DLL that can enjoy all of the MFC classes your application enjoys.

**Example 85    Resource Libraries**  We create an MFC DLL that's devoid of all functionality and exists simply as a repository of resources, such as text strings and dialog box templates. Resource libraries can be used to share the same dialog box template or icon or bitmap among several applications. Applications requiring a mutlilingual interface can also use resource libraries to house all of your application's language-specific resources.

# Example 82 Statically Linked C/C++ Libraries

## Objective

You would like to package your C or C++ functions in a library that you then statically link with your application. Static linking will make your final application executable larger than dynamic linking, but you won't have to worry about including additional DLL files when installing your application.

---

NOTE: This example assumes you will not be using any MFC classes in this library. To create a library of functions that will be using the MFC class library, see Example 84.

---

## Strategy

We will be using the Developer Studio to create our project workspace. The Studio creates a workspace with the correct build settings, but with no source files or even an outline of a source file. Therefore, we will have to either create or import source files into this project. We will also configure our C library so that it can be used directly by our C++ application by using the __cplusplus compiler directive.

## Steps

### Create a Static Library

1. Click the New menu command to open your Developer Studio's New dialog. Select the Projects tab and then Win32 Static Library. Give your project a name and then click the OK button. The Studio will now create an empty project with the correct settings to build either a C or C++ static library.

Example 82   Statically Linked C/C++ Libraries   **481**

A static library project creates either a release or debug version of a .lib file. The release version is optimized for performance, while the debug version is bloated with debug symbols. It's a good idea to change the name of the debug version of your .lib file so that it won't be confused with its release version when including it in your application. The debug version of a library will not correctly link or run with the release version of your application, and vice versa.

2. To give your library's debug version a new name, click on the Studio's Project/Settings menu commands to open the Project Settings dialog. Find the Settings for combo box and select Win32 Debug. Then, select the Library tab and change the name that appears in the Output File Name edit box. Typically, you would append the name listed there with the letter "d" to signify debug.

   The StdAfx.cpp and StdAfx.h files define every run-time library and MFC class your application could ever want. Rather than having to compile this monster every time one of your source modules needs compiling, these files can be precompiled by your project so that just the original code is compiled. This feature is automatically assumed for any C++ file. However, the C++ files in this project will not be using the MFC class library, so we need to manually turn this feature off, as follows.

3. For C++ files, you must change their settings in your project *not* to use the precompiled header. Open the Project Settings dialog box as before, change the Settings for to All Configurations, select your C++ files in the file tree, and then select the C/C++ tab. From the Category combo box, choose Precompiled headers and click the Not using precompiled headers for each of your C++ files.

## Add Functions to a Static Library

1. Either import or create your .c or .cpp files.
2. To allow your library functions to use the standard run-time C library, include the following file.

```
#include <stdlib.h>
```

3. To allow your library functions to use the Windows API, include the following file.

```
#include <windows.h>
```

**IV** 15

4. If you are creating a C library, you will need to include the following at the start and end of your C include file, so that this include file can be used in either a C or C++ file. We need to add these directives because the C++ compiler gives the exact same function name a different internal symbol name than the C compiler. This symbol name is used by the linker to link your application's objects together. To force the C++ compiler to assign a C symbol name to a function, we need to use the extern "C" directive. However, since the extern "C" directive would cause our C compiler to generate a syntax error, we can only conditionally process it if the _cplusplus macro is defined. We use the __cplusplus directive because it's TRUE when compiling a C++ file, but FALSE when compiling a C file. The end result is that when this file is included in a C file, it's business as usual. However, when compiled in a C++ file, the prototypes are forced to be C prototypes.

```
/* at the start of your C .h file */
#ifdef __cplusplus
extern "C" {
#else
#endif  /* __cplusplus */
:    :    :
your C function prototypes
:    :    :
/* at the end of your C .h file */
#ifdef __cplusplus
}
#endif
```

5. For example C and C++ static library functions, see "Listings — Example Static C Library" on page 483 and "Listings — Example Static C++ Library" on page 484.

## Use Your New Static Library

1. You can include your static .lib file in another application by opening the target application's Project Settings dialog and selecting the Link tab. Then add your .lib name to Object/library modules edit box. Make sure you add the debug version of your library to the debug version of

Example 82   Statically Linked C/C++ Libraries   **483**

the target application and release version to release version — otherwise, the target application won't link.

## Notes

- The classic use of a static C or C++ library with your MFC application is porting legacy code from an old application to your new MFC application. If your legacy code is written in C, however, you might be better served to convert each function to static C++ functions, instead. That way, your legacy code can also easily call your MFC functions. (There is no easy way for a C routine to call a C++ routine). How do you convert C source to C++? Simply by changing it's filename from .c to .cpp. Then, compile and fix any syntax errors which are usually caused by your C routine using a reserved C++ keyword (e.g., class, new, etc.).
- See Appendix D for more tips on working with libraries.

## Listings — Example Static C Library

```
/* WzdStatic.h : NonMFC Static Library
 *
 ***************************************************************/

#if !defined WZDSTATIC_H
#define WZDSTATIC_H

#ifdef __cplusplus
extern "C" {
#else
#endif  /* __cplusplus */

void WzdMessageBox(LPSTR pszString);
void DestroyWzdWindow(HWND hWnd);
void WzdFunc3(BOOL b);

#ifdef __cplusplus
}
#endif


#endif
```

IV  15

```c
/*
 * WzdStatic.c : NonMFC "C" Static Library Using the Win32 API directly
 */

#include <stdlib.h>
#include <windows.h>
#include "WzdStatic.h"

BOOL flag=FALSE;

// must call Win32 API directly
void WzdMessageBox(LPSTR pszString)
{
    MessageBox(NULL,pszString,"Wzd Static Library",MB_OK);
}

void DestroyWzdWindow(HWND hWnd)
{
    DestroyWindow(hWnd);
}

void WzdFunc3(BOOL b)
{
    flag=b;
}
```

## Listings — Example Static C++ Library

```cpp
// WzdCpp.h : C++ Static Library
//
#if !defined WZDCPP_H
#define WZDCPP_H

class CWzdDllCpp
{
    BOOL  m_bFlag;

public:
    CWzdCpp();
    ~CWzdCpp();
```

Example 82   Statically Linked C/C++ Libraries   **485**

```
        void WzdMessageBox(LPSTR pszString);
        void DestroyWzdWindow(HWND hWnd);
        void WzdFunc3(BOOL b);

};


#endif

// WzdCpp.cpp : C++ Static library using the Win32 API directly
//

#include <windows.h>
#include <stdlib.h>
#include "WzdCpp.h"

CWzdCpp::CWzdCpp()
{
    m_bFlag=FALSE;
}


CWzdCpp::~CWzdCpp()
{
}


// must call the Win32 API directly
void CWzdCpp::WzdMessageBox(LPSTR pszString)
{
    MessageBox(NULL,pszString,"Wzd DLL Cpp",MB_OK);
}

void CWzdCpp::DestroyWzdWindow(HWND hWnd)
{
    DestroyWindow(hWnd);
}

void CWzdCpp::WzdFunc3(BOOL b)
{
    m_bFlag=b;
}
```

IV 15

# Example 83   Dynamically Linked C/C++ Libraries

## Objective

You would like to package your C or C++ functions in a dynamically linked library, which will allow other applications to share these functions and potentially reduce overall memory requirements. You do not need any MFC classes in these functions.

---

NOTE: This example assumes you will not be using any MFC classes in this library. To create a library of functions that will be using the MFC class library, see Example 84.

---

## Strategy

We will be using the Developer Studio to create our project workspace. The Studio creates a workspace with the correct build settings, but with no source files or even an outline of a source file. Therefore, we will have to either create or import source files into this project. We will also configure our C library so that it can be used directly by our C++ application by using the __cplusplus compiler directive.

## Steps

### Create a Dynamically Linked Library

1. Click the New menu command to open your Developer Studio's New dialog. Select the Projects tab and then Win32 Dynamic-Linked Library. Give your project a name and then click the OK button. The Studio will now create an empty project with the correct settings to build a dynamically linked.

   A dynamically linked library project creates either a release or debug version of a .lib and a .dll file. The release version is optimized for performance, while the debug version is bloated with debug symbols. It's a good idea to change the name of the debug version of your .lib and .dll files so that they won't be confused with their release versions when including them in your application. The debug version of a library will not correctly link or run with the release version of your application, and vice versa.

Example 83    Dynamically Linked C/C++ Libraries    **487**

2. To give your library's debug version a new name, click on the Studio's Project/Settings menu commands to open the Project Settings dialog. Find the Settings for combo box and select Win32 Debug. Then, select the Library tab and change the name that appears in the Output File Name edit box. Typically, you would append the name listed there with the letter "d" to signify debug.

The `StdAfx.cpp` and `StdAfx.h` files define every run-time library and MFC class your application could ever want. Rather than having to compile this monster every time one of your source modules needs compiling, these files can be precompiled by your project, so that just the original code is compiled. This feature is automatically assumed for any C++ file. However, the C++ files in this project will not be using the MFC class library, so we need to manually turn this feature off, as follows.

3. For C++ files, you must change their settings in your project *not* to use the precompiled header. Open the Project Settings dialog box as before, change the Settings for to All Configurations, select your C++ files in the file tree, and then select the C/C++ tab. From the Category combo box, choose Precompiled headers and click the Not using precompiled headers for each of your C++ files.

## Add Functions to a Dynamically Linked Library

1. Either import or create your `.c` or `.cpp` files.
2. To allow your library functions to use the standard run-time C library, include the following file.

```
#include <stdlib.h>
```

3. To allow your library functions to use the Windows API, include the following file.

```
#include <windows.h>
```

A dynamically linked library has functions that can be accessed from outside the library and functions that can only be internally accessed. To tell the compiler and linker to make a function available to the outside world (external), you need to declare it with the `__declspec(dllexport)` function type. However, when including the prototypes for your library in your application, you don't want to include this function type. To resolve this problem, we will be creating a macro that will conditionally mean

**IV**  **15**

__declspec(dllexport) when creating the library or nothing when including these prototypes in your application.

4. Add the following macro to the include that defines the prototypes for the functions in this library.

```
#ifdef WZDDLL_BLD
#define DLL __declspec(dllexport)
#else
#define DLL
#endif
```

5. To make this DLL macro work correctly, you now need to define the WZDDLL_BLD symbol in your DLL project settings. To do this, click on your Developer Studio's Project/Settings menu commands to open the Project Settings dialog box. Find the Settings for combo box and select All Configurations. Then select the C/C++ tab and add WZDDLL_BLD to the Preprocessor definitions edit box.

6. To use this DLL macro to declare a C function as external, stick it into the definition and implementation of the function.

```
void DLL WzdMessageBox(LPSTR pszString);
void DLL WzdMessageBox(LPSTR pszString)
{
        :     :     :
}
```

7. To use this DLL macro to declare a C++ class and all of its functions as external, simply stick it into the declaration of the class.

```
class DLL CWzdDllCpp
{

};
```

8. If you are creating a C library, you will need to include the following at the start and end of your C include file so that this include file can be included used in both a C or C++ file. We need to add these directives because the C++ compiler gives the exact same function name a different internal symbol name than the C compiler. This symbol name is what's used by the linker when it links your application's objects together. To force the C++ compiler to assign a C symbol name to a function, we need

Example 83 Dynamically Linked C/C++ Libraries **489**

to use the `extern` "C" directive. However, since the `extern` "C" directive would cause our C compiler to generate a syntax error, we can only conditionally process it if the `__cplusplus` macro is defined. We use the `__cplusplus` directive because it's `TRUE` when compiling a C++ file, but `FALSE` when compiling a C file. The end result is that when this file is included in a C file, it's business as usual. However, when compiled in a C++ file, the prototypes are forced to be C prototypes.

```
/* at the start of your C .h file */
#ifdef __cplusplus
extern "C" {
#else
#endif  /* __cplusplus */
:     :     :
your C function prototypes
:     :     :
/* at the end of your C .h file */
#ifdef __cplusplus
}
#endif
```

9. For example C and C++ dynamically linked library functions, see "Listings — Example Dynamically Linked C Library" on page 490 and "Listings — Example Dynamically Linked C++ Library" on page 492.

## Use Your New Dynamically Linked Library

1. You can include your dynamically linked `.lib` file in another application by opening that target application's Project Settings dialog and selecting the Link tab. Then, add your `.lib` name to Object/library modules edit box. Make sure you add the debug version of your library to the debug version of the target application and release version to release version — otherwise, the target application won't link. To run your application, the `.dll` file created by your library project must be included in your system's execution path. You can stick it in the same directory as your application's executable, or in one of the directories specified in your system's PATH environment variable. For more tips on working with DLLs, see Appendix D.

**IV 15**

## Notes

- A dynamically linked library allows the functions that it contains to be shared among concurrently executing applications and, therefore, can potentially save on overall memory requirements if each application would have otherwise had to contain the same functions privately. However, since the Windows operating system makes use of virtual memory anyway, the true advantage to DLLs is really in the fact that the functions your application is craving are more likely to be in main memory than out on the disk in the swap file.
- See Appendix D for more tips on working with libraries.

## CD Notes

- Use the Testdll project to test this project.

## Listings — Example Dynamically Linked C Library

```
/* WzdDll.h : NonMFC Dll
 *
 ************************************************************/

#if !defined WZDDLL_H
#define WZDDLL_H

#ifdef __cplusplus
extern "C" {
#else
#endif  /* __cplusplus */



#ifdef WZDDLL_BLD
#define DLL __declspec(dllexport)
#else
#define DLL
#endif

void DLL WzdMessageBox(LPSTR pszString);
void DLL DestroyWzdWindow(HWND hWnd);
void DLL WzdFunc3(BOOL b);
```

Example 83   Dynamically Linked C/C++ Libraries   **491**

```c
#ifdef __cplusplus
}
#endif

#endif


/*
 * WzdDll.c : NonMFC "C" Dll Using the Win32 API directly
 */

#include <windows.h>
#include <stdlib.h>
#include "WzdDll.h"

BOOL flag=FALSE;

// must call Win32 API directly
void DLL WzdMessageBox(LPSTR pszString)
{
    MessageBox(NULL,pszString,"Wzd DLL",MB_OK);
}


void DLL DestroyWzdWindow(HWND hWnd)
{
    DestroyWindow(hWnd);
}


void DLL WzdFunc3(BOOL b)
{
    flag=b;
}
```

IV 15

# Listings — Example Dynamically Linked C++ Library

```cpp
// WzdDllCpp.h : C++ NonMFC Dll
//


#if !defined WZDDLLCPP_H
#define WZDDLLCPP_H

#ifdef WZDDLL_BLD
#define DLL __declspec(dllexport)
#else
#define DLL
#endif

class DLL CWzdDllCpp
{
    BOOL  m_bFlag;

public:
    CWzdDllCpp();
    ~CWzdDllCpp();

    void WzdMessageBox(LPSTR pszString);
    void DestroyWzdWindow(HWND hWnd);
    void WzdFunc3(BOOL b);

};


#endif
```

Example 83 Dynamically Linked C/C++ Libraries **493**

```
// WzdDllCpp.cpp : NonMFC "C++" Dll Using the Win32 API directly
//

#include <windows.h>
#include <stdlib.h>
#include "WzdDllCpp.h"

CWzdDllCpp::CWzdDllCpp()
{
    m_bFlag=FALSE;
}


CWzdDllCpp::~CWzdDllCpp()
{
}


// must call the Win32 API directly
void CWzdDllCpp::WzdMessageBox(LPSTR pszString)
{
    MessageBox(NULL,pszString,"Wzd DLL Cpp",MB_OK);
}


void CWzdDllCpp::DestroyWzdWindow(HWND hWnd)
{
    DestroyWindow(hWnd);
}


void CWzdDllCpp::WzdFunc3(BOOL b)
{
    m_bFlag=b;
}
```

IV 15

# Example 84 Dynamically Linked MFC Extension Class Libraries

## Objective

You would like to package your C++ functions in a dynamically linked library. You would also like to be able to use MFC classes in these functions.

## Strategy

We will be using the Developer Studio to create an MFC DLL project workspace. Then, we will create or import our function files. Each function we intend to export will receive a __declspec(dllexport) function declare.

## Steps

### Create an MFC Extension Library

1. Click the New menu command to open your Developer Studio's New dialog. Select the Projects tab and then MFC AppWizard (dll). Give your project a name and then click the OK button.

2. The AppWizard will prompt you to create one of three types of DLLs. A Regular DLL is one that can be used in an application that does not already use MFC. An MFC Extension DLL must be used with an MFC application. Your third choice is a Regular DLL that is statically linked with the MFC library. Choosing to statically link with MFC will cause your DLL to be much larger, but eliminates one extra file you need to distribute with your application. For this example, choose MFC Extension DLL.

---

**NOTE:** If your application is statically linked with the MFC library, you cannot use an MFC Extension DLL. You must change your application to dynamically link with MFC. The symptom for this problem is hundreds of link errors caused by duplicate MFC symbols.

---

A dynamically linked library project creates either a release or debug version of a .lib and a .dll file. The release version is optimized for

**Example 84   Dynamically Linked MFC Extension Class Libraries   495**

performance, while the debug version is bloated with debug symbols. It's a good idea to change the name of the debug version of your `.lib` and `.dll` files so that they won't be confused with their release versions when including them in your application. The debug version of a library will not correctly link or run with the release version of your application, and vice versa.

3. To give your library's debug version a new name, click on the Studio's Project/Settings menu commands to open the Project Settings dialog. Find the Settings for combo box and select Win32 Debug. Then select the Library tab and change the name that appears in the Output File Name edit box. Typically, you would append the name listed there with the letter "d" to signify debug.

## Add Functions to Your MFC Extension Library

1. The files that the AppWizard creates can be ignored for now. Any functionality you need to add to this DLL must be created or imported. You can also use the ClassWizard to create a new class from an MFC base class — thus the name MFC Extension Library. See "Listings — Example Extension Class" on page 497 for an example of manually added functionality.

   A dynamically linked library has functions that can be accessed from outside the library and functions that can only be internally accessed. To tell the compiler and linker to make a function available to the outside world (external), you need to declare it with the `__declspec(dllexport)` function type. However, when including the prototypes for your library in your application, you don't want to include this function type. To resolve this problem, we will be creating a macro that will conditionally mean `__declspec(dllexport)` when creating the library or nothing when including these prototypes in your application.

2. Add the following macro to the include that defines the prototypes for the functions in this library.

```
#ifdef _WINDLL
#define DLL __declspec(dllexport)
#else
#define DLL
#endif
```

**IV** **15**

The _WINDLL symbol is already defined in the settings of the DLL project that the AppWizard created for you. Therefore, unlike the previous two examples, you need not add your own symbol.

3. To use this DLL macro to declare a C++ class and all of its functions as external, simply stick it into the declaration of the class.

```
class DLL CWzdDllCpp
{


};
```

## Use Your New Dynamically Linked Library

1. You can include your dynamically linked .lib file in another application by opening that target application's Project Settings dialog and selecting the Link tab. Then, add your .lib name to the Object/library modules edit box. Make sure you add the debug version of your library to the debug version of the target application and release version to release version — otherwise, the target application won't link.

2. To run your application, the .dll file created by your library project must be included in your system's execution path. You can stick it in the same directory as your application's executable, or in one of the directories specified in your system's PATH environment variable. For more tips on working with DLL's, see Appendix D

## Notes

- If you create a Regular DLL with Shared MFC DLL, you must add the AFX_MANAGE_STATE() macro to any functions that access an MFC class. This macro must be the first line in that function.

```
void CWzdDllMFC::DestroyWzdWindow(HWND hWnd)
{
    AFX_MANAGE_STATE(AfxGetStaticModuleState());
    // rest of function here
}
```

- See Appendix D for more tips on working with libraries.

Example 84   Dynamically Linked MFC Extension Class Libraries   **497**

## CD Notes

- Use the Testdll project to test this project.

## Listings — Example Extension Class

```
// WzdDllMFC.h : MFC Dll
//

#if !defined WZDDLLMFC_H
#define WZDDLLMFC_H

#ifdef _WINDLL
#define DLL __declspec(dllexport)
#else
#define DLL
#endif

class DLL CWzdDllMFC
{
    BOOL  m_bFlag;

public:
    CWzdDllMFC();
    ~CWzdDllMFC();

    void WzdMessageBox(LPSTR pszString);
    void DestroyWzdWindow(HWND hWnd);
    void WzdFunc3(BOOL b);

};


#endif

// WzdDllMFC.cpp : MFC Dll
//

#include "stdafx.h"
#include "WzdDllMFCx.h"
```

```
CWzdDllMFC::CWzdDllMFC()
{
    m_bFlag=FALSE;
}


CWzdDllMFC::~CWzdDllMFC()
{
}

// can use MFC classes and static functions
void CWzdDllMFC::WzdMessageBox(LPSTR pszString)
{
    AfxMessageBox(pszString);
}


void CWzdDllMFC::DestroyWzdWindow(HWND hWnd)
{
    CWnd wnd;
    wnd.Attach(hWnd);
    wnd.DestroyWindow();
}


void CWzdDllMFC::WzdFunc3(BOOL b)
{
    m_bFlag=b;
}
```

# Example 85   Resource Libraries

## Objective

You would like to package all of your resources in a DLL so that they can be shared among your other applications. You would like to segregate all of your strings and dialog templates into a DLL so that different versions can be created to support different languages (e.g., French, Italian, Esperanto, etc.).

Example 85    Resource Libraries    **499**

# Strategy

We will be creating an MFC Extension DLL that will only contain resources. Your application can access the resources in an MFC Extension DLL the same way it accesses its own resources through resource IDs. Your only concern when using resources from a DLL is that the resource IDs in your application don't conflict with those in the DLL.

# Steps

## Create a Resource Library

1. Follow the steps in the previous example to create an MFC Extension DLL.
2. Add at least one class to this DLL, which can optionally be devoid of functions or variables.
3. Use the Developer Studio editors to add resources to this project. Make sure the resource IDs don't conflict with any other DLL or your main application. You may want to assign ID ranges or possibly use text IDs instead of numeric ones. To create a text ID, simply enclose the ID in double quotes.

```
"MYDLL_RESOURCE1"
```

4. Cut these new resource IDs from your resource.h file and paste them into a new .h file that you can include in a project that will be using this resource DLL.
5. To make your application multilingual, you should never use text strings directly in your code. Instead, you should put all text in a string table in a resource DLL and access that string as seen below. You should also put all dialog box templates in your resource DLL.

```
CString str;
str.LoadString(id);
AfxMessageBox(str);
```

IV 15

# Notes

- It may be more prudent not to worry about adding multilingual support to your application until after the main development effort is done.

Constantly updating a string table, especially when several developers are involved, can make for a sloppy resource file. Mixing this task in with your main development effort can also be an unnecessary distraction that can just as easily be accomplished at the end of the project.

## CD Notes

- Use the Testdll project to test this project.

# Control Window Styles

There are several ways to create a control window using MFC and the Windows API. You can use the Dialog Editor to add a control to a dialog box template, which will then be created by the Windows API when the dialog box is created. Or you can use an MFC control class, such as CButton, to manually create the control window yourself. If MFC hasn't yet wrapped a common control in a class, you can still create the control using the generic CWnd class and the name of the common control's Window Class.

Each control comes loaded with different visual features that you can access by simply changing its window style. The only way you can see the full range of styles available to you is by experimenting with the Dialog Editor, and even then, the Dialog Editor doesn't support every window style.

This appendix documents the more significant windows styles available to you for each control. In particular, the window styles that can affect the appearance of a control are listed here along with a picture of its effect.

# Control Windows Provided by Windows 3.1 and Above

## Button Controls

### Create Button Controls with the Windows API

```
HWND CreateWindowEx(dwExStyle, "BUTTON", "Text",
    WS_CHILD|WS_VISIBLE|dwStyle x, y, width, height,
    hWndParent, (HMENU) id, hInstance, NULL);
```

### Create Button Controls with MFC

```
CButton m_button;                 // usually embedded in parent class
m_button.Create(
    "Text", WS_VISIBLE|WS_CHILD|dwStyle, rect, pParentWnd, id);
```

## Visual Styles

## Figure A.1    Button Control Styles

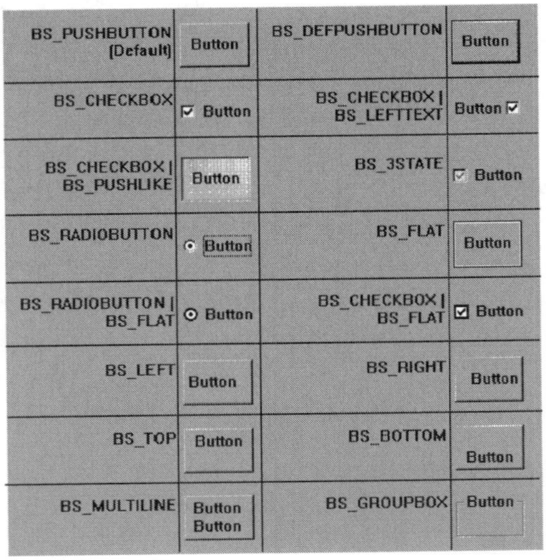

## Other Styles

| | |
|---|---|
| BS_DEFPUSHBUTTON | A button with this style is "clicked" when the user presses the Enter key. This is only true, however, if its parent window (e.g., a dialog box) has the input focus. Only one button in a parent window can have this style at one time. |
| BS_AUTOCHECKBOX | Automatically toggles its state between checked and unchecked when the user clicks on it. |
| BS_AUTO3STATE | Automatically toggles between three states of checked, unchecked, and indeterminate when the user clicks on it. |
| BS_AUTORADIOBUTTON | Automatically unselects other radio buttons in its same group. |
| BS_OWNERDRAW | Parent must draw the button. |
| BS_ICON | The button will draw an icon on its face, which you must specify in another step. |
| BS_BITMAP | The button will draw a bitmap on its face, which you must specify in another step |

### Notes

- Notice that a group box is actually a button control that's been told to draw a box, put its name in the upper-left corner, and then ignore any input. This piling up of nonsimilar features in a single control is possibly due to the days when there were only seven common controls and no room to add more.
- Additional button effects can be achieved by using the border styles shown in "Plain Window Styles" on page 533.

# Static Controls

## Create Static Controls with the Windows API

```
HWND CreateWindowEx(dwExStyle, "STATIC", "Text",
    WS_CHILD|WS_VISIBLE|dwStyle x, y, width, height,
    hWndParent, (HMENU) id, hInstance, NULL);
```

IV

A

## Create Static Controls with MFC

```
CStatic m_static;               // usually embedded in parent class
m_static.Create(
    "&Text", WS_VISIBLE|WS_CHILD|dwStyle, rect, pParentWnd,
    id);
```

## Visual Styles

## Figure A.2    Static Control Styles

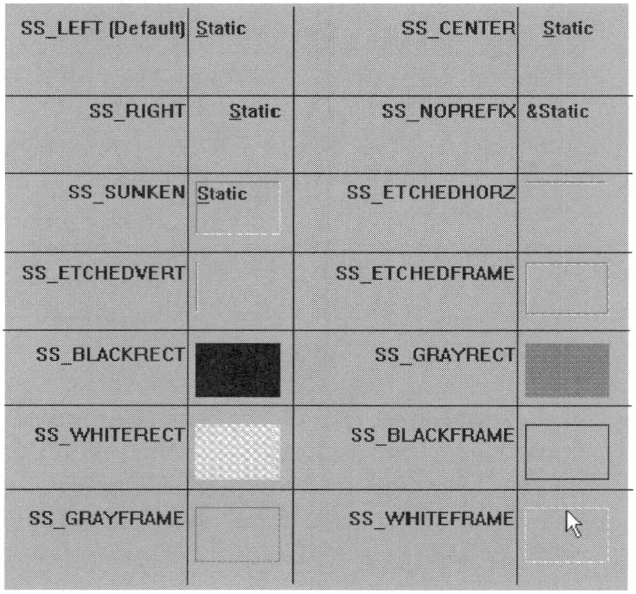

## Other Styles

| | |
|---|---|
| SS_OWNERDRAW | Parent window draws control. |
| SS_NOTIFY | Static controls normally pass any mouse clicks directly to the parent window. This style forces the control to process these messages itself. |
| SS_ICON | An icon, which is set in another step, is drawn in the control. |

| SS_BITMAP | A bitmap, which is specified in another step, is drawn in the control. |
|---|---|
| SS_ENHMETAFILE | An enhanced metafile, which is specified in another step, is drawn in this control. |
| SS_CENTERIMAGE | If this static is displaying an icon, bitmap, or enhanced metafile, this style causes it to be centered in the control. |

### Notes

- If you specify an ampersand character (&) before a letter in your static text, that character will be underlined. When the user presses that letter, the control that comes next in the tab order will receive input focus. The tab order is simply the order of control entries in the dialog resource.

- A static control with a SS_ETCHEDVERT or SS_ETCHEDHORZ style will draw a simple etched line for you. How can you do this with the Dialog Editor? Where is this option? There is none. However, there is a way ... see Example 69 for how. The text equivalent (text characters seemingly etched into the window) can be achieved with CDC::DrawState() with the disabled option. DrawState() is used to draw disabled menu items.

# Edit Controls

## Create Edit Controls with the Windows API

```
HWND CreateWindowEx(dwExStyle, "EDIT", "Text",
    WS_CHILD|WS_VISIBLE|dwStyle x, y, width, height,
    hWndParent, (HMENU) id, hInstance, NULL);
```

## Create Edit Controls with MFC

```
CEdit m_edit;                    // usually embedded in parent class
m_edit.CreateEx(
    WS_VISIBLE|WS_CHILD|dwStyle, rect, pParentWnd, id);
```

**IV**

**A**

## Visual Styles

### Figure A.3 Edit Control Styles

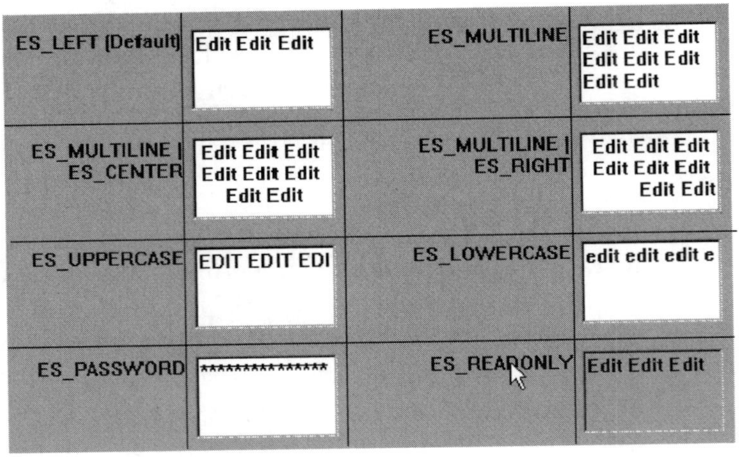

## Other Styles

| ES_AUTOVSCROLL | Causes the entry to automatically scroll left as new text is entered. |
|---|---|
| ES_AUTOHSCROLL | Causes the text to scroll up as new text is entered. Only valid with the ES_MULTILINE style. |
| ES_NOHIDESEL | Causes selected text to remain selected even if the control loses input focus. |
| ES_WANTRETURN | On a multiline control, allows the enter key to be passed to the control rather than acting as if you had pushed the default button on the parent window. |

## Notes

- The edit controls shown here have had an additional border style added to them. By default, an edit control window has no border.
- All of the text in an edit control will have the same font (which you can set with SetFont()). If you would like to have some text underlined and some boldface, you should use the Rich Edit Control, instead.

# List Box Controls

## Create List Box Controls with the Windows API

```
HWND CreateWindowEx(dwExStyle, "LISTBOX", "",
    WS_CHILD|WS_VISIBLE|dwStyle x, y, width, height,
    hWndParent, (HMENU) id, hInstance, NULL);
```

## Create List Box Controls with MFC

```
CListBox m_listbox;              // usually embedded in parent class
m_listbox.Create(
    WS_VISIBLE|WS_CHILD|dwStyle, rect, pParentWnd, id);
```

### Visual Styles

**Figure A.4  Listbox Control Styles**

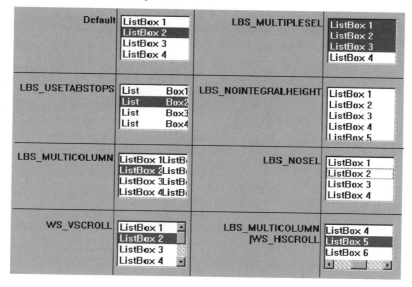

## Initialize Control

- To reset the contents, use

```
m_listbox.ResetContent();
```

IV

A

- To add a string, use

```
m_listbox.AddString("ListBox1");
```

- To set the width of each column when LBS_MULTICOLUMN is set, use

```
m_listbox.SetColumnWidth(60);
```

- To set tab stops when LBS_USETABSTOPS is set, use

```
int tabs[NUM]={10,20,30};
m_listbox.SetTabStops(NUM,tabs);
```

## Other Styles

| | |
|---|---|
| LBS_NOTIFY | Causes listbox control to process mouse clicks rather than passing them onto the parent window. |
| LBS_SORT | Strings are sorted alphabetically. |
| LBS_OWNERDRAWFIXED | The parent window must draw the control, and each entry is the same size vertically. |
| LBS_OWNERDRAWVARIABLE | The parent window must draw the control, and each entry can have a different size. |
| LBS_HASSTRINGS | For owner-drawn controls, the control will still maintain strings internally so that you can still set and get strings from the control. |
| LBS_EXTENDEDSEL | Like LBS_MULTIPLESEL, several items can be selected at once. But with LBS_EXTENDEDSEL, the user must push Shift and Control keys to make multiple selects. With LBS_MULTIPLESEL, the user just clicks on an entry. |

## Notes

- The WS_HSCROLL style will only work when the LBS_MULTICOLUMN style set — but then the WS_VSCROLL style doesn't work.
- Normally, a listbox will override the rectangle size you assign it so that the last line is not partially obscured by the bottom of the control. To turn this off, use the LBS_NOINTERGRALHEIGHT style.

- The listbox controls shown here have had an additional border style added to them. By default, a list box has no border.

# Scroll Bars

## Create Scroll Bars with the Windows API

```
HWND CreateWindowEx(dwExStyle, "SCROLLBAR", "",
    WS_CHILD|WS_VISIBLE|dwStyle x, y, width, height,
    hWndParent, (HMENU) id, hInstance, NULL);
```

## Create Scroll Bars with MFC

```
CScrollBar m_scrollbar;      // usually embedded in parent class
m_scrollbar.Create(
    WS_VISIBLE|WS_CHILD|dwStyle, rect, pParentWnd, id);
```

## Visual Styles

**Figure A.5    Scrollbar Control Styles**

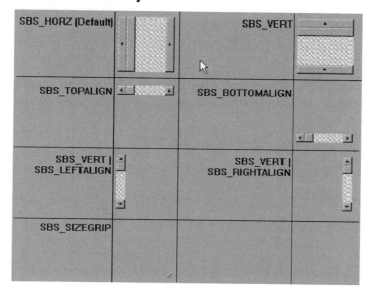

## Notes

- Unlike the scrollbars that Windows will draw in the nonclient area of any window, these scrollbars are, in fact, entire child windows unto themselves. A.ent window can have its own scroll bars and any number of scrollbar control windows. Unfortunately, both nonclient scrollbars and control window scrollbars both report back to the parent using the same `WM_HSCROLL` and `WM_VSCROLL` windows messages. To distinguish them in the parent window, you need to look at the scrollbar handle in the `lParam` of the message, which MFC wraps for you in a `CScrollBar` class when you add a handler for this message.

- When `SBS_HORZ` and `SBS_VERT` are specified alone, the width and height of the scrollbar can be any size. When specified along with `SBS_TOPALIGN`, `SBS_BOTTOMALIGN`, etc., the scrollbar assumes a standard width, as seen in Figure A.5.

- Creating a scrollbar control with the `SBS_SIZEGRIP` style will allow the user to grab that control with the mouse to resize the window in which the control is sitting, no matter where the control is located in that window.

# Combo Boxes

## Create Combo Boxes with the Windows API

```
HWND CreateWindowEx(dwExStyle, "COMBOBOX", "",
    WS_CHILD|WS_VISIBLE|dwStyle x, y, width, height,
    hWndParent, (HMENU) id, hInstance, NULL);
```

## Create Combo Boxes with MFC

```
CComboBox m_combobox;           // usually embedded in parent class
m_combobox.Create(
    WS_VISIBLE|WS_CHILD|dwStyle, rect, pParentWnd, id);
```

## Visual Styles

### Figure A.6    Combo Box Control Styles

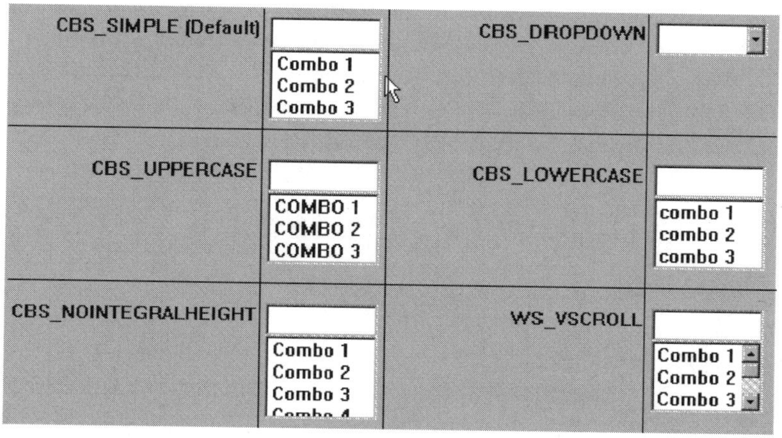

## Initialize Control

- To add a string to a combo box, use

```
m_combobox.AddString("Combo 1");
```

## Other Styles

| | |
|---|---|
| CBS_OWNERDRAWFIXED | The parent draws the control, and each entry has the same height. |
| CBS_OWNERDRAWVARIABLE | The parent draws the control and each entry can have a different height. |
| CBS_HASSTRINGS | Even if the control is drawn by the parent, the control still maintains a string for each entry. |
| CBS_AUTOHSCROLL | If the edit control is enabled, entered text will automatically scroll left when entered past the end of the window. |
| CBS_SORT | The list will be alphabetically sorted. |

**IV**

**A**

## Notes

- Combo boxes actually control two of their very own controls windows: an EDIT control for the edit box and a ComboLBox control for the list box portion. Because some window messages, such as mouse clicks, go to one of these control windows and not the combo box, you will occasionally need to subclass them directly. Just search the child windows under the combo box. If you use the dropdown style, however, the list box control will not be available until it drops down.

# Control Windows Provided by Windows 95/NT and Above

## Rich Edit Controls

### Create Rich Edit Controls with the Windows API

```
HWND CreateWindowEx(dwExStyle, "RICHEDIT", "",
    WS_CHILD|WS_VISIBLE|dwStyle x, y, width, height,
    hWndParent, (HMENU) id, hInstance, NULL);
```

### Create Rich Edit Controls with MFC

```
CRichEditCtrl m_richeditctrl;           // usually embedded in
                                        // parent class

m_richeditctrl.Create(
    WS_VISIBLE|WS_CHILD|dwStyle, rect, pParentWnd, id);
```

## Notes

- Rich Edit Controls use most of the same window styles as Edit Boxes. The major difference is that Rich Edit Controls can process Rich Text Formatted files, which allow text to be several colors, styles, and font sizes, all within the same sentence. To get this support, however, you can't use SetWindowText() or GetWindowText() to put text in the window. Instead, you need to use the member functions StreamIn() and StreamOut(). In fact, you can still use SetWindowText() and GetWindowText() — but any text sequences the RTF format uses to change text format will become visible.

# List Controls

## Create List Controls with the Windows API

```
HWND CreateWindowEx(dwExStyle, "SysListView32", "",
    WS_CHILD|WS_VISIBLE|dwStyle x, y, width, height,
    hWndParent, (HMENU) id, hInstance, NULL);
```

## Create List Controls with MFC

```
CListCtrl m_listctrl;          // usually embedded in parent class
m_listctrl.Create(
    WS_VISIBLE|WS_CHILD|dwStyle, rect, pParentWnd, id);
```

## Visual Styles

## Figure A.7    List Control Styles

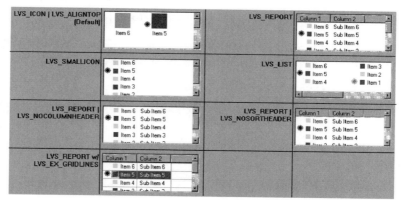

## Initialize Control

- To add columns to a list control, use

```
m_listctrl.InsertColumn(0,"Column 1",LVCFMT_LEFT,width,0);
m_listctrl.InsertColumn(1,"Column 1",LVCFMT_RIGHT,width,1);
```

IV

A

- To add images that can be used by the list control, use

```
m_listctrl.SetImageList(&m_imageLarge, LVSIL_NORMAL);
m_listctrl.SetImageList(&m_imageSmall, LVSIL_SMALL);
m_listctrl.SetImageList(&m_imageState, LVSIL_STATE);
```

- To add a new line, use

```
LV_ITEM lvi;
lvi.mask = LVIF_TEXT | LVIF_IMAGE;
lvi.iItem = 0;
lvi.iSubItem = 0;
lvi.pszText = "Item 1";
lvi.iImage = 1;
int inx=m_listctrl.InsertItem(&lvi);
```

- To add an item to a column, use

```
m_listctrl.SetItemText(inx, 1,  "Sub Item 1");
```

- To change the state image of a line, use

```
m_listctrl.SetItemState(inx, INDEXTOSTATEIMAGEMASK(1),
    LVIS_STATEIMAGEMASK);
```

- To set an extended style, such as LVS_EX_GRIDLINES, use

```
m_listctrl.SendMessage(LVM_SETEXTENDEDLISTVIEWSTYLE,
    0,LVS_EX_GRIDLINES|LVS_EX_FULLROWSELECT);
```

## Other Styles

| | |
|---|---|
| LVS_SINGLESEL | Allows only one item to be selected at a time. |
| LVS_SHOWSELALWAYS | Causes selection(s) to stay selected, even if the window loses focus. |
| LVS_SORTASCENDING | Sorts list entries in ascending order. |
| LVS_SORTDESCENDING | Sorts list entries in descending order. |
| LVS_EDITLABELS | Allows list entries to be edited in place. When the user is done editing, the parent will receive an LVN_ENDLABELEDIT message. |
| LVS_NOSCROLL | Disables scrolling. |
| LVS_OWNERDRAWFIXED | The parent is responsible for drawing an entry. |

## Notes

- Column widths are required in pixels. A quick way to convert character widths into pixels is

```
CDC* dc = GetDC();

TEXTMETRIC tm;
dc->GetTextMetrics(&tm);
ReleaseDC(dc);
```

Then, when specifying the column width, you can use the number of characters desired * `tm.tmAveCharWidth`.

- The image list that you specify for use with a control must stay around after you set the control; otherwise, no images will appear and the control will act like nothing is wrong. The best way to keep an image list around is to embed it in the parent window class of the control.

```
class CDialog
{
    :    :    :
CImageList m_imagelist;
};
```

# Extended Combo Boxes

## Create Extended Combo Boxes with the Windows API

```
HWND CreateWindowEx(dwExStyle, "ComboBoxEx32", "Text",
    WS_CHILD|WS_VISIBLE|dwStyle x, y, width, height,
    hWndParent, (HMENU) id, hInstance, NULL);
```

## Create Extended Combo Boxes with MFC

```
CComboBoxEx m_comboex;    // usually embedded in parent class
m_comboex.Create(
    WS_VISIBLE|WS_CHILD|dwStyle, rect, pParentWnd, id);
```

**IV**

**A**

## Visual Styles

### Figure A.8  Extended Combo Box Control Styles

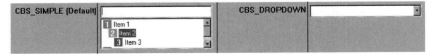

### Initialize Control

- To add images that can be used by the control, use

```
m_comboex.SetImageList(&m_imageList);
```

- To add items to the control, use

```
COMBOBOXEXITEM cbei;
cbei.mask = CBEIF_INDENT | CBEIF_TEXT |CBEIF_IMAGE;
cbei.iItem        = 0;
cbei.pszText      = pszItem1;
cbei.cchTextMax   = sizeof(pszItem1);
cbei.iImage       = 0;           //IMAGE TO DISPLAY
cbei.iSelectedImage = 1;         //IMAGE TO DISPLAY WHEN SELECTED
cbei.iIndent      = 0;           //# OF PIXELS TO INDENT
m_comboex.InsertItem(&cbei);
```

### Notes

- The extended combo box uses most of the same window styles as the regular combo box, except that you can specify an image to be place next to an entry and entries can be indented by a pixel amount.
- The extended combo box is relatively new and some versions of MFC have yet to support it. You can still access it, however, by using the Windows API call and communicating with the control using windows messages.
- You are responsible for maintaining the image list object even after you've set it into the control. Failure to do so will cause images not to appear, but *no* errors will be sent from MFC or windows.

# Animation Controls

## Create Animation Controls with the Windows API

```
HWND CreateWindowEx(dwExStyle, "SysAnimate32", "",
    WS_CHILD|WS_VISIBLE|dwStyle x, y, width, height,
    hWndParent, (HMENU) id, hInstance, NULL);
```

## Create Animation Controls with MFC

```
CAnimateCtrl m_animate;     // usually embedded in parent class
m_animate.Create(
    WS_VISIBLE|WS_CHILD|dwStyle, rect, pParentWnd, id);
```

## Visual Styles

| | |
|---|---|
| ACS_CENTER | Centers the .avi image in the control. |
| ACS_TRANSPARENT | Draws the control without a background color to give the appearance the control is being drawn directly on the parent window. |
| ACS_AUTOPLAY | Automatically starts playing when the .avi file is opened. |
| ACS_TIMER | Normally, the .avi file is played back by another thread. This style causes it to be played back without a thread. |

## Initialize Control

- To open an .avi file and play it, use

```
m_animate.Open("filename");
m_animate.Play(0,                // from frame (0=start)
    -1,                          // to frame (-1=end)
    1);                          // number of replays (-1=forever)
```

- To stop and close an .avi file, use

```
m_animate.Stop();
m_animate.Close();
```

IV

A

### Notes

- See Examples 9 and 57.

# Slider Controls

## Create Slider Controls with the Windows API

```
HWND CreateWindowEx(dwExStyle, "msctls_trackbar32", "",
    WS_CHILD|WS_VISIBLE|dwStyle x, y, width, height,
    hWndParent, (HMENU) id, hInstance, NULL);
```

## Create Slider Controls with MFC

```
CSliderCtrl m_slider;          // usually embedded in parent class
m_slider.Create(
    WS_VISIBLE|WS_CHILD|dwStyle, rect, pParentWnd, id);
```

## Visual Styles

## Figure A.9   Slider Control Styles

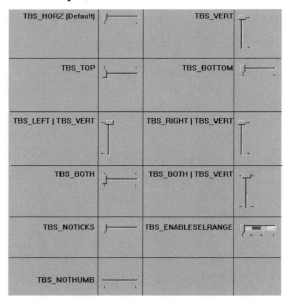

## Other Styles

| TBS_AUTOTICKS | Tick marks will be created for each increment in the control's range. |
|---|---|
| TBS_TOOLTIPS | Causes a popup to appear that displays the bar's current position when the mouse cursor is held over the control. |

## Notes

- The Slider (a.k.a., Trackbar) control reports back to its parent window using the same WM_VSCROLL and WM_HSCROLL messages that scrollbar control windows use. The OnHScroll() and OnVScroll() message handlers wrap the control in a CScrollBar class, but it's really wrapping a Slider control and you may need to type cast this class to a CSliderCtrl class.

# Tree View Controls

## Create Tree View Controls with the Windows API

```
HWND CreateWindowEx(dwExStyle, "SysTreeView32", "Text",
    WS_CHILD|WS_VISIBLE|dwStyle x, y, width, height, hWndParent,
    (HMENU) id, hInstance, NULL);
```

## Create Tree View Controls with MFC

```
CTreeCtrl m_treectrl;      // usually embedded in parent class
m_treectrl.Create(
    WS_VISIBLE|WS_CHILD|dwStyle, rect, pParentWnd, id);
```

**IV**

**A**

## Visual Styles

### Figure A.10  Tree Control Styles

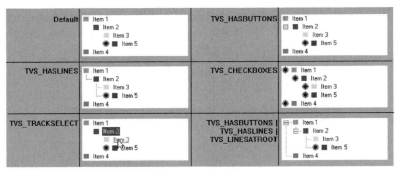

## Initialize Control

- To add a list of images that can be displayed in the control (the blocks seen in Figure A.10 are LVSIL_NORMAL images and the stars are LVSIL_STATE images), use

```
m_treectrl.SetImageList(&m_imageSmall, LVSIL_NORMAL);
m_treectrl.SetImageList(&m_imageState, LVSIL_STATE);
```

- To add an item to the list, use

```
TV_INSERTSTRUCT    tvi;
tvi.item.mask = TVIF_IMAGE | TVIF_TEXT |TVIF_SELECTEDIMAGE;
tvi.hParent = TVI_ROOT;
tvi.hInsertAfter = TVI_LAST;
tvi.item.iImage = tvi.item.iSelectedImage =0;
tvi.item.pszText = pszItem1;
tvi.item.stateMask = TVIS_STATEIMAGEMASK;
tvi.item.state = INDEXTOSTATEIMAGEMASK(1);
HTREEITEM hTreeRoot = m_treectrl.InsertItem(&tvi);
```

## Other Styles

| | |
|---|---|
| TVS_EDITLABELS | Allows the user to directly edit the entries in the control. |
| TVS_DISABLEDRAGDROP | Prevents entries from being dragged from the control. |
| TVS_SHOWSELALWAYS | Causes selections to remain selected, even when the control loses input focus. |
| TVS_TRACKSELECT | Causes the control to react visually when the mouse cursor moves over it. |

## Notes

- As can be seen in the previous examples, the tree control, by default, has no border. You need to add one yourself by specifying it with a border window style. See the last examples in this Appendix for border window styles.

# Spin Button Controls

## Create Spin Button Controls with the Windows API

```
HWND CreateWindowEx(dwExStyle, "msctls_updown32", "Text",
    WS_CHILD|WS_VISIBLE|dwStyle x, y, width, height, hWndParent,
    (HMENU) id, hInstance, NULL);
```

## Create Spin Button Controls with MFC

```
CSpinButtonCtrl m_spin;     // usually embedded in parent class
m_spin.Create(
    WS_VISIBLE|WS_CHILD|dwStyle, rect, pParentWnd, id);
```

## Visual Styles

**Figure A.11  Spin Control Styles**

## Other Styles

| | |
|---|---|
| UDS_WRAP | Causes the "buddy" window's value to wrap to the start or end if the value goes past the end or start of the control's range. |
| DS_NOTHOUSANDS | Doesn't insert comma or period between thousands when formatting the value. |
| UDS_AUTOBUDDY | Automatically selects the previous window in the parent window as its "buddy" window. Otherwise, you must use the SetBuddy() function of CSpinButtonCtrl. The Spin control increments or decrements the value in the "buddy" window. |
| UDS_SETBUDDYINT | Causes the control to update its "buddy" window with a formatted text string version of its current numeric value. |
| UDS_ALIGNRIGHT | Causes the control to be drawn flush with the right side of its "buddy" window. |
| UDS_ALIGNLEFT | Causes the control to be drawn flush with the left side of its "buddy" window. |
| UDS_ARROWKEYS | Causes the arrow keys to also increment or decrement this control. |
| UDS_HOTTRACK | Causes the control to react visually when the mouse cursor moves over it. |

## Notes

- The main use of the Spin or Up/Down Button control is to increment an edit control. In this scenario, the button is called a "buddy" button.
- With the default window style, the button can be any height, but only a standard width. With the UDS_HORZ style, however, the button is any size you make it, as seen in Figure A.11.

## Progress Control Styles

### Create Progress Control Styles with the Windows API

```
HWND CreateWindowEx(dwExStyle, "msctls_progress32", "Text",
    WS_CHILD|WS_VISIBLE|dwStyle x, y, width, height, hWndParent,
    (HMENU) id, hInstance, NULL);
```

### Create Progress Control Styles with MFC

```
CProgressCtrl m_progress;     // usually embedded in parent class
m_progress.Create(
    WS_VISIBLE|WS_CHILD|dwStyle, rect, pParentWnd, id);
```

### Visual Styles

### Figure A.12  Progress Control Styles

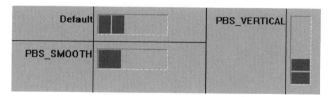

### Initialize Control

- To set the range over which the control represents, use

```
m_progress.SetRange(0,100);
```

- To set the actual progress of the control, use

```
m_progress.SetPos(33);
```

### Notes

- You might have seen progress controls that also display a number indicating the progress in the center of the control. There isn't a windows style that will do this for you automatically. You can instead simply create a plain window and then override the WM_PAINT message to draw the bar with Rectangle() and the percent number with TextOut().

**IV**

**A**

# Header Control Styles

## Create Header Control Styles with the Windows API

```
HWND CreateWindowEx(dwExStyle, "SysHeader32", "",
    WS_CHILD|WS_VISIBLE|dwStyle x, y, width, height,
    hWndParent, (HMENU) id, hInstance, NULL);
```

## Create Header Control Styles with MFC

```
CHeaderCtrl m_header;        // usually embedded in parent class
m_header.Create(
    WS_VISIBLE|WS_CHILD|dwStyle, rect, pParentWnd, id);
```

## Visual Styles

### Figure A.13  Header Control Styles

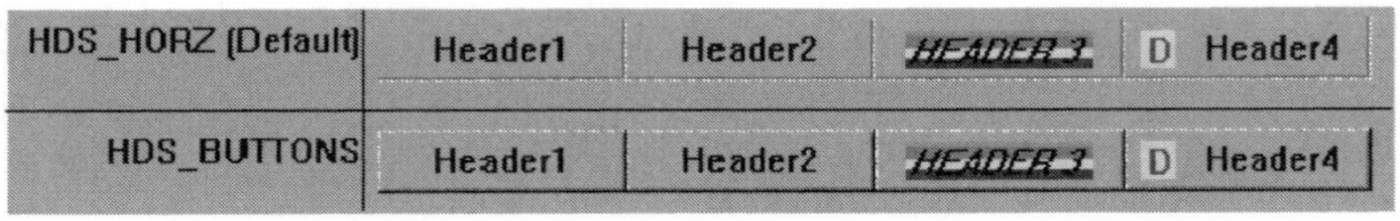

## Initialize Control

- To add a text column header (as seen in columns 1 and 2 of Figure A.13), use

```
HD_ITEM hdi;
hdi.mask=HDI_TEXT|HDI_WIDTH|HDI_FORMAT;
hdi.fmt=HDF_CENTER|HDF_STRING;
hdi.cxy=100;                         // column width in pixels
hdi.pszText=pszHeader1;              // "text"
hdi.cchTextMax=sizeof(pszHeader1);
m_header.InsertItem(0,&hdi);         // 0=column 1
```

- To add a bitmap for a header (as seen in column 3 of Figure A.13), use

```
hdi.mask = HDI_FORMAT | HDI_WIDTH | HDI_BITMAP;
hdi.fmt = HDF_CENTER|HDF_BITMAP;
hdi.cxy = 100;                              //column width in pixels
hdi.hbm = HBITMAP(m_bitmap);
m_header.InsertItem(2,&hdi);
```

- To add a text column with an icon (as seen in column 4 of Figure A.13), use

```
m_header.SetImageList(&m_imageList);
hdi.mask=HDI_IMAGE| HDI_FORMAT| HDI_TEXT;
hdi.fmt=HDF_LEFT |HDF_IMAGE | HDF_STRING;
hdi.pszText=pszHeader4;                  // "text"
hdi.cchTextMax=sizeof(pszHeader4);
hdi.iImage= 1;                           // image number in image list
hdi.cxy=100;                             // column width in pixels
m_header.InsertItem(3,&hdi);
```

## Other Styles

| | |
|---|---|
| HDS_DRAGDROP | Allows the user to change which header is above which column. |
| HDS_FULLDRAG | Causes the control to display its full title, even when the user is resizing it. |
| HDS_HOTTRACK | Causes the control to react visually when the mouse cursor moves over it. |

## Notes

- The header control by itself is fairly worthless. Its real importance is that it's used as the header of a list control when in report mode. Once the list control has been created, you can wrap a CHeaderCtrl class around the header control used by that list control with

```
CListCtrl m_listctrl;
m_listctrl.Create(…);
CHeaderCtrl *pHeader=m_listctrl.GetHeaderCtrl();
```

IV

A

Then you can change the style of that list control's header using `ModifyStyle()` and the styles listed previously.

• The `GetHeaderCtrl()` is a new member function of `CListCtrl` and may not be available with your version of MFC. If so, you can still locate the handle of the header control yourself by searching the child windows of a list control for the `SysHeader32` Windows Class, and then attaching it to a `CHeaderCtrl` class yourself.

• Even after you have set an image in a header control, you are still responsible for keeping that image in memory. Usually, you can accomplish this by making the image list a member variable of the parent window.

# Tab Controls

## Create Tab Controls with the Windows API

```
HWND CreateWindowEx(dwExStyle, "SysTabControl32", "",
    WS_CHILD|WS_VISIBLE|dwStyle x, y, width, height,
    hWndParent, (HMENU) id, hInstance, NULL);
```

## Create Tab Controls with MFC

```
CTabCtrl m_tab;              // usually embedded in parent class
m_tab.Create(
    WS_VISIBLE|WS_CHILD|dwStyle, rect, pParentWnd, id);
```

## Visual Styles

## Figure A.14  Tab Control Styles.

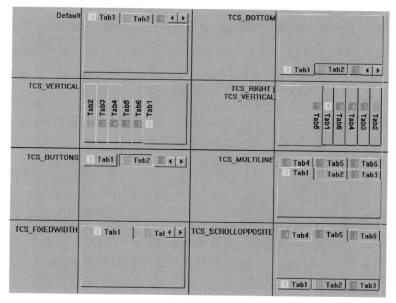

## Initialize Control

- To add a list of images that can be used by the tab control, use

```
m_tab.SetImageList(&m_imageList);
```

IV

A

- To add a text item with an image to a tab control, use

```
TC_ITEM tc;
tc.mask=TCIF_TEXT|TCIF_IMAGE;
tc.pszText=pszTab1;
tc.iImage=0;
m_tab.InsertItem(0,&tc);
```

## Other Styles

| | |
|---|---|
| TCS_MULTISELECT | Multiple tabs can be selected at once by using the Ctrl key; however, this only applies when using the TCS_BUTTONS style. |
| TCS_FORCEICONLEFT | Forces icon to the left of the tab, but applies only when also using the TCS_FIXEDWIDTH style. |
| TCS_FORCELABELLEFT | Forces label to the left of the tab, but applies only when also using the TCS_FIXEDWIDTH style. |
| TCS_OWNERDRAWFIXED | Allows the parent to draw the control. |
| TCS_HOTTRACK | Causes the control to react visually when the mouse cursor moves over it. |

## Notes

- The tab control by itself is fairly worthless — just like the header control is nothing without the list control. If you like the tab paradigm, you are better served by creating a property sheet window with CPropertySheet and then adding pages to it using the CPropertyPage class. If you then want to change the style of the tab control that the property sheet is using, you can gain access to it by using GetTabControl().

```
CPropertySheet ps;
ps.Create(…);
CTabCtrl *pTab=ps.GetTabControl();
```

Then, use ModifyStyle() to change the style.

## Month Calendar Controls

### Create Month Calendar Controls with the Windows API

```
HWND CreateWindowEx(dwExStyle, "SysMonthCal32", "",
    WS_CHILD|WS_VISIBLE|dwStyle x, y, width, height,
    hWndParent, (HMENU) id, hInstance, NULL);
```

### Create Month Calendar Controls with MFC

```
CMonthCalCtrl m_month;    // usually embedded in parent class
m_month.Create(
    WS_VISIBLE|WS_CHILD|dwStyle, rect, pParentWnd, id);
```

### Visual Styles

## Figure A.15  Month Calendar Control Styles

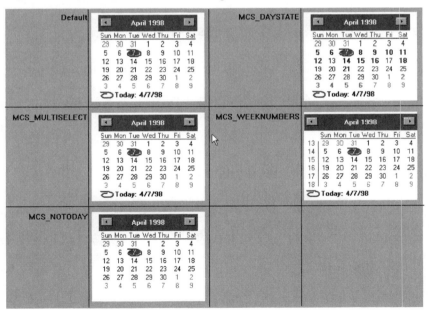

IV

A

### Notes

- The `CMonthCalCtrl` class is new to MFC and may not be available with your version. You can still create this control using the Windows API call by specifying the `SysMonthCal32` Window Class. To then manipulate this control, you can send windows messages to it.
- As you can see in the examples, the Month/Calendar control has no border, by default. And although you can make it any size, the calendar itself will remain the same size in the center — you will only be increasing or decreasing the size of the white border around it by changing its size.

# Date/Time Controls

## Create Date/Time Controls with the Windows API

```
HWND CreateWindowEx(dwExStyle, "SysDateTimePick32", "",
    WS_CHILD|WS_VISIBLE|dwStyle x, y, width, height,
    hWndParent, (HMENU) id, hInstance, NULL);
```

## Create Date/Time Controls with MFC

```
CDateTimeCtrl m_datetime;      // usually embedded in parent class
m_datetime.Create(
    WS_VISIBLE|WS_CHILD|dwStyle, rect, pParentWnd, id);
```

### Visual Styles

### Figure A.16  Date/Time Control Styles

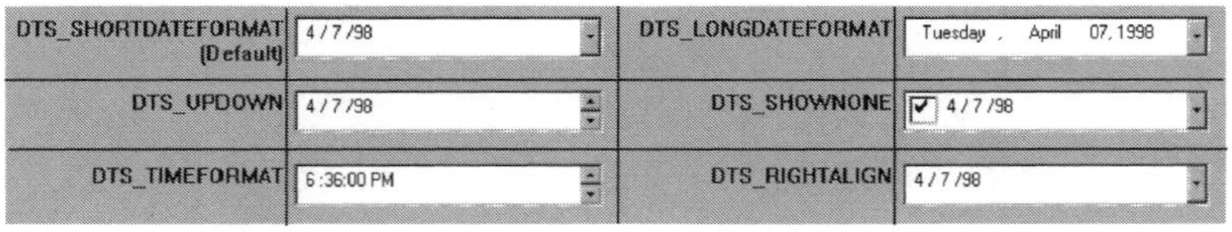

| DTS_SHORTDATEFORMAT [Default] | 4 / 7 /98 | DTS_LONGDATEFORMAT | Tuesday , April 07, 1998 |
| --- | --- | --- | --- |
| DTS_UPDOWN | 4 / 7 /98 | DTS_SHOWNONE | ☑ 4 / 7 /98 |
| DTS_TIMEFORMAT | 6 :36:00 PM | DTS_RIGHTALIGN | 4 / 7 /98 |

## Other Styles

| DTS_SHOWNONE | Unless checkmarked, the control will not return a date when your application requests it. |
|---|---|
| DTS_APPCANPARSE | Allows the user to edit the date when they push the F2 key. When the user is done, the control sends a DTN_USERSTRING message to the parent. |
| DTS_RIGHTALIGN | The Month/Calendar control will be right-aligned with the Date/Time Control when it drops down, instead of left-aligned. |

### Notes

- The CDateTimeCtrl class is new to MFC and may not be available with your version. You can still use this control using the Windows API call by specifying the SysDateTimePick32 Window Class. To then manipulate this control, you can send windows messages to it.
- When a Date/Time control drops down, it in fact opens a Month/Calendar control. You can access that control by calling the GetMonthCal-Ctrl() member function of CDateTimeCtrl.

# Dialog Window Styles

## Dialog Windows

### Create Dialog Windows with the Windows API

```
::CreateDialogIndirect(hInst, lpDialogTemplate,
    pParentWnd->GetSafeHwnd(), AfxDlgProc);
```

### Create Dialog Windows with MFC

```
CDialog m_dialog;              // usually embedded in parent class
m_dialog.Create(
    "Text", WS_VISIBLE|WS_CHILD|dwStyle, rect, pParentWnd, id);
```

**IV**

**A**

## Other Styles

| | |
|---|---|
| DS_ABSALIGN | Causes the dialog box to be created with screen coordinates, instead of client coordinates. In other words, the dialog will be created relative to the entire screen and not your application's main window. |
| DS_SYSMODAL | Causes the dialog to become a top-most window until the user dismisses it. Useful for system-wide failures. |
| DS_SETFONT | Causes CreateDialogIndirect() to also send a WM_FONT message to the window it creates to define a new font. |
| DS_NOIDLEMSG | Modal dialog boxes normally send a WM_ENTERIDLE to its owner when the dialog is idle. This style turns off this feature. |
| DS_SETFOREGROUND | Causes the dialog to be initially brought to the foreground. |
| DS_3DLOOK | Causes all control windows created by the dialog to have 3-D borders. Also makes all text nonbold. |
| DS_FIXEDSYS | Font defaults to SYSTEM_FIXED_FONT instead of SYSTEM_FONT. |
| DS_CONTROL | Sets all the styles necessary for this dialog to become a property page in a property sheet (WS_CHILD, no border, etc.). |
| DS_CENTER | Centers the dialog in the working area. This can be the parent window, or if DS_ABSALIGN is also specified, the center of the screen. |
| DS_CENTERMOUSE | When created, the mouse cursor is centered in the middle of the dialog. |
| DS_CONTEXTHELP | Puts a question mark in the upper corner of the dialog. If the user clicks on it and then a control, the control will receive a WM_HELP message. |

## Notes

- A dialog box is a popup window that's been specialized to create and manipulate control windows in bulk. In fact, CreateDialogIndirect() simply calls CreateWindowEx() and then sends a few extra window

messages to this window. The dialog template resource then provides a
list of controls to create.

# Plain Window Styles

So far we have only listed those styles that pertain to a specific control. We
will now review the window styles common to all windows.

### Creating a Plain With the Windows API

```
HWND CreateWindowEx(dwExStyle, "AfxWnd", "Text",
    WS_CHILD|WS_VISIBLE|dwStyle x, y, width, height,
    hWndParent, (HMENU) id, hInstance, NULL);
```

### Creating a Plain Window With MFC

```
CWnd m_wnd;                     // usually embedded in parent class
m_wnd.CreateEx(
    dwExStyle,"AfxWnd","Title",dwStyle, x, y, width,
    height, hwndParent, nIDorHMenu, lpParam /*= NULL*/);
```

IV

A

# Visual Styles

## Figure A.17 Plain Window Styles

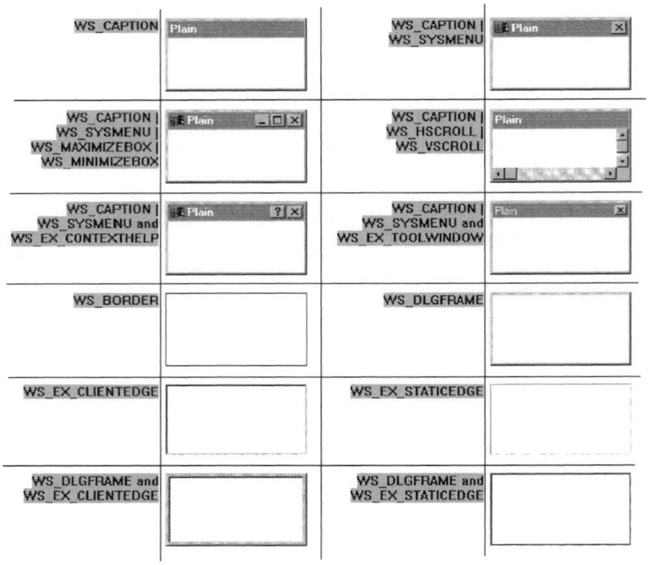

| | | | |
|---|---|---|---|
| WS_CAPTION | Plain | WS_CAPTION \| WS_SYSMENU | Plain ✕ |
| WS_CAPTION \| WS_SYSMENU \| WS_MAXIMIZEBOX \| WS_MINIMIZEBOX | Plain ▁☐✕ | WS_CAPTION \| WS_HSCROLL \| WS_VSCROLL | Plain |
| WS_CAPTION \| WS_SYSMENU and WS_EX_CONTEXTHELP | Plain ?✕ | WS_CAPTION \| WS_SYSMENU and WS_EX_TOOLWINDOW | Plain ✕ |
| WS_BORDER | | WS_DLGFRAME | |
| WS_EX_CLIENTEDGE | | WS_EX_STATICEDGE | |
| WS_DLGFRAME and WS_EX_CLIENTEDGE | | WS_DLGFRAME and WS_EX_STATICEDGE | |

# Visual Styles 2—Interesting Combinations

### Figure A.18   Mixed Control Styles

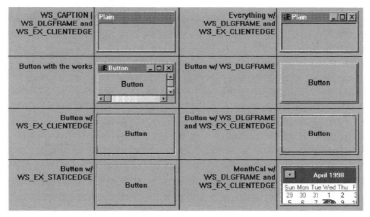

## Other Styles

| | |
|---|---|
| WS_CHILD | Usually entirely drawn by its parent and contained entirely within its parent window on the screen (e.g., control windows and views). |
| WS_OVERLAPPED | Usually the main window of an application (e.g., the frame windows). |
| WS_OVERLAPPEDWINDOW | Contains all the other styles normally associated with a frame window (e.g., border) |
| WS_POPUP | Usually an informational or data gathering window (e.g., dialog and message boxes). |
| WS_POPUPWINDOW | Contains all of the other styles normally associated with a popup, such as border. |
| WS_CLIPCHILDREN | Prevents a parent window from drawing over its child windows. |
| WS_CLIPSIBLINGS | Prevents a child window from drawing over its sibling windows (other child windows). |

IV

A

| | |
|---|---|
| WS_GROUP | Used in conjunction with radio buttons to delineate groups. When a radio button is clicked, the window unclicks other radio buttons in that group. This just marks the first entry in a group — the lost entry is defined as the one that occurs before the next WS_GROUP style. |
| WS_TABSTOP | A marker that Windows looks for to know that a control window is eligible to receive input focus when the user is tabbing around a parent window. |
| WS_MAXIMIZE | Causes a window to be maximized upon creation to the size of the screen, no matter what is specified for the window's width and height. |
| WS_MINIMIZE | Causes a window upon creation to immediately be minimized. |
| WS_VISIBLE | Causes a window to be visible when first created. The default action is for the window to be invisible until the ShowWindow(SW_SHOW) function is called. |
| WS_DISABLED | Grays a window and passes all mouse clicks to its parent. |
| WS_EX_ACCEPTFILES | Windows with this style will receive a WM_DROPFILES window message when the user releases the mouse over it while in drag/drop mode. |
| WS_EX_CONTEXTHELP | Puts a question mark in the upper corner of the window. If the user clicks on it and then a control, the control will receive a WM_HELP message. |
| WS_EX_CONTROLPARENT | Allows the user to move the focus around the child window of this window by using the tab key. Each child window must have the WS_TABSTOP style. |

| WS_EX_LEFTSCROLLBAR | The vertical scrollbar will appear on the left side of the window, instead of the right. This refers to the nonclient area scrollbar. Scrollbar control windows are not affected. |
|---|---|
| WS_EX_MDICHILD | Creates an MDI child window. In an MDI application, the main window fills its client area with a window created with the MDICLIENT window class. This window then maintains the windows created within. This style is a marker for that window to indicate which of its child windows it must maintain. |
| WS_EX_NOPARENTNOTIFY | A child window will normally send a WM_PARENTNOTIFY message to its parent window when it's created, destroyed, or the mouse clicks on it. This style turns off this feature. |
| WS_EX_TOPMOST | This window becomes a top-most window. The Window Manager maintains two lists, one of regular windows and one of top-most windows. The Manager always causes the top-most windows to draw last, which keeps them on top of the desktop. Among themselves, however, top-most windows must still fight for top spot. |
| WS_EX_TRANSPARENT | A window with this style doesn't erase its background and any window(s) underneath, thus allowing any windows underneath to show through. This can also be achieved by not specifying a background color in the Window Class with which this window is created. Moving this window however will then cause the background to move with it. This style is intended for modal operation, usually when drawing a highlight around another window, as when the Dialog Editor highlights a control. |

**IV**

**A**

# B

# Messages, Notifications, and Message Map Macros

See the tables on the following pages.

# Some Common Windows Messages

| WM Command | Description | Argument Notes |
|---|---|---|
| WM_CREATE | Sent to tell a window to initialize itself. | lParam points to a CREATESTRUCT, which contains the CreateWindow() arguments. Should return zero (0) to indicate success. |
| WM_INITDIALOG | Sent to tell a dialog box to initialize itself. Sent after the WM_CREATE message and after all of the control windows in the dialog have been created. | wParam is handle of a Control Window that should receive initial focus. lParam is an optional parameter passed by CreateDialogParam(). |
| WM_GETMINMAXINFO | Sent to tell a window that it's about to be resized to solicit a minimum or maximum size within which the window must stay. | lParam points to a MINMAXINFO structure that allows the window to set its minimum and maximum size. |
| WM_SIZE | Sent to tell a window that it's being resized. | wParam is a flag telling whether the window is being minimized, maximized, or just resized. lParam tells the size of the new window. |
| WM_PAINT | Sent to tell a window to draw its client area. | wParam is the device context with which to draw. |
| WM_DESTROY | Sent to tell a window to deallocate anything on the global heap. | wParam and lParam are unused. |
| WM_NCCREATE | Sent to tell a window that the nonclient area is being initialized. | lParam points to the same CREATESTRUCT used with WM_CREATE. |
| WM_NCDESTROY | Sent to tell a window that the nonclient area has just been destroyed. | wParam and lParam are unused. |
| WM_NCPAINT | Sent to tell a window to draw its nonclient area. | wParam has the handle of the Region to paint. Regions are described in Chapter 4. |
| WM_ERASEBKGND | Sent to tell a window to erase its background. | wParam specifies the device context with which to erase (a rectangle is drawn and filled). |
| WM_SHOWWINDOW | Sent to tell a window to hide or show itself. | If wParam is nonzero, the window is about to be shown. |
| WM_ENABLE | Sent to tell a window to enable or disable itself. | If wParam is nonzero, the window has been enabled. |
| WM_INITMENUPOPUP | Sent to tell a window that either its system menu or main menu is about to open. | wParam holds the handle of the menu object. The low word of lParam indicates the submenu position. If the high word of lParam is nonzero, it's the system menu about to open. |

| WM Command | Description | Argument Notes |
| --- | --- | --- |
| WM_HELP | Sent to tell a window that the F1 key has been pressed. | lParam contains a pointer to a HELPINFO structure that contains information needed to process context-sensitive help. |
| WM_MOUSEMOVE | Posted to a window when the mouse has moved in its client area. | wParam indicates whether any keyboard Shift-Ctrl-Alt keys were depressed and lParam indicates the new mouse cursor position relative to the client area. |
| WM_LBUTTONDOWN | Posted to a window when the left mouse button has been pressed in its client area. | Same as WM_MOUSEMOVE. |
| WM_LBUTTONUP | Posted to a window when the left mouse button has been lifted in its client area. | Same as WM_MOUSEMOVE. |
| WM_LBUTTONDBLCLK | Posted to a window when the left mouse button has been double-clicked in its client area. | Same as WM_MOUSEMOVE. Note that windows without the CS_DBLCLKS Window Class style will not receive this Window Message. |
| WM_RBUTTONDOWN WM_RBUTTONUP WM_RBUTTONDBLCLK | Same as WM_LBUTTONDBLCLK, except for right mouse button. | Same as WM_MOUSEMOVE, except for right mouse button. |
| WM_KEYDOWN | Posted to a window that has the keyboard focus when a key has been pressed — unless the Alt key is also pressed without the Ctrl key. | wParam contains the virtual key code and lParam contains additional key data. |
| WM_KEYUP | Posted to a window that has keyboard focus when a key has been released — unless the Alt key is currently pressed without the Ctrl key. | Same as WM_KEYDOWN. |
| WM_CHAR | Posted to a window after a WM_KEYDOWN message has been translated into a key character. | wParam contains the character and lParam is the same as WM_KEYDOWN. |

IV

B

# Control Notification Examples

## WM_XXX Control Notifications

| Control Window | Notification Sent |
| --- | --- |
| All child windows created without the WS_EX_NOPARENTNOTIFY window style. | Notification WM_PARENTNOTIFY is sent to tell a parent window when a control or any child window has been created or destroyed or when the mouse has been clicked on it. |
| Button, Combo Box, List Box | Notifications WM_CTLCOLOR, WM_DRAWITEM, WM_MEASUREITEM, WM_DELETEITEM, WM_CHARTOITEM, WM_VKEYTOITEM, and WM_COMPAREITEM are sent to parent windows to tell it went to draw this control. |
| Scroll Bars | The notifications are: WM_HSCROLL, WM_VSCROLL. |

## WM_COMMAND Control Notifications

| Control Window | Notification Sent |
| --- | --- |
| Static | Notifications include: STN_CLICKED, STN_DBLCLK, STN_ENABLE, STN_DISABLE. |
| Button | Notifications include: BN_CLICKED, BN_PAINT, BN_DISABLE, BN_PUSHED, BN_UNPUSHED, BN_DBLCLK, BN_SETFOCUS, BN_KILLFOCUS |
| Edit | Notifications include: EN_SETFOCUS, EN_KILLFOCUS, EN_CHANGE, EN_UPDATE, EN_ERRSPACE, EN_MAXTEXT, EN_HSCROLL, EN_VSCROLL |
| List Box | Notifications include: LBN_SELCHANGE, LBN_DBLCLK, LBN_SELCANCEL, LBN_SETFOCUS, LBN_KILLFOCUS |
| Combo Box | Notifications include: CBN_SELCHANGE, CBN_DBLCLK, CBN_SETFOCUS, CBN_KILLFOCUS, CBN_EDITCHANGE, CBN_EDITUPDATE, CBN_DROPDOWN, CBN_CLOSEUP, CBN_SELENDOK, CBN_SELENDCANCEL |

## WM_NOTIFY **Control Notifications**

| Control Window | Notification Sent |
|---|---|
| Animation, Progress, Slider | Only provides: NM_OUTOFMEMORY |
| Spin | Provides: NM_OUTOFMEMORY and UDN_DELTAPOS. For a notification of UDN_DELTAPOS, the NMHDR structure is a header to the NMUPDOWN structure. |
| Tab | Provides NM_OUTOFMEMORY, TCN_KEYDOWN, TCN_SELCHANGE, TCN_SELCHANGING. **When the notification is** TCN_KEYDOWN, the NMHDR structure is a header to the NMTCKEYDOWN structure. |
| Tree View Control | Provides TVN_SELCHANGING, TVN_SELCHANGED, TVN_GETDISPINFO, TVN_SETDISPINFO, TVN_ITEMEXPANDING, TVN_ITEMEXPANDED, TVN_BEGINDRAG, TVN_BEGINRDRAG, TVN_DELETEITEM, TVN_BEGINLABELEDIT, TVN_ENDLABELEDIT. Also provided are the generic notification messages: NM_OUTOFMEMORY, NM_CLICK, NM_DBLCLK, NM_RETURN, NM_RCLICK, NM_RDBLCLK, NM_SETFOCUS, NM_KILLFOCUS, NM_CUSTOMDRAW, NM_HOVER. The NMHDR structure can be a header to one of the following structures depending on the notification: NMTVDISPINFO, NMTVKEYDOWN, NMTVCUSTOMDRAW |
| List Control | Provides LVN_ITEMCHANGING, LVN_ITEMCHANGED, LVN_INSERTITEM, LVN_DELETEITEM, LVN_DELETEALLITEMS, LVN_COLUMNCLICK, LVN_BEGINDRAG, LVN_BEGINRDRAG, LVN_ODCACHEHINT, LVN_ITEMACTIVATE, LVN_ODSTATECHANGED, LVN_ODFINDITEM, LVN_BEGINLABELEDIT, LVN_ENDLABELEDIT, LVN_GETDISPINFO, LVN_SETDISPINFO, LVN_KEYDOWN, LVN_MARQUEEBEGIN. Also provides the generic notification messages listed above. The NMHDR structure can be a header to one of the following structures depending on the notification: NMLVDISPINFO, NMLVKEYDOWN |

## **Control Window With Both** WM_COMMAND **and** WM_NOTIFY **Control Notifications**

| Control Window | Notification Sent |
|---|---|
| Rich Edit | The WM_COMMAND notifications are seen above with the Edit control. The WM_NOTIFY control notifications are: EN_MSGFILTER, EN_REQUESTRESIZE, EN_SELCHANGE, EN_DROPFILES, EN_PROTECTED, EN_CORRECTTEXT, EN_STOPNOUNDO, EN_IMECHANGE, EN_SAVECLIPBOARD, EN_OLEOPFAILED, EN_OBJECTPOSITIONS, EN_LINK, EN_DRAGDROPDONE |

IV

B

# Message Map Macros For Window Messages

| Macro | Comments | Message Handler Calling Arguments |
|---|---|---|
| `ON_MESSAGE(WM_XXX, Handler)`<br>Examples:<br>`ON_MESSAGE(WM_CREATE,MyHandler)`<br>`ON_MESSAGE(WM_PAINT,MyHandler)` | Can be used to process any window message where WM_xxx is any Windows Message value and Handler can be any name. | `LRESULT Handler(WPARAM wParam,`<br>`                 LPARAM lParam)` |
| `ON_WM_XXX`<br>Examples:<br>`ON_WM_CREATE()`<br>`ON_WM_DESTROY()` | Every standard window message also has its very own macro that takes no arguments and whose name and use is based on its WM_XXX designation. | `OnXxx(message specific)`<br>(since the arguments are message specific, it's best to have the ClassWizard add these macros to your Message Maps) |

# Reflected Messages

| Macro | Comments | Message Handler Calling Arguments |
|---|---|---|
| WM_CTLCOLOR_REFLECT()<br>WM_DRAWITEM_REFLECT()<br>WM_MEASUREITEM_REFLECT()<br>WM_DELETEITEM_REFLECT()<br>WM_CHARTOITEM_REFLECT()<br>WM_VKEYTOITEM_REFLECT()<br>WM_COMPAREITEM_REFLECT()<br>WM_HSCROLL_REFLECT()<br>WM_VSCROLL_REFLECT()<br>WM_PARENTNOTIFY_REFLECT() | These macros allow Control Windows to handle WM_XXX type Control Notifications that are reflected back to it by the parent window. | HBRUSH CtlColor(CDC*,UINT)<br>void DrawItem(LPDRAWITEMSTRUCT)<br>void MeasureItem(LPMEASUREITEMSTRUCT)<br>void DeleteItem(LPDELETEITEMSTRUCT)<br>int CharToItem(UINT,UINT)<br>int VKeyToItem(UINT,UINT)<br>int CompareItem(LPCOMPAREITEMSTRUCT)<br>void HScroll(UINT,UINT)<br>void VScroll(UINT,UINT)<br>void ParentNotify(UINT,LPARAM) |
| ON_CONTROL_REFLECT(nCode,Handler) | This macro allows Control Windows to handle WM_COMMAND type Control Notifications that are reflected back to it by the parent window, where nCode is the notification code and Handler is any name you choose. | Void Handler() |
| ON_CONTROL_REFLECT_EX(nCode, Handler) | Same as above except now your handler returns a Boolean, which if TRUE allows the Parent to process the message too. | BOOL Handler() |
| ON_NOTIFY_REFLECT(nCode,Handler) | This macro allows Control Windows to handle WM_NOTIFY type Control Notifications that are reflected back to it by the Parent. Where nCode is the notification code and Handler is any name you choose. | Void Handler(NMHDR * pNotifyStruct, LRESULT* result) |
| ON_NOTIFY_REFLECT_EX(nCode, Handler) | Same as above except now your handler returns a Boolean, which if TRUE allows the Parent to process the message too. | BOOL Handler(NMHDR * pNotifyStruct, LRESULT* result) |

IV

B

# Command and Notification Messages

| Macro | Comments | Message Handler Calling Arguments |
|---|---|---|
| `ON_COMMAND(id, Handler)` | This macro directs WM_COMMAND messages to your handler where the id is the command id. | `void Handler()` |
| `ON_COMMAND_RANGE(id, idLast, Handler)` | Same as above except your handler is allowed to handle a contiguous range of command id's. | `void Handler(UINT id)` |
| `ON_COMMAND_EX(id, Handler)` | Same as above however now your handler can return a Boolean value which if FALSE will now allow OnCmdMsg() to continue to scan Message Maps for another handler. | `BOOL Handler(UINT id)` |
| `ON_COMMAND_EX_RANGE(id, idLast, Handler)` | Same as above except as a range. | `BOOL Handler(UINT)` |
| `ON_CONTROL(nCode, id, Handler)` | This macro directs WM_COMMAND control notifications to your handler. | `void Handler()` |
| `ON_CONTROL_RANGE(nCode, id, idLast, Handler)` | Same as above except a range of control IDs are allowed. | `void Handler(UINT id)` |
| `ON_XXX_XXXX(id, Handler)` Example: `ON_BN_DOUBLECLICKED(id, Handler)` | Each WM_XXX control notification has its very own macro which simply feeds the nCode to ON_CONTROL(). | `void Handler()` |
| `ON_NOTIFY(nCode, id, Handler)` | This macro directs WM_NOTIFY control notifications to your handler. | `void Handler(NMHDR *, LRESULT *)` |
| `ON_NOTIFY_RANGE(nCode, id, idLast, Handler)` | Same as above except a range of control IDs are allowed. | `void Handler(UINT id, NMHDR *, LRESULT *)` |
| `ON_NOTIFY_EX(nCode, id, Handler)` | Same as above however now your handler can return a Boolean value which if FALSE will now allow OnCmdMsg() to continue to scan Message Maps for another handler. | `BOOL Handler(UINT id, NMHDR *, LRESULT *)` |
| `ON_NOTIFY_EX_RANGE(nCode, id, idLast, Handler)` | Same as above except as a range. | `BOOL Handler(UINT id, NMHDR *, LRESULT *)` |

# Accessing Other Application Classes

## Table of Pointers

See "Table of Pointers" on page 549.

## Notes

- Other than the GetDocument() function from your CView class, you should use these functions sparingly. If you find that a function is constantly referring to another class, move the function into that class. See Example 1 for ideas on how to divide your application among the various application classes.

- Notice that because the Document Class can have more than one view, it gets the start of a list of views that it can sift though. You can access this list directly as m_ViewList, which is a CPtrList member variable of CDocument. To access only the currently active view, the Document can call the following for an SDI application

```
GetAfxMainWnd()->GetActiveView();
```

or for an MDI application

```
GetAfxMainWnd()->MDIGetActive()->GetActiveView();
```

- Notice that you can only use theApp to access the application class if you include the following in application class's include file past the declaration.

```
class CWzdApp : public CWinApp
{
};
extern theApp;
```

However, using theApp is frowned upon in an encapsulated C++ environment when you can use AfxGetApp(), instead.

# Table of Pointers

**From this class**   — you can access these classes, using the functions shown.

| | CWinApp | CMainFrame | CChildFrame | CDocument | CView |
|---|---|---|---|---|---|
| **CWinApp** | | AfxGetMainWnd()<br>or<br>m_pMainWnd | GetAfxMainWnd()-><br>MDIGetActive() | GetAfxMainWnd()-><br>GetActiveView()-><br>GetDocument() | GetAfxMainWnd()-><br>GetActiveView() |
| **CMainFrame** | AfxGetApp()<br>or<br>theApp | | MDIGetActive()<br>or<br>GetActiveFrame() | **if SDI:**<br>GetActiveView()-><br>GetDocument()<br>**or if MDI:**<br>MDIGetActive()-><br>GetActiveView()-><br>GetDocument() | **if SDI:**<br>GetActiveView()<br>**else if MDI:**<br>MDIGetActive()-><br>GetActiveView() |
| **CChildFrame** | AfxGetApp()<br>or<br>theApp() | GetParentFrame() | | GetActiveView()-><br>GetDocument() | GetActiveView() |
| **CDocument** | AfxGetApp()<br>or<br>theApp | AfxGetMainWnd() | GetAfxMainWnd()-><br>MDIGetActive() | | POSITION pos=<br>GetFirst\<br>ViewPosition();<br>GetNextView(pos); |
| **CView** | AfxGetApp()<br>or<br>theApp | AfxGetMainWnd() | GetParentFrame() | GetDocument() | |
| **any other class** | AfxGetApp() | AfxGetMainWnd() | AfxGetMainWnd()-><br>MDIGetActive()<br>or<br>AfxGetMainWnd()-><br>GetActiveFrame() | **if SDI:**<br>AfxGetMainWnd()-><br>GetActiveView()-><br>GetDocument()<br>**or if MDI:**<br>AfxGetMainWnd()-><br>MDIGetActive()-><br>GetActiveView()-><br>GetDocument() | **if SDI:**<br>AfxGetMainWnd()-><br>GetActiveView()<br>**else if MDI:**<br>AfxGetMainWnd()-><br>MDIGetActive()-><br>GetActiveView() |

IV

C

# Development Notes

There are several issues left to cover that, although unrelated to VC++ or the MFC class library, are never the less important when creating an application using the Developer Studio.

**Setting up the Developer Studio**  The Studio can very easily be configured to conform to anyone's standards for developing an application. Here we look at some of those options.

**Debugging**  There are several resources available to you for debugging your application. Besides the obvious interactive debugger that's invoked when you run the debug version of your application, there are also framework, Developer Studio, and even MFC class functions you can call, as well as third party run-time debuggers.

**Organizing and Building Projects**  If your application will be a single executable file with no additional DLL files, you will probably remain within the confines of your project's workspace. However, if your application will span several libraries and possibly multiple executables, you will need to consider some strategies to build and maintain your project.

**Microsoft Hieroglyphs**  The somewhat informal and totally voluntary variable naming convention that Microsoft adheres to is reviewed here.

**Spying**  The SPY.EXE utility is examined here as a potential source not only of debugging your own application, but also as a way of "spying" on another application to figure out how someone else implemented their interface.

**Other Example Resources**  Besides this book, we review other sources you can turn to that might have one or more examples of what you are attempting to implement in your application using MFC VC++.

# Setting up the Developer Studio

## Invoking Other Utilities

The Developer Studio allows you to run other utilities from within the Studio by adding custom menu commands to the Tools menu.

1. Click on the Tools/Customize menu commands to open the Customize Property Sheet.
2. Select the Tools tab and scroll to the bottom of the menu list (the Studio should already have several commands preinstalled in this menu).
3. Select the empty focus rectangle on the bottom and enter the name you want to appear in the menu. Pressing Enter should cause three edit boxes to open up below for Command, Arguments, and Initial Directory.
4. In Command, you enter the path name and executable you want to run.
5. In Arguments, you enter the arguments you want to pass to the executable.
6. In Initial Directory, you enter the directory the Studio should change to before running this command.

You might consider adding one of the following commands to your Tools menu.

**An Explorer Command** to bring up the Windows Explorer initially opened to your project's directory. Type in Explorer for the menu name, c:\windows\explorer.exe or c:\winnt\explorer.exe for the command (depending on your operating system), /e for the arguments, and leave the initial directory blank.

**A DOS Command** to open an MS-DOS window initially to your project's directory. Type in MS-DOS for the menu command, c:\winnt\system32\cmd.exe or c:\windows\command.exe for the command (depending on your operating system), nothing for arguments, and $(CurDir) for the initial directory.

**An Editor Command** to opens a third party text editor and then pass the name of the currently selected file to that editor. Although the text editor provided with the Studio is fairly efficient, there are several editors on the market with more bells and whistles. Type in Editor for the menu command, the path of the editor for the Command, $(FileName)$(FileExt) for the arguments, and $(WkspDir) for the initial directory. When you execute any menu command, the Studio saves all of your files so that even if you modified a file before you call up the external editor, you will be editing the latest version when it opens.

## Toolbar Buttons

The Developer's Studio also allows you to configure the buttons in your toolbars.

1. Click on the Tools/Customize menu commands to open the Customize Property Sheet and select the Commands tab.
2. The buttons shown in the Buttons group box can be dragged into an existing toolbar or into a clear space to create a new toolbar.
3. The Category combo box changes the selection of buttons from which you can choose.
4. To delete a button, drag it back into the Property Sheet.

**IV**

**D**

Clicking on each icon in the Commands page will display a description of the command it represents. You can also discover what predefined icons go with what menu command by looking at the menu itself. The icons that appear next to some menu commands depict their toolbar button.

Possible candidates for their own toolbar button include the following.

- Any command you add to the Tools menu, which will typically start at seven. Therefore, look in the Tools category of this page for a button icon with a small hammer and a seven (7) in the lower-right corner for the first command.

- The ClassWizard, which can be found under the View category as a triangle with a wizard's wand.

- The Open Workspace... command, which is one of the most commonly used commands, yet has no default toolbar button. You can create a button for this command by selecting the All Commands category. There, you will find a complete list of all Developer Studio commands. Scroll down until you find WorkspaceOpen nearly at the bottom. Drag this selection to a spot on the toolbar. Since this command does not have a default toolbar icon associated with it, the Studio will then prompt you with a list of generic icons you can assign to this button. You can also choose to include the name of the menu command in the button.

---

**NOTE:** After you get the Studio customized as you like it, exit immediately. The changes you make to the toolbars and menus aren't saved until you exit. If the Studio crashes before you exit normally, your changes will be lost! Also, be careful when running more than one instance of the Studio at one time. Make sure the last Studio to exit is the one to which you made the changes. The last Studio to exit gets final dibs on your Studio's configuration.

---

# Debugging

Most of your debugging effort will be done through the Studio's Text Editor interface, which is quite intuitive. See your MFC documentation for more. However, there are several other resources available to you for debugging code.

## TRACE()

If your application is time- or sequence-sensitive, you might need to print debug messages rather than step through a critical section using the interactive debugger. Historically, programmers have printed these messages to a printer or disk file using printf() statements. The framework provides a function called TRACE() that is equivalent to a printf() command, except that it prints directly to your Studio's Debug window. It cannot exceed 512 characters. An example follows.

---

**NOTE:** Make sure you end your statement with a new line character (\n)!

---

```
TRACE("Integer = %d, String = %s\n", i, sz);
afxTraceEnabled=FALSE; // turns off all TRACE() statements
```

---

**NOTE:** The TRACE() statement does nothing in the release version of your application. If the problem only appears in the release version, you must revert to printf()s. You could also use AfxMessageBox(), as discussed next.

---

## AfxMessageBox()

If a problem only appears in your application on a system without an installed Studio or if the problem only occurs in the release version of your application, you can use printf()s to a file. You might also consider using AfxMessageBox(), instead. You can format a message or create a static message indicating the progress of your application through a critical section and use AfxMessageBox() to display these messages. AfxMessageBox() doesn't even have to be in a class that controls a window — or in a class at all.

## ASSERT(), VERIFY()

Two other global functions you can use to check your work are ASSERT() and VERIFY(). These macros are typically used to check whether the arguments passed to a function are valid. If these arguments originated with your user, you would probably want to display some sort of error message

**IV**

**D**

with `AfxMessageBox()`. However, if the arguments originated with another part of your application, `ASSERT()` and `VERIFY()` can be used with a lot less overhead. The MFC library uses `ASSERT()`s and `VERIFY()`s extensively to prevent any bonehead mistakes from causing strange results.

In the debug version of your application, both `ASSERT()` and `VERIFY()` evaluate a valid C or C++ expression and, if not `TRUE`, display an error message and abort the application. In the release version of your application, `ASSERT()` does nothing — it doesn't even evaluate the expression within its parentheses. `VERIFY()`, on the other hand, does evaluate its expression, but it still doesn't halt the release version of your application. An example of each follows.

```
ASSERT(this)            // if false, debug application aborts
                        // statement not evaluate on release
                        // version
VERIFY(GetFile (...))   // if false, debug application aborts
                        // statement still evaluated in release
                        // version
```

## Dump()

Every class you derive from `CObject` inherits a `Dump()` member function. This function is similar in concept to `Serialize()`, except that it can output all the members your class in a human-readable format. You can dump a single object or you can dump all the objects in a document, in a method similar to serializing all data objects to disk using `Serialize()` (Example 67). However, instead of passing `Dump()` the instance of an archive class, as is the case with `Serialize()`, you pass it an instance of a dump context. You can use the default dump context of `afxDump`, which outputs to the Studio's debug window, or you can use a file.

```
CFile file;
file.Open("dump.txt", CFile::modeCreate | CFile::modeWrite);
CDumpContext dc(&file);
```

You can then dump an object as follows.

```
pObject->Dump(&dc);
```

If you are implementing a full dumping system, you might call the first `Dump()` function using a menu command. All other `Dump()` functions in your document could then be called in a descending pattern.

An example `Dump()` follows.

```
void CMyData::Dump(CDumpContext &dc) const
{
    CObject::Dump(dc);
    dc << "Item 1 = " << m_nItem1;
    dc << "Item 2 = " << m_nItem2;
    dc << "Item 3 = " << m_nItem3;
        :     :     :
}
```

Again, this facility is only available in the debug version of your application.

# Exceptions

You might occasionally see the following type of error in the debug window of the Studio.

```
First-chance exception in xxx.exe (KERNEL32.DLL):
    0xE06D7363: Microsoft C++ Exception.
```

This indicates that an exception was thrown in a `try{} catch{}` configuration in the MFC library and eventually handled. If you would like to see what caused the exception yourself, you can tell the debugger to stop on this type of exception, rather than letting it be handled automatically in the `catch{}`.

1. Start by running your application in the debug mode to expose the Debug menu command.
2. Then, click the Debug/Exceptions menu commands to open the Exceptions dialog box.
3. In the Exceptions dialog box, you will find a list of all exceptions. Select Microsoft C++ Exceptions, which is at the bottom of the list, and then click on Stop Always.

Now whenever an MFC C++ exception occurs, the debugger will stop, positioned over the exception.

**IV**

**D**

## GetLastError()

A lot of the Windows API calls that MFC makes or that you make directly will fail with no explanation other than what you expected to happen didn't happen. However, you can access the error for a select few API calls (refer to your documentation for which) by calling ::GetLastError(). The error code returned can be found in your MFC documentation.

## AfxCheckMemory()

For sporadic problems that cause your heap memory to become corrupted, you can use AfxCheckMemory() to try to pinpoint the problem. AfxCheck-Memory() checks the validity of your heap memory. This function returns TRUE if everything is okay, making it a natural for the following syntax.

```
ASSERT(AfxCheckMemory())
```

Insert this line around the trouble area and hope it catches something.

## CMemoryState

MFC's CMemoryState class helps track down memory leaks by taking a series of snapshots of the global heap for later comparison. A typical use follows.

```
CMemoryState msBefore, msAfter, msDif;
msBefore.Checkpoint();

    // allocate and deallocate memory using "new" and "delete"

msAfter.Checkpoint();
msDif.Difference(msBefore, msAfter);
msDif.DumpStatistics();     // displays leaks
```

Note that we needed three instances of the class to do the comparison.

## Third Party Error Checkers

The Developer Studio's debugger does some limited memory leak checking, but no resource leak checks. A resource leak occurs when you create a device context or pen object and fail to destroy it after its use. To really do a thorough job of checking your program for leaks, you should consider one of the third party leak checkers such as BoundsChecker™ or Purify™. Not only do these utilities check for leaks, but they can also spot other types of bugs that might take you days to find otherwise, such as a `memcpy()` writing where it shouldn't

I personally prefer BoundsChecker for its simplicity of operation. You run the program right out of the box and it spits out a few error messages, one of which might be an actual problem. Yes, the other messages might be bogus. An inherent problem with this type of utility is that it continually comes up with false leads. One of the major jobs of BoundsChecker, for instance, is just to filter out all the known error messages coming out of MFC code that you can't change. My experience with Purify in this regard has been similar. However, it does seem to work sometimes when BoundsChecker doesn't. Both seem to find problems that the other doesn't, so in an ideal world, you would have both.

## Dr. Watson

Last and very least as a resource for debugging your applications is the program error debugger built right into your operating system. This is the small window that appears on your screen when an application has had a page fault or tried to access forbidden memory and tells you that you must click Yes to destroy hours of work. On a system without the Developer Studio installed, only the registers and stack are displayed. These values won't help you to decipher anything, in most cases. When the Studio is installed, you are given the opportunity to put the application into a debugging session. However, if the application is a release version, doing this will also be close to useless, unless you understand assembler and have a few hours to kill.

# Organizing and Building Projects

If your application can be contained within a single executable, you can ignore this section. Building the application for you is a simple matter of clicking the Build All button. If your application will involve several executables, however, you will need to organize how your application is built so that, for example, application or library A gets built before B and that C can find the includes and libraries that D generates.

When planning a large application spanning more than one project directory, you are presented with the logistics problem of how one project will be able to access the include files and libraries of another project. Some enterprising souls have solved this problem by simply copying the necessary files

**IV**

**D**

into the necessary directories. However this presents another logistics nightmare of trying to determine which files in a directory belong to that project or not. And if not, are they the latest version. Also, if several projects need to access the same libraries, will there now be several possible versions of each file lying around?

## Use \Include and \Lib Directories

A more reasonable approach might be to create two common directories that can be shared by your projects: \include and \lib. The \include directory, obviously, would contain include files (.h files). The \lib file, besides containing libraries, would also contain any DLL files and the debug symbol tables of debug libraries (.lib, .dll, and .pdb files).

When you run an executable that requires .dlls, your operating system first looks for these .dll files in the same directory as the executable. If it can't find them there, it then looks in your system's normal execution path. Therefore, you must also add the \lib directory to your system's normal path. With Windows 95 and 98, this means editing the autoexec.bat file in your system's root directory and adding the \lib directory to the PATH= statement. With a Windows NT system, this means running the system configurator in the Control Panel and adding it to the PATH= statement listed in the top box.

---

**WARNING:** In both cases, make sure to reboot your system — otherwise, the changes won't take effect!

---

## Rename Debug Versions

Which brings up the next point: because libraries would now be sharing the same directory, you must also give debug libraries a name different from release libraries. Typically, you can accomplish this by appending the letter d to the end of the debug library filename. For example, make the debug version of mylib.lib equal to mylibd.lib. You can do this through your Project Settings dialog box. You will also need to create an additional .def file in your project to contain the name of the debug version of your .dll. Simply copy the existing .def file to a new file whose name is appended with d (e.g. mydll.def is copied into mydlld.def). Then append a d to the names contained in that new .def file. When you include this new .def file in your project, you will notice that Studio squawks at you that you can't have more then one `.def` file per project. Using the Project Settings, select the old `.def` file and exclude it from the debug version of your project. Then, include this new `.def` file in the debug version.

## Post Build Step

You can get the appropriate files into the `\include` and `\lib` directories automatically by adding a DOS command to the Post Build Step tab of your Project Settings dialog. This command can be a simple `COPY` statement, copying from your project's `\Release` or `\Debug` directory into the `\include` or `\lib` directory. For DLL projects, copy both the `.lib` and `.dll` files. For the debug version, also include the `.pdb` file, which contains what the debugger needs to do a source level debugging session with the DLL.

## Developer Studio Directories

You can tell the Developer Studio where to find the `\include` and `\lib` files through its Options Property Sheet, which you can open by clicking on Tools/Options. Then, select the Directories tab. Under the Show Directories for combo box you will find entries for both Include files and Library files. Add your `\include` and `\lib` directories to these lists and exit the Studio (closing all other instances of Studio first) to save these additions. Now when you build your application, the Studio's compiler and linker will also search those directories. This assumes, of course, that you added these libraries to your project's settings (Example 84).

## Version Control Software

A version control utility is also highly recommended, not only to safeguard the source, but also as a way of easily creating older versions of your application. Two good version control utilities are Visual Source Safe and PVCS.

## Build Machine and Batch File

An impartial machine is usually designated as a "build" machine. This is typically also a file server and the machine in which the version control utility resides. Because a build can take forever, especially as your application matures and takes on more features, a batch file is highly recommended for building your entire application, libraries, and executables. A batch file eliminates human error when a long process requires several brief but crucial steps. A batch file ensures that your application is created the same way

**IV**

**D**

each time and eliminates the build itself as a suspect when your application is acting strangely. A build batch file typically has three steps.

1. Clean up the old build — delete all files from the \include and \lib directories, etc.

2. Extract the correct software versions using the version control utility. With Visual Source Safe, you can accomplish this with the following. You can also pull out other versions to create an earlier version of your application.

```
cd \develop\project1
ss Get $/Project1 -R
ss Get $/Project1 -R -V2.3 -- gets version 2.3
```

3. Build each project in the required sequence as follows, outputting any errors to a text file you can review later. Be aware that the string assigned in the cfg="" declare must be identical to that in the .mak file.

```
nmake -fProject1.mak -xProject1.err cfg="Project1 - Win32
    Release">>err.err
```

Which brings up the question, where the heck do I get a .mak file? Earlier versions of the Studio generated them automatically. However, the latest versions make it an option you have to set.

4. Click the Studio's Tools/Options commands and select the Build tab.

5. Then, make sure you set the option Export makefile when saving project.

You probably don't want to do this for larger projects because this export slows down a simple project save. You might do better to simply create a new makefile whenever the project changes. To create a makefile on demand, simply click Project/Update Makefile... .

When you build in batch mode, outside of the Studio, the directories you setup inside no longer apply. The compiler and linker are now using the paths defined in your operating system's environment. Therefore, to once again make the compiler and linker aware of where the communal \include and \lib directories are, you must add the following to your environment.

```
lib=\lib
include=\lib
path=\lib (so that dll files can be found)
```

With Windows NT, you must make these changes to the System Variables using the System utility in the Control Panel. You must also reboot afterwards.

## Build Machine and the Developer Workspace

As mentioned previously, newer versions of the Studio don't automatically generate a .mak file. Instead, they use two new files: a .dsp file, which keeps track of individual projects, and a .dsw file, which keeps track of a workspace. A workspace can contain several projects, which can even have dependencies on each other. Therefore, if you are using the newer studios, you might also want to consider using these facilities to build large projects in the following steps.

1. Assuming you already have several .dll or .exe project subdirectories, use the Developer Studio to open the .dsw file of the main .exe project.

2. In the Studio's Project menu, use the Insert Project into Workspace command to add all of the other projects to this project's workspace.

3. Again under the Project menu, use the Dependencies... command to set up all of the interdependencies of your entire project. For example, select an .exe project and then click off all of the .dll projects that must be built before this .exe can be built.

4. You can use the Projects/Set Active command to determine which project will be built when you tell the Studio to Build or Build all. If you were to make the main .exe project the active project, you can rebuild your entire project just by hitting Build or Build all.

5. If you want to build both the Debug and Release versions of your entire project without having to come back after one has completed, you can use the Developer Studio's Build menu to select the Batch Build command and select everything to rebuild everything.

6. This workspace configuration is saved in the .dsw file. For a particularly large project, you might consider having a .dsw file that encompasses the entire project just on your build machine.

## Build Kit

The last step to organizing a large application is creating a Build Kit. A Build Kit is simply a selection of files that anyone can use to create a new version of the application. This includes the files in the \include and \lib

**IV**

**D**

directories, and, optionally, the source files with which the libraries were built. The Build Kit should reside on the Build Machine in a separate directory from the build. Making the kit separate allows it to be accessed for new development, even in the middle of a new build. And if the new build doesn't function correctly, new work isn't postponed for everyone else while the build is fixed. Once a build is verified to work, another batch file can be used to copy it into the Build Kit directory.

# Microsoft Hieroglyphs

Microsoft has instituted a somewhat informal, voluntary naming convention for program variables used in conjunction with MFC VC++ programming. While this convention is not required in any way to create an MFC application, and is not always applicable, it does have some merit. In addition to the usual descriptive name you can give a variable, using this convention you can also discern a variable's type and whether or not it's a member variable of a class. The convention is as follows.

**Variables that start with**

| | |
|---|---|
| m_ | are member variables of a class |
| m_b | are Boolean |
| m_n | are integer |
| m_p | are pointers |
| m_dw | are DWORDs |
| m_classname | are of the class type specified in classname |

Critics of this convention point out that it goes against the flexibility of the language. For example, after having named your variable m_bFlag, it would now be difficult to turn it into an integer without either editing all occurrences or leaving it alone and possibly confusing future software developers. However, from personal experience, I have found that you rarely need to change the type of a variable. The advantage of being able to quickly figure out what a variable does far outweighing this disadvantage.

# Spying

A possibly overlooked utility in your VC++ development package is SPY.EXE. This little utility cannot only help you figure out what's going on in your own application, but as the name implies, it can help you figure out how someone else implemented windows in their application's interface.

## Finding Messages

You can read all you want to about which windows messages are available to which window in your application, but when all your application's interface components are slapped together, it's anyone's guess just what messages will be available to you. To find out for sure, use the SPY.EXE utility to intercept and display all messages to any visible window.

1. Start by executing your application.
2. Then execute SPY.EXE (which can be found in your MFC C++'s \bin directory).
3. Click on Spy's Spy/Messages menu commands to open the Message Options Property Sheet.
4. Find the Finder Tool and drag it over the window on which you want to spy.

A new window will open within Spy and start displaying windows messages to and from that window. From there, you might spot a window message you hadn't thought of to process.

## Finding Windows

Using that same Finder Tool, you can sometimes figure out how someone else was able to achieve an effect with their interface. You can, for instance, determine whether the effect is a window style you hadn't used before or the creators of the other application drew the effect themselves. Grab the Finder Tool as before, except now drag it over the effect in question. The dimensions of the underlying window will be highlighted, its class and styles revealed. You'll be surprised sometimes by what you'll find.

# Other Example Resources

If you can't find an example of what you want to do in this book, there's still a good chance it exists somewhere out there. Your VC++ distribution kit CD came with its own directory of samples. Another excellent, albeit unusual, source of samples is the collection of source files from which your MFC class libraries were created, which can be found in your VC++'s \mfc\src directory. This is also the place to look for undocumented virtual MFC functions.

IV

D

Another major repository of examples can be found on the Internet at:
`ftp://ftp.microsoft.com/Softlib/MSLFILES/`

although you may have to access it at 3 A.M. to beat the 2,500 users limit.

# Index

# Why Do Serious Embedded Developers Read *Dr. Dobb's Journal?*

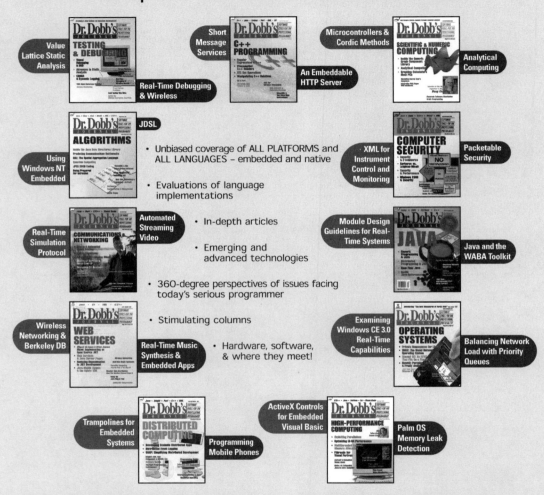

Value Lattice Static Analysis

Real-Time Debugging & Wireless

Short Message Services

Microcontrollers & Cordic Methods

An Embeddable HTTP Server

Analytical Computing

JDSL

Using Windows NT Embedded

- Unbiased coverage of ALL PLATFORMS and ALL LANGUAGES – embedded and native

- Evaluations of language implementations

XML for Instrument Control and Monitoring

Packetable Security

Real-Time Simulation Protocol

Automated Streaming Video

- In-depth articles

- Emerging and advanced technologies

Module Design Guidelines for Real-Time Systems

Java and the WABA Toolkit

- 360-degree perspectives of issues facing today's serious programmer

Wireless Networking & Berkeley DB

- Stimulating columns

Real-Time Music Synthesis & Embedded Apps

- Hardware, software, & where they meet!

Examining Windows CE 3.0 Real-Time Capabilities

Balancing Network Load with Priority Queues

Trampolines for Embedded Systems

Programming Mobile Phones

ActiveX Controls for Embedded Visual Basic

Palm OS Memory Leak Detection

In each issue, serious developers depend on Dr. Dobb's Journal for an environment that is relevant, exciting and helpful to their jobs of creating unique and powerful software programs.

If your job demands a knowledge of emerging or advanced software technologies and tools, regardless of language or platform, embedded or native, you need to add Dr. Dobb's Journal to your toolbox.

To subscribe online using your special embedded rate, go to:

## www.ddj.com/sub/

and type code: 2DCK

CMP
United Business Media

Dr. Dobb's
JOURNAL

SOFTWARE
TOOLS FOR THE
PROFESSIONAL
PROGRAMMER

## Debugging Visual C++ Windows

### by Keith Bugg

**Understand and control the Bug cycle!** This detailed reference provides tutorial-based examples and a conceptual model for preventing and eliminating bugs during the design cycle that includes suggestions on identification, prevention, and correction for each of the four types of bugs: compile-time errors, run-time errors, logic and design errors, and machine errors. Learn how compiler tools such as TRACE, Spy, and Stress work and get a critical review of commercial debuggers (including BoundsChecker and Code Wizard). Disk included, 224pp, ISBN 0-87930-545-2

**RD2985**　　　**$29.95**

**RD3056**　　　**$49.95**

## Supercharge MFC
### by Jeffrey Galbraith

**Achieve User Interface functionality not yet offered by Microsoft!** Extend MFC with the author's revolutionary use of C++ subclassing to achieve pseudo-multiple inheritance. Get a sophisticated message handler that can be used with any window — eliminating the need for complicated MFC message maps. If MFC objects should do the work, it lets the message pass through; if the developer wants customized class extensions to do the work, then the C++ wrappers handle the messages themselves. CD-ROM included, 544pp, ISBN 0-87930-569-X

# What's on the CD-ROM?

**Visual C++ MFC Programming by Example** is accompanied by the companion CD-ROM which includes all of the examples described in the book as well as an MFC Quick Reference Guide. Each example is contained in a Visual C++ v5.0 project subdirectory with a name that corresponds to the example in the book. Each project subdirectory also includes a \Wizard subdirectory that contains just the source necessary to add that example to your application.

Also on the CD is the author's **SampleWizard**™, a utility that uses these \Wizard subdirectories to add the source from an example directly to your application. You can even add your own examples to the **SampleWizard**'s collection.

## Using SampleWizard™

- Pick the section, chapter, and example in the book that contains the type of example you're looking for.
- Determine how the example source will be inserted into your application.
- View the example and insert by one of the following three methods: (a) copy and paste sections of the file into your source, (b) copy the entire displayed file to the clipboard, or (c) copy any of the example files to a target directory.

---

For more information on using **SampleWizard**™, see the Readme file on the CD. For the most up-to-date CD-ROM data and an added v6.0 project, download the files from the publisher's ftp site: ftp://ftp.cmpbooks.com/pub/errata/visualC.zip.

---